The Phylogenetic Handbook

The Phylogenetic Handbook is a broad introduction to the theory and practice of nucleotide and amino-acid phylogenetic analysis. As a unique feature of this book, each chapter contains an extensive practical section, in which step-by-step exercises on real data sets introduce the reader to the most widely used phylogeny software, including CLUSTAL, PHYLIP, PAUP*, DAMBE, PAML, TREE-PUZZLE, TREECON, SplitsTree, TreeView, MEGA2, and SimPlot. Chapters 1 through 10 provide a strong background in basic topics such as the use of sequence databases, alignment algorithms, tree-building methods, estimation of genetic distances, and testing models of evolution. Additional chapters briefly survey special topics in evolution; for example, modeling evolution with networks, studying recombination, testing for positive selection, and methods in population genetics. The book will be an invaluable resource for advanced-level undergraduate and graduate students, as well as for professionals working in the fields of molecular biology and evolution.

Marco Salemi is a postdoctoral scientist at the Medical Faculty, Katholieke Universiteit Leuven, Belgium. He currently holds a research position at the University of California, Irvine, where he works on epidemiological virology and phylogenetic analysis, focusing on the molecular epidemiology of human pathogenic viruses such as HIV and HCV. He is supported by the Fund for Scientific Research – Flanders, Belgium.

Anne-Mieke Vandamme is a Full Professor in the Medical Faculty at the Katholieke Universiteit Leuven, Belgium, working in the field of clinical and epidemiological virology. Her laboratory investigates treatment responses in HIV-infected patients and is respected for its scientific and clinical contributions on virus-drug resistance. Her laboratory also studies the evolution and molecular epidemiology of human viruses such as HIV and HTLV.

For many years, the editors have organized a widely recognized annual workshop on virus evolution and molecular epidemiology. This book has arisen in response to requests for tutorials on phylogenetics at these workshops, and as a result of the extensive teaching experience of the authors and editors.

The Phylogenetic Handbook

A Practical Approach to DNA and Protein Phylogeny

Edited by

Marco Salemi
University of California, Irvine and Katholieke Universiteit Leuven, Belgium

and

Anne-Mieke Vandamme
Rega Institute for Medical Research, Katholieke Universiteit Leuven, Belgium

 CAMBRIDGE
UNIVERSITY PRESS

PUBLISHED BY THE PRESS SYNDICATE OF THE UNIVERSITY OF CAMBRIDGE
The Pitt Building, Trumpington Street, Cambridge, United Kingdom

CAMBRIDGE UNIVERSITY PRESS
The Edinburgh Building, Cambridge CB2 2RU, UK
40 West 20th Street, New York, NY 10011-4211, USA
477 Williamstown Road, Port Melbourne, VIC 3207, Australia
Ruiz de Alarcón 13, 28014 Madrid, Spain
Dock House, The Waterfront, Cape Town 8001, South Africa

http://www.cambridge.org

First published 2003

Printed in the United Kingdom at the University Press, Cambridge

Typefaces Minion 10.5/14 pt. and Formata   *System* LaTeX 2$_\varepsilon$ [TB]

A catalog record for this book is available from the British Library.

Library of Congress Cataloging in Publication Data

The phylogenetic handbook : a practical approach to DNA and protein phylogeny /
   edited by Marco Salemi, Anne-Mieke Vandamme.

      p.   cm.

   Includes bibliographical references and index.

   ISBN 0-521-80390-X

   1. DNA – Analysis – Handbooks, manuals, etc.   2. Proteins – Analysis – Handbooks,
manuals, etc.   3. Cladistic analysis – Handbooks, manuals, etc.   I. Salemi, Marco, 1968–II.
Vandamme, Anne-Mieke, 1960–

   QP624 .P485   2003
   572.8′633 – dc21   2002073927

ISBN 0 521 80390 X hardback

To Luisa, Simona, Barbara, and Martha

Contents

vii

Foreword

Theodosius Dobzhansky (1973) wisely said, "Nothing in biology makes sense except in the light of Evolution." This truism is so often repeated that it is nearly a mantra and, with the complete genomes of many organisms being completed nearly daily, all kinds of people, but especially molecular biologists and informaticists, are rediscovering that truth. And with that discovery they are coming to need to know the tools of the trade that have been under development for nearly forty years. This book is for them in particular but it has much that, except for polymaths, may be useful even to the cognoscenti.

The book has grown out of Drs. Vandamme's and Salemi's annual course in these methods at the Katholieke Universiteit in Leuven, Belgium, where they have produced an exceptional workshop for eight years that does for Europe what a similar outstanding and long-running workshop at Woods Hole did for the United States and Canada. But the latter has not created a book like this.

The coverage is comprehensive. Topics touched upon include databases, multiple alignments, nucleotide substitution models, phylogeny inference methods (such as distance, maximum likelihood, and maximum parsimony), post-phylogenetic information (such as molecular clocks and selection), and useful subsidiary statistical techniques (such as bootstrapping and likelihood ratio tests).

Each of the major sections is written by an expert in the field, and each such section is divided into two major subsections, theory and practice. This permits the novice to proceed with his analysis without having to master the theory. That is, of course, very dangerous in this field where so many methods have different assumptions and the failure of any one of those assumptions (clocklike behavior, all sites equally mutable, all substitutions neutral) can reduce your analysis to rubbish, if untrue, which they frequently are. Still, there are people like that and we may hope that a good text such as this, with its many caveats and generally simple prose, will reduce the published trash.

The material is enhanced by the use of specific examples from which you can see what to expect, and see if you can get the same answer, and then try your own data to see if anything strange has happened. The examples also aid in locating what you need to find in the text.

Another aspect of the book that enhances its utility for the reader is the repeated use of the same three data sets, even by different authors, to illustrate the methods. This increases immensely the value of the exercises. This is especially true when the results from different methods are ostensibly for the same desired end, and one gets to see how they differ and why (or at least to worry about it).

The example data sets used in the book can be downloaded from the book's website [http://www.kuleuven.ac.be/aidslab/phylogenybook.htm]. On the website the reader can also find useful links to the major phylogeny resources on the internet, as well as the results of all the analyses discussed in the text, including phylogenetic trees, unaligned and aligned sequences, and so forth.

It is appropriate to compare this work with others in the general area. The first two are by Weir (1990) and by Waterman (1995). They are both highly theoretical and quite capable of turning off many biologists quickly (although Waterman's book can be highly engaging as in his recounting of the efforts of George Gamow to predict that the genetic code was a commaless code). At the other extreme, Hall (2001) is really simple-minded enough (intentionally so) that a bright senior could easily master the methods. However, the Hall book lacks the comprehensiveness of the Salemi and Vandamme work. Two other good books, Li (1998), and Page and Holmes (1998), are largely theoretical although they make a great effort to make the subject palatable to the biologist who is mathematically challenged. In sum, there is no other book even trying to occupy the niche of this one.

In conclusion, this is a relatively easy-to-use workbook for phylogenetics, especially if the index is properly looked to (I haven't seen it). However, I have to present a strongly worded negative comment. Although tables and figures in the book have titles, many have no legends and many of the remainder have poor legends. For example, numbers normally have dimensions, (such as nucleotide differences per hundred nucleotide positions), that should have been given. Figures and tables should be as self-sufficient as is reasonable. This is not true here. Let us hope this is corrected in the next printing, which I am sure this book will achieve.

Walter M. Fitch
December 27, 2002

REFERENCES

Dobzhansky, Theodosius (1973). Nothing in biology makes sense except in the light of evolution. *American Biology Teacher*, 125–129.

Weir, Bruce S. (1990). *Genetic Data Analysis*, Sunderland, MA: Sinauer Associates.

Waterman, Michael S. (1995). *Introduction to Computational Biolog: Maps, sequences and genomes*. New York: Chapman and Hall.

Hall, Barry G. (2001). *Phylogenetic Trees Made Easy*, Sunderland, MA: Sinauer Associates.

Li, Wen-Hsiung (1997). *Molecular Evolution*, Sunderland, MA: Sinauer Associates.

Page, Roderic D. M. and Edward C., Holmes (1998). *Molecular Evolution: A Phylogenetic Approach*. London: Blackwell Science.

Note: During the writing of this book the alpha release of the new version of PHYLIP, PHYLIP 3.6 has been made available on the PHYLIP web page. All the exercises with PHYLIP refer to version 3.5, but additional exercises covering PHYLIP v3.6 can be found on the book website.

Acknowledgments

We would like to thank the talented researchers who took the time to contribute to the present book and the Flemish "Fonds voor Wetenschappelijk Onderzoek", for their financial support of our research. We are also grateful to Philippe Lemey, Dimitrios Paraskevis, Sonia Van Dooren, Tulio de Oliveira who, on several occasions, gave us helpful suggestions on how to improve the readibility of the manuscript for the non-specialist reader. We thank Professor Walter M. Fitch, who has been very supportive of our work, for his critical reading and help to revise carefully the final version of the book. Finally, we owe a special thanks to all the students of the Workshop on Virus Evolution and Molecular Epidemiology that we have organized at the University of Leuven over the last eight years. Their questions and their suggestions stimulated our writing of this manual. We are also grateful to the Katholieke Universiteit Leuven, Belgium, and to the European Community for financial support to these workshops.

Contributors

Guy Bottu
Belgian EMBnet Node
Brussels, Belgium

Walter Fitch (Foreword)
Ecology and Evolutionary Biology
University of California, Irvine
California, USA

Takashi Gojobori
Center for Information Biology
and DNA Data Bank of Japan
National Institute of Genetics
Mishima, Japan

Arndt von Haeseler
Heinrich-Heine-Universitat Düsseldorf
Institut für Bioinformatik
Düsseldorf, Germany

Des Higgins
Department of Biochemistry
University College
Cork, Ireland

Mary K. Kuhner
Department of Genome Sciences
University of Washington
Washington, USA

Vincent Moulton
Physics and Mathematics Department
Mid Sweden University
Sundsvall, Sweden

Fred R. Opperdoes
C. de Duve Institute of Cellular Pathology
Universite Catholique de Louvain
Brussels, Belgium

Yves Van de Peer
Department of Plant Systems Biology
Flanders Interuniversity Institute for
    Biotechnology (VIB)
Ghent University
Ghent, Belgium

David Posada
Departamento de Bioquímica, Xenética e
    Inmunoloxía
Facultade de Ciencias
Universidade de Vigo
Vigo, Spain

Marc Van Ranst
Rega Institute for Medical Research
Katholieke Universiteit Leuven
Leuven, Belgium

Marco Salemi
Rega Institute for Medical Research
Katholieke Universiteit Leuven
Leuven, Belgium
and
Ecology and Evolutionary Biology
University of California, Irvine
California, USA

Mika Salminen
HIV Laboratory
National Public Health Institute
Department of Infectious Disease
   Epidemiology
Helsinki, Finland

Korbinian Strimmer
Department of Statistics
University of München
München, Germany

Jack Sullivan
Department of Biological Science
University of Idaho
Idaho, USA

Yoshiyuki Suzuki
Center for Information Biology
and DNA Data Bank of Japan
National Institute of Genetics
Mishima, Japan

David L. Swofford
School of Computational Science and
   Information Technology
and Department of Biological Science
Florida State University
Florida, USA

Anne-Mieke Vandamme
Rega Institute for Medical Research
Katholieke Universiteit Leuven
Leuven, Belgium

Xuhua Xia
Biology Department
University of Ottawa
Ottawa, Ontario
Canada

Zheng Xie
Institute of Environmental Protection
Hunan University
China

Basic concepts of molecular evolution

Anne-Mieke Vandamme

1.1 Genetic information

The phenotype of living organisms is always a result of the genetic information
that they carry and pass on to the next generation and of the interaction with the
environment. The genome, carrier of this genetic information, is in most organisms
deoxyribonucleic acid (**DNA**), whereas some viruses have a ribonucleic acid (**RNA**)
genome. Part of the genetic information in DNA is transcribed into RNA, either
mRNA, which acts as a template for **protein** synthesis; rRNA, which together with
ribosomal proteins constitutes the protein translation machinery; or tRNA, which
offers the encoded **amino acid**. The genomic DNA also contains elements, such
as **promotors** and **enhancers**, that orchestrate the proper transcription into RNA.
A large part of the genomic DNA of eukaryotes consists of genetic elements, such
as introns, alu-repeats, the function of which is still not entirely clear. Proteins,
RNA, and to some extent DNA, through their interaction with the environment,
constitute the phenotype of an organism.

DNA is a double helix in which the two **polynucleotide** strands are antiparallelly
oriented, whereas RNA is a single-stranded polynucleotide. The backbone in each
DNA strand consists of deoxyriboses with a phosphodiester linking each 5′ carbon
with the 3′ carbon of the next sugar. In RNA, the sugar moiety is ribose. On
each sugar, one of the following four bases is linked to the 1′ carbon in DNA:
the **purines**, **adenine** (**A**) or **guanine** (**G**); or the **pyrimidines**, **thymine** (**T**), or
cytosine (**C**); in RNA, thymine is replaced by **uracil** (**U**). Hydrogen bonds and
base stacking result in binding of the two DNA strands, with strong (triple) bonds
between G and C, and weak (double) bonds between T/U and A (Figure 1.1).
These hydrogen-bonded pairs are called **complementary**. During DNA duplication
or RNA transcription, DNA or RNA polymerase synthesizes a complementary
5′–3′ strand starting with the lower 3′–5′ DNA strand as template, such that the
genetic information is preserved. This genetic information is represented by a one-
letter code, indicating the 5′–3′ sequential order of the bases in the DNA or RNA

5' - A   C   G   T   G   T - 3'

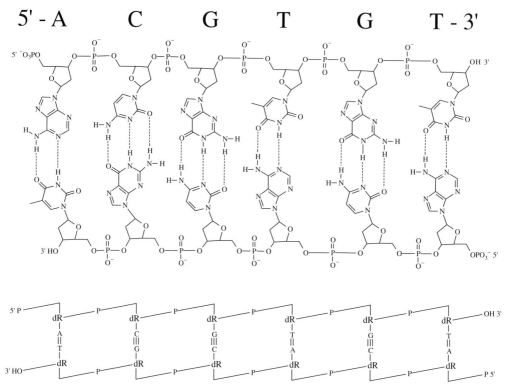

Figure 1.1   Chemical structure of double-stranded DNA. The chemical moieties are indicated as follows: dR, deoxyribose; P, phosphate; G, guanine; T, thymine; A, adenine; and C, cytosine. The strand orientation is represented in a standard way: in the upper strand 5'–3', indicating that the chain starts at the 5' carbon of the first dR, and ends at the 3' carbon of the last dR. The one-letter code of the corresponding genetic information is given on top, and only takes into account the 5'–3' upper strand. (Courtesy of Christophe Pannecouqe.)

(Figure 1.1). A nucleotide sequence is thus represented by a contiguous stretch of the four letters A, G, C, and T/U.

In the RNA strands that encode a protein, each triplet of bases is recognized by the ribosomes as a code for a specific amino acid. This translation results in polymerization of the encoded amino acids into a protein. Amino acids can be represented by a three- or one-letter abbreviation (Table 1.1). An amino-acid sequence is represented by a contiguous stretch of the 21 letters of the one-letter amino-acid abbreviation.

The **genetic code** is universal for all organisms, with only a few exceptions such as the mitochondrial code, and is usually represented as an RNA code because the RNA is the direct template for protein synthesis (Table 1.2). The corresponding DNA code can be easily reconstructed by replacing the U with a T. Each position of the triplet code can be one of four bases; hence, 4^3 or 64 possible triplets encode 20 amino

Table 1.1 Three- and one-letter abbreviations of 20 naturally encoded amino acids

Amino acid	Three-letter abbreviation	One-letter abbreviation
Alanine	Ala	A
Arginine	Arg	R
Asparagine	Asn	N
Aspartic acid	Asp	D
Cysteine	Cys	C
Glutamic acid	Glu	E
Glutamine	Gln	Q
Glycine	Gly	G
Histidine	His	H
Isoleucine	Ile	I
Leucine	Leu	L
Lysine	Lys	K
Methionine	Met	M
Phenylalanine	Phe	F
Proline	Pro	P
Serine	Ser	S
Threonine	Thr	T
Tryptophan	Trp	W
Tyrosine	Tyr	Y
Valine	Val	V

acids (61 *sense* codes) and 3 stop codons (3 *non-sense* codes). The genetic code is said to be degenerated, or redundant, because all amino acids except methionine have more than one possible code. The first codon for methionine *downstream* (or 3′) of the ribosome entry site also acts as the start codon for the translation of a protein. As a result of the triplet code, each contiguous nucleotide stretch has three reading frames in the 5′–3′ direction. The complementary strand encodes three other reading frames. A reading frame that is able to encode a protein starts with a codon for methionine and ends with a stop codon. These reading frames are called *open reading frames* (*ORFs*).

During duplication of the genetic information, the DNA or RNA polymerase can occasionally incorporate a noncomplementary nucleotide. In addition, bases in a DNA strand can be chemically modified due to environmental factors such as UV light or chemical substances. These modified bases can potentially interfere with the synthesis of the complementary strand and thereby also result in a nucleotide incorporation that is not complementary to the original nucleotide. When these changes escape the cellular repair mechanisms, the genetic information is altered, resulting in what is called a *point mutation*. The genetic code has evolved in such a

Table 1.2 Universal codon table

Codon	Amino acid[a]	Codon	Amino acid	Codon	Amino acid	Codon	Amino acid
UUU	Phe	UCU	Ser	UAU	Tyr	UGU	Cys
UUC	Phe	UCC	Ser	UAC	Tyr	UGC	Cys
UUA	Leu	UCA	Ser	UAA	Ter	UGA	Ter
UUG	Leu	UCG	Ser	UAG	Ter	UGG	Trp
CUU	Leu	CCU	Pro	CAU	His	CGU	Arg
CUC	Leu	CCC	Pro	CAC	His	CGC	Arg
CUA	Leu	CCA	Pro	CAA	Gln	CGA	Arg
CUG	Leu	CCG	Pro	CAG	Gln	CGG	Arg
AUU	Ile	ACU	Thr	AAU	Asn	AGU	Ser
AUC	Ile	ACC	Thr	AAC	Asn	AGC	Ser
AUA	Ile	ACA	Thr	AAA	Lys	AGA	Arg
AUG	Met	ACG	Thr	AAG	Lys	AGG	Arg
GUU	Val	GCU	Ala	GAU	Asp	GGU	Gly
GUC	Val	GCC	Ala	GAC	Asp	GGC	Gly
GUA	Val	GCA	Ala	GAA	Glu	GGA	Gly
GUG	Val	GCG	Ala	GAG	Glu	GGG	Gly

[a] Amino acids are indicated by three-letter codes as indicated in Table 1.1.

way that a point mutation at the 3rd position rarely results in an amino-acid change (only in 30% of possible changes). A change at the second position always, and at the 1st position usually (96%), results in an amino-acid change. Mutations that do not result in amino-acid changes are called **silent** or **synonymous mutations.** When a mutation results in the incorporation of a different amino acid, it is called nonsilent or **nonsynonymous.** A site within a coding triplet is said to be **fourfold degenerate** when all possible changes at that site are synonymous; **twofold degenerate** when only two different amino acids are encoded by the four possible nucleotides at that position; and **nondegenerate** when all possible changes alter the encoded amino acid.

Incorporation errors replacing a purine with a purine and a pyrimidine with a pyrimidine are for steric reasons more easily made. The resulting mutations are called **transitions. Transversions,** purine to pyrimidine changes and the reverse, are less likely. When resulting in an amino-acid change, transversions often have a larger impact on the protein than transitions. There are four possible transition errors $(A \rightleftarrows G, C \rightleftarrows T)$ and eight possible transversion errors $(A \rightleftarrows C, A \rightleftarrows T, G \rightleftarrows C, G \rightleftarrows T)$; therefore, if a mutation would occur randomly, a transversion would be two times more likely than a transition. However, in many genes, transitions are twice as more likely to occur than transversions, which is used as default substitution parameter in substitution models that can score transitions and transversions differently (see also Chapter 4).

Single nucleotide changes in a single codon often result in an amino acid with similar properties (e.g., hydrophobic), such that the tertiary structure of the encoded protein is not altered dramatically. Living organisms can therefore tolerate a limited number of nucleotide point mutations in their coding regions. Point mutations in noncoding regions are subject to other constraints, such as conservation of binding places for proteins or conservation of base pairing in RNA tertiary structures.

Errors in duplication of genetic information can also result in the **deletion** or **insertion** of one or more nucleotides, called **indels**. When multiples of three nucleotides are inserted or deleted in coding regions, the reading frame remains intact, but one or more amino acids are inserted or deleted. When a single nucleotide or two nucleotides are inserted or deleted, the reading frame is disturbed and the resulting gene generally codes for an entirely different protein, with different amino acids and a different length than the original gene. Insertions or deletions are therefore rare in coding regions, but rather frequent in noncoding regions. When occurring in coding regions, indels can occasionally change the reading frame of a gene and make another ORF of the same gene accessible. Such mutations can lead to acquisition of new gene functions. Viruses make extensive use of this possibility. They often encode several proteins from a single gene by using overlapping ORFs.

When parts of two different DNA strands are recombined into a single strand, the mutation is called a **recombination**. Recombinations have major effects on the affected gene. **Splicing** is the most common form of recombination. Eukaryotic genes are encoded by coding gene fragments called **exons**, which are separated from each other by **introns**. Joining of the introns occurs in the nucleus at the pre-mRNA level in dedicated spliceosomes. Mutations can result in altered splicing patterns. These usually destroy the gene function, but can occasionally result in the acquisition of a new gene function. Viruses have again used these possibilities extensively. By alternative splicing, sometimes in combination with the use of different reading frames, viruses are able to encode multiple proteins by a single gene. For example, human immunodeficiency virus (HIV) is able to encode two additional regulatory proteins using part of the coding region of the *env* gene by alternative splicing and overlapping reading frames. Another common form of recombination happens during **meiosis**, when recombination occurs between **homologous chromosomes**, shuffling the **alleles** for the next generation. Consequently, recombination has a major contribution to evolution of diploid organisms. In general, these recombinations occur in-between genes. However, if they occur within genes, they have deleterious effects on the affected gene, although sometimes genes with entirely different functions can be created.

A special case of recombination is ***gene duplication***. Gene duplication results in genome enlargement and can involve a single gene, or large genome sections (e.g., chromosome duplication or ***aneuploidy***). They can be partial, involving only gene fragments, or complete, whereby entire genes are duplicated. Genes in which a partial duplication took place, such as domain duplication, can potentially have a greatly altered function. An entirely duplicated gene can evolve independently. After a long history of independent evolution, duplicated genes can eventually acquire a new function. Duplication events have played a major role in the evolution of species. For example, complex body plans were possible due to separate evolution of duplications of the homeobox genes (Carroll, 1995).

1.2   Population dynamics

Mutations in a gene that are passed on to the offspring and that coexist with the original gene result in ***polymorphisms***. At a polymorphic site, two or more variants of a gene circulate in the population simultaneously. Population geneticists deal with the dynamics of the frequency of these polymorphic sites over time. In population dynamics, the evolution of these frequencies is investigated on a small time scale, covering a number of generations. The location in the genome where two variants coexist is called the ***locus***. The different variants are each called an allele. Virus genomes are flexible to genetic changes; RNA viruses can contain many polymorphic sites simultaneously in a single population. HIV, for example, has no single genome, but consists of a swarm of variants called a ***quasispecies*** (Eigen and Biebricher, 1988; Domingo et al., 1997). This is due to the rapid and error-prone replication of RNA viruses. ***Diploid*** organisms always carry two alleles. When both alleles are identical, the organism is ***homozygous*** at that locus; when the organism carries two different alleles, it is ***heterozygous*** at that locus. Heterozygous positions are polymorphic.

Evolution is always a result of changes in ***allele frequencies***, also called ***gene frequencies***. Whereby some alleles are lost over time and other alleles increase their frequency to 100 percent, they become ***fixed*** in the population (Figure 1.2). For RNA viruses, this evolution is reflected in the frequency of a variant in the quasi-species distribution. The long-term evolution of a species results from the successive fixation of particular alleles, which reflects fixation of mutations. The rate at which these mutations are fixed in the population is called the ***evolutionary rate***, or ***fixation rate***, and it is usually expressed as number of nucleotide (or amino acid) changes per site per year. This rate is dependent on the ***mutation rate***, the rate at which mutations arise at the DNA level, usually expressed as number of nucleotide (or amino acid) changes per site per replication cycle, on the ***generation***

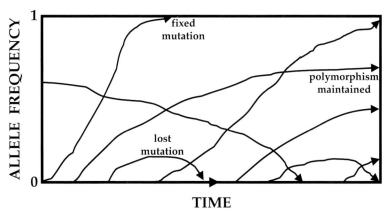

Figure 1.2   Loss or fixation of an allele in a population.

time, the time separating two generations, and on *evolutionary forces*, such as the fitness of the organism carrying the allele or variant, positive and negative selective pressure, population size, genetic drift, reproductive potential, and competition of alleles.

If a particular allele is more fit than its polymorphic allele in a particular environment, it will be subjected to *positive selective pressure*; if it is less fit, it will be subjected to *negative selective pressure*. An allele can be less fit when it is homozygous, but have an advantage as heterozygote. In this case, polymorphism is advantageous and can be selected; this is called *balancing selection*. For example, humans who carry the hemoglobin S allele on both chromosomes suffer from sickle-cell anemia, whereas heterozygotes are to some extent protected against malaria (Allison, 1956). Fitness of a variant is always the result of a particular phenotype of the organism; therefore, in coding regions, selective pressure always acts on mutations that alter function or stability of a gene or the amino-acid sequence encoded by the gene. Synonymous mutations could at first sight be expected to be neutral because they do not result in amino-acid changes. However, this is not always true. For example, synonymous changes can change RNA secondary structure and influence RNA stability; also, they result in the usage of a different tRNA, which may be less abundant. Still, most synonymous substitutions can be considered to be selectively neutral.

Whether a mutation becomes fixed through *deterministic* or *stochastic* forces depends on the *effective population size* (*Ne*) of the organism. This can be defined as the size of an ideal, randomly making population that has the same gene frequency changes as the population being studied. The effective population size can be smaller than the overall *population size* (*N*) because some members of a population may produce no offspring and there may be some level of inbreeding. It is the effective population size that determines the allele frequencies over time. When

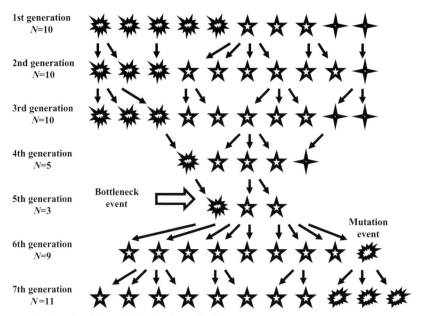

Figure 1.3   Population size (*N*), and the bottleneck effect.

the effective population size varies over multiple generations, the rates of evolution are notably influenced by generations with the smallest effective population sizes. This may be particularly true when population sizes are greatly reduced due to catastrophes or during migrations, etc. (Figure 1.3). These are called bottlenecks.

A ***deterministic model*** assumes that changes in allele frequencies or quasispecies distributions depend solely on the reproductive fitness of the variants in a particular environment and on the environmental conditions. In such a model, the gene frequencies can be predicted if the fitness and environmental conditions are known. In deterministic evolution, changes other than environmental conditions (e.g., chance events) do not influence allele frequencies or quasispecies distributions; therefore, this can only be true if the effective population size is infinite. ***Natural selection***, the effect of positive and negative selective pressure, accounts entirely for the changes in frequencies. When random fluctuations determine in part the allele frequencies, chance events play a role and allele frequencies or quasispecies distributions cannot be entirely predicted. In such a ***stochastic model***, one can only determine the probability of frequencies in the next generation. These probabilities still depend on the reproductive fitness of the variants in a particular environment and on the environmental conditions; however, in this case, chance events – due to the limited ***population size*** – also play a role. Allele frequencies or quasispecies distributions can only be predicted approximately. ***Random genetic drift***, therefore, contributes significantly to changes in frequencies under the stochastic model. The smaller the effective population size, the larger the effect of chance events and the more the mutation rate is determined by ***genetic drift*** rather than by selective pressure.

Evolution is never entirely deterministic or entirely stochastic. Depending on the effective population size, allele frequencies and quasispecies distributions evolve more due to natural selection or random genetic drift. Although genetic changes are always random, an ***adaptive change*** under positive selective pressure will increase its frequency and become fixed after fewer generations than a neutral change, provided the effective population size is large enough. A mutation under negative selective pressure can become fixed due to random genetic drift when it is not entirely deleterious, but this requires more generations than for a neutral change. ***Nonsynonymous*** mutations result in a change in the phenotype of an organism, changing the interaction of that organism with its environment, and are thus subject to selective pressure. Synonymous substitutions, when not under constraints other than their coding potential, are neutral and therefore only become fixed due to random genetic drift. The effect of positive and negative selective pressure can be investigated by comparing the ***synonymous and nonsynonymous substitution rate*** (see also Chapter 11).

Darwin realized that the factors that shaped evolution were an environment with limited resources, inheritable variations among organisms that influenced fitness, competition between organisms, and natural selection. In his view, the survival of the fittest was the result of these factors and the major force behind the origin of species (Darwin, 1859). Only in the twentieth century, after the rediscovery of Mendelian laws, was it realized that the sources of variation on which selection could act were random mutations. In ***neo-Darwinism***, random mutations result in genetic variation, on which natural selection acts as the dominant force in evolution. Advantageous changes become fixed due to positive selective pressure, changes that result in a disadvantage are eliminated, and neutral changes result in polymorphisms that are maintained in a population. Changes in the environment can change the fate of neutral changes into advantageous or disadvantageous changes, resulting in subsequent fixation or elimination. Polymorphism also can be selected through balancing selection. Neo-Darwinism corresponds to a rather deterministic approach. In neo-Darwinism, a gene substitution is always the result of a positive adaptive process. The surviving organisms increase their fitness and become increasingly more adapted to the environment. This is called ***adaptive evolution.***

The ***neutral theory of evolution*** follows a more stochastic approach. Kimura (1983) advocated that the majority of gene substitutions were the result of random fixation of neutral or nearly neutral mutations. Positive selection does operate, but the effective population size is generally so small in comparison with the magnitude of the selective forces that the contribution of positive selection to evolution is too weak to shape the genome. According to the neutral theory, only a small minority of mutations become fixed because of positive selection. Organisms are generally so well adapted to the environment that many nonsynonymous changes are deleterious and, therefore, quickly removed from the population by negative selection. Stochastic events predominate and substitutions, which are fixed mutations, are mainly the results of random genetic drift.

To what extent natural selection or neutral evolution acts on an organism or gene can be investigated with specific tools that are explained in detail in Chapter 11.

1.3   Data used for molecular phylogenetic analysis

To investigate the evolution and relationships among genes and organisms, different kinds of data can be used. The classical way of estimating the relationship between species is by comparing their morphological characters (Linnaeus, 1758). Taxonomy is still mainly based on morphology. The molecular information that is increasingly becoming available, such as nucleotide or amino-acid sequences and restriction fragment length polymorphism (RFLP), also can be used to infer phylogenetic relationships, based on the concepts of natural selection and neutral evolution. Whether the morphological or molecular approach is preferable for any particular evolutionary question has been hotly debated during the last *decennia* (Patterson, 1987). However, the use of molecular data for inferring *phylogenetic trees* has now gained considerable interest among biologists of different disciplines, and it is often used in addition to morphological data to study relationships in further detail. For extinct species, it is difficult or impossible to obtain molecular data, and using morphological characteristics of mummies or fossils is usually the only way to estimate their relationships. However, organisms such as viruses do not leave fossil records. The only way to study their past is through the phylogenetic relationships of existing viruses. In this book, we introduce the concepts, mathematics, and techniques to infer phylogenetic trees from molecular data and, in particular, from nucleotide and amino-acid sequences. Therefore, all applications described in this book restrict themselves to the use of sequence data.

According to the evolutionary theory, all organisms evolved from one common ancestor, going back to the origin of life. Different mechanisms of acquiring variation have led to today's biodiversity. These mechanisms include mutations, duplication of genes, reorganization of genomes, and recombination. Of all these sources, only mutations (i.e., point mutations, insertions, and deletions) are used by the different molecular phylogenetic methods to infer relationships between genes. To perform these evaluations, the similarity of the genes is considered, assuming that they are homologous (i.e., they share a common ancestor). Although it is assumed that all organisms share a common ancestor, over time the similarity in two genes can be eroded such that the sequence data themselves do not carry enough information on the relationship between the two genes and they have accumulated too much variation. Therefore, the term *homology* is used only when the common ancestor is recent enough such that sequence information has retained enough similarity to be used in phylogenetic analysis. Thus, genes are either homologous or they are not. Consequently, there does not exist such an expression as 95% homology; rather, one should speak of 95% *similarity*.

When two sequences are compared, one can always calculate the percentage similarity by counting the amount of identical nucleotides or amino acids, relative to the length of the sequence. This can be done even if the sequences are not homologous. DNA is composed of four different types of residues: A, G, C, and T. If gaps are *not* allowed, on average, 25% of the residues in two randomly chosen aligned sequences would be identical. If gaps *are* allowed, as much as 50% of the residues in two randomly chosen aligned sequences can be identical, resulting in a 50% similarity. For proteins, with 21 different types of codons (i.e., twenty amino acids and one terminator), it can be expected that two random protein sequences – after allowing gaps – can have up to 20% identical residues. In general, the higher the similarity, the more likely that the sequences are homologous.

Taxonomic comparisons show that the genes of closely related species usually only differ from one another by point mutations. These are usually found in the third (i.e., redundant) codon positions of ORFs such that the 3rd codon position has a faster evolutionary rate than the 1st and 2nd codon positions. The redundancy of the genetic code ensures that nucleotide sequences usually evolve more quickly than the proteins they encode. The sequences also may have a few inserted or deleted nucleotides (i.e., indels). Genes of more distantly related species differ by a greater number of changes of the same type. Some genes are conserved more than others, especially those parts encoding, for example, catalytic sites or the core of proteins. Other genes may have little or no similarity. Distantly related species often have discernible sequence relatedness only in the genes that encode enzymes or structural proteins. These similarities, when found, can be very distant and involve only short segments (i.e., motifs) interspersed with large regions with no similarity and of variable length, which indicates that many mutations and indels have occurred since they evolved from their common ancestor. Some of the proteins from distantly related species may have no significant sequence similarity but clearly have similar secondary and tertiary structures. Primary structure is lost more quickly than secondary and tertiary structure during evolutionary change. Thus, differences between closely related species are assessed most sensitively by analysis of their nucleotide sequences. More distant relationships, between families and genera, are best analyzed by comparing amino-acid sequences and may be revealed only by parts of some genes and their encoded proteins (see also Chapter 8).

```
VI557     TTAAATGCATGGGTAAAAGTAGTAGAAGAGAAGGCTTTTAGCCCAGAAGT 50

VI69      TTAAATGCATGGGTAAAGGTGATAGAAGAGAAGGCTTTTAGTCCAGAAGT 50

BZ162     TTAAATGCATGGGTAAAGGTGATAGAAGAGAAGGCTTTTAGCCCAGAAGT 50

VI313     TTGAATGCGTGGGTAAAAGTAATAGAGGAGAAGGCTTTCAGCCCGGAGGT 50

UG268     TTGAATGCATGGGTAAAAGTAATAGAGGAAAAGGCTTTCAGCCCAGAGGT 50

DJ259     TTGAATGCATGGGTAAAAGTAATAGAGGAGAAGGCTTTCAGCCCAGAAGT 50

K112      TTGAATGCATGGGTAAAAGTAATAGAAGAAAAGGCTTTCAGCCCAGAAGT 50

CI20      TTGAATGCATGGGTGAAGGTAATAGAGGAAAAGGCTTTCAGCCCAGAAGT 50

GAG46     TTAAATGCATGGGTAAAAGTAGTAGAAGAAAAGGCTTTCAGCCCAGAAGT 50

LAV       TTAAATGCATGGGTAAAAGTAGTAGAAGAGAAGGCTTTCAGCCCAGAAGT 50

HAN       TTAAATGCATGGGTAAAAGTAGTGGAAGAGAAGGCTTTCAGCCCAGAAGT 50

BZ121     TTAAATGCATGGGTCAAAGTAGTAGAAGAGAAGGCTTTCAGCCCAGAAGT 50

LBV217    TTAAATGCATGGGTAAAAGTAGTAGAAGAAAAGGCCTTCAGTCCAGAAGT 50

HIVBL     TTGAATGCATGGGTAAAAGTAGTAGAAGAAAAGGCCTTCAGTCCAGAAGT 50

VI191     TTGAATGCATGGGTAAAAGTAATAGAAGAAAAGAACTTCAGTCCAGAAGT 50

VI174     TTAAATGCATGGGTAAAGGTGATAGAAGAGAAAGCTTTTAGCCCAGAAGT 50

VI525     TTAAATGCATGGGTAAAAGTAGTAGAAGAAAAGGCTTTTAGCCCAGAAGT 50

Z2Z6      TTGAACGCATGGGTAAAAGTAATAGAAGAAAAGGCTTTCAGCCCAGAAGT 50

NDK       TTGAACGCATGGGTAAAAGTAATAGAAGAAAAGGCCTTCAGCCCGGAAGT 50

VI203     TTGAACGCATGGGTAAAAGTAATAGAGGAAAAGGCTTTCAATCCAGAAGT 50
```

Figure 1.4   Nucleotide sequence alignment of a fragment of the HIV-1 *gag* sequence. This alignment is part of the alignment used to draw the tree in Debyser et al. (1998).

Phylogenetic analysis estimates the relationship between genes or gene fragments by inferring the common history of the genes or gene fragments. To do this, it is essential that homologous sites be compared with each other (i.e., **positional homology**). For this reason, the homologous sequences under investigation are aligned such that homologous sites form columns in the alignment (Figure 1.4). Obtaining the correct alignment is easy for closely related species and can even be done manually using a word processor. The more distantly related the sequences are, the trickier it is to find the best alignment. Therefore, alignments are usually constructed with specific software packages using particular algorithms. This topic is extensively discussed in Chapter 3.

Most algorithms start by comparing the sequence similarity of all sequence pairs, aligning first the two sequences with the highest similarity. The other sequences, in order of similarity, are added progressively. The alignment continues in an iterative fashion, adding gaps where required to achieve positional homology, but gaps are introduced at the same position for all members of each growing cluster. Alignments obtained in this way are optimal for clusters of sequences, as there is no global optimization of the total alignment. When several gaps have been added to clusters of sequences, the total alignment often can be improved by manual editing. Obtaining a good alignment is one of the most crucial steps toward a good phylogenetic tree. When the sequence similarity is so low that an alignment becomes too ambiguous to be confident that homologous sites are aligned correctly, it is better to delete that particular gene fragment from the alignment so as not to distort the phylogenetic tree. Gaps at the beginning and end of a sequence, representing missing sequence data for the shorter sequences, have to be removed to consider equal amounts of data for all sequences. Often, columns in the middle of the sequence with deletions and insertions for the majority of the sequences are also removed from the analysis (see Chapter 3). The best alignment possible is the data that phylogenetic software packages use to construct phylogenetic trees.

For a reliable estimate of the phylogenetic relationship between genes, the entire gene under investigation must have the same history. Therefore, recombination events within the fragment under investigation, which distort this common history, also will distort a phylogenetic tree. Recombination outside the fragment of interest does not disturb the tree; however, knowledge of the recombination event is necessary when the two fragments are both investigated.

Genes originating from a duplication event recent enough to reveal their common ancestry at the nucleotide or amino-acid level are called *paralogous*. Comparing such genes by phylogenetic analysis will result in information on the duplication event. Homologous genes in different species that have started a separate evolution because of the speciation are called *orthologous*. Comparing such genes by phylogenetic analysis will result in information on the speciation event. Therefore, when performing phylogenetic analysis on homologous genes, it is important to know whether the genes are orthologous or paralogous. This prevents making the wrong conclusions on speciation events by comparing paralogous genes instead of orthologous genes.

Evolution of nonhomologous genes under similar selective pressures can result in *parallel* or *convergent evolution*. When two enzymes evolve to have a similar function, the similar functional requirements can result in a similar active site consisting of the same or similar amino acids. This effect can result in the two sequences having higher than expected similarity, which can be mistaken for homology. Other events can result in a higher similarity of two sequences than the similarity expected from their evolutionary history. *Sequence reversals* occur when a mutation reverts back to the original nucleotide; *multiple hits* when a mutation has occurred several times at the same nucleotide, resulting in the same nucleotide at homologous positions in two divergent sequences; and *parallel sub-stitutions* when the same substitution happened in two different lineages. All these events disturb the linear relationship between the time of evolution and sequence divergence. This effect is called *homoplasy* (see Chapter 4).

Presently, sequence information is stored in databases such as the National Center for Biotechnology Information (NCBI), the National Library of Medicine (NLM), the European Molecular Biology Organization (EMBO), and the DNA Database of Japan (DDJ). A search for homologous sequences in individual databases can be done in various ways, based on scoring the similarity between sequences. Some organizations provide a search service via the international computer network (e.g., BLAST). However, no search method is perfect and related sequences may be missed. Information on search engines is provided in Chapter 2.

1.4   What is a phylogenetic tree?

Evolutionary relationships among genes and organisms can be elegantly illustrated by a phylogenetic tree, comparable to a pedigree showing which genes or organisms are most closely related. Phylogenetic trees are described this way because the various diagrams used for depicting these relationships resemble the structure of a tree (Figure 1.5), and the terms referring to the various parts of these diagrams (i.e., *root, stem, branch, node,* and *leaf*) are also reminiscent of trees. *External (terminal) nodes,* the *extant (existing) taxa,* are often called *operational taxonomic units (OTUs)*, a generic term that can represent many types of comparable *taxa* (e.g., a family of organisms, individuals, or virus strains of a single species; a set of related genes; or even gene regions). Similarly, *internal nodes* may be called *hypothetical taxonomic units (HTUs)* to emphasize that they are the hypothetical progenitors of OTUs. A group of taxa that belong to the same branch have a *monophyletic* origin and is called a *cluster.* In Figure 1.5, the taxa A, B, and C form a cluster, have a common ancestor H, and, therefore, are of monophyletic origin. C, D, and E do not form a cluster without including additional strains; thus, they are not of monophyletic origin. The branching pattern – that is, the order of the nodes – is called the *topology* of the tree.

An *unrooted* tree only positions the individual taxa relative to each other without indicating the direction of the evolutionary process. In an unrooted tree, there is no indication of which node represents the ancestor of all OTUs. To indicate the direction of evolution in a tree, it must have a root that leads to the common ancestor of all the OTUs in it (see Figure 1.5). The tree can be *rooted* if one or more of the OTUs form an *outgroup* because they are known as, or are believed to be, the most distantly related of the OTUs (i.e., *outgroup rooting*). The remainder then forms the *ingroup.* The root node is the node that joins the ingroup and the

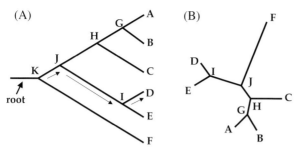

Figure 1.5   Structure of rooted (A) and unrooted (B) phylogenetic trees. Both trees have the same
topology. A rooted tree is usually drawn with the root to the left. A, B, C, D, E, and F are external nodes or OTUs. G, H, I, J, and K are internal nodes or HTUs, with K as root node. The
unrooted tree does not have a root node. The lines between the nodes are branches. The
arrow indicates the direction of evolution in the rooted tree (e.g., from root K to external
node D). The direction of evolution is not known in an unrooted tree.

outgroup; therefore, it must represent the common ancestor of both the outgroup
and the ingroup. It is still possible to assign a root even when it is not known which
OTU to use as the outgroup. Assuming that the rate of evolution in the different
lineages is similar, the root will then lie either at the midpoint of the path joining
the two most dissimilar OTUs, or at the mean point of the paths that join the most
dissimilar OTUs connected through a single edge (i.e., **midpoint rooting**).

When trying to root a tree, do not choose an outgroup that is distantly related
to the ingroup taxa. This may result in serious topological errors because sites may
have become saturated with multiple mutations, by which information may have
been erased. Also, do not choose an outgroup that is too closely related to the taxa
in question; in this case, it may not be a true outgroup. The use of more than one
outgroup generally improves the estimate of tree topology. As noticed previously,
midpoint rooting could be a good alternative when no outgroups are available, but
only in case of approximately equal evolutionary rates over all branches of the tree.

Various styles are used to depict phylogenetic trees. Figure 1.6 demonstrates the
same tree as in Figure 1.5, but in a different style. Branches at internal nodes can be
rotated without altering the topology of a tree. Both trees in Figure 1.6 have identical
topologies. Compared with tree (A), tree (B) was rotated at nodes J and H.

Phylogenetic trees illustrate the relationship among the sequences aligned;
therefore, they are always **gene trees**. Whether these gene trees can be interpreted as
representing the relationship among species depends on whether the genes provided
to the alignment are orthologous or paralogous genes. When the ancestral gene A
is duplicated into A1 and A2 within the same species, then the relationship between
A1 and A2 will give information on the duplication event. Suppose the speciation
into species C and D – with C1 and D1 being the descendant of gene A1, and C2
and D2 descendant from A2 – occurs after the gene duplication. Comparing C1

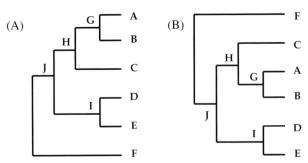

Figure 1.6 This is the same tree as in Figure 1.5, but in a different style. Both trees (A) and (B) have identical topologies, with some of the internal nodes rotated.

with D2 (or C2 with D1) will give information on the duplication event, whereas comparing C1 with D1 (or C2 with D2) will give information on the speciation event and the tree can be considered a ***species tree***. When dealing with a gene that has polymorphic sites in that species, the nodes in the gene tree never really indicate the speciation event. Because some sequence variation existed before speciation – represented by the different alleles (or quasispecies variation for RNA viruses) – the gene tree is a population tree and the nodes represent the separation of the different alleles, which precede speciation. Alternatively, some alleles may have become extinct after speciation and the separation of the different alleles may follow speciation.

The ***coalescence time*** is the time when the ***most recent common ancestor*** (***MRCA***) of the extant alleles still existed. When trying to acquire information on the origin of a species population by analyzing the sequence variability of different alleles of a particular gene, the coalescence time depends on the extinction of alleles after speciation. In Figure 1.3, individuals of the 7th generation have one common ancestor in the 4th generation; therefore, the coalescence time is later than the first generation. Thus, when the effective population size is small and alleles are lost after speciation, the coalescence time for the different alleles within a species in a population tree is later than the speciation time. For example, the coalescence time of human mitochondrial DNA, which is inherited through the female line, is calculated to be around two hundred thousand years ago (Vigilant et al., 1991; Ingman et al., 2000). The coalescence time for the Y chromosome is around seventy thousand years ago (Dorit et al., 1995; Thomson et al., 2000), yet human speciation was not at a different time for women than for men. The estimated dates are the coalescence times for the two different genes analyzed, whose polymorphic origins do not necessarily have to be simultaneous. When the origin of polymorphism predated speciation, the coalescence time of the existing alleles of a species can even precede speciation. Whether the coalescence time of existing alleles precedes or follows speciation is dependent on the effective population size.

To estimate coalescence times of genes, alleles, or quasispecies variants, a specific assumption – that sequence divergence increases over time – always has to be made. Time runs only in one evolutionary direction; therefore, even if the morphology of a species has not changed, its sequence divergence will almost always have increased. The time since speciation is related to the extent of sequence divergence in a species. The easiest way to calculate divergence times is to assume that sequence divergence accumulates linearly over time; this is called a ***molecular clock***. When the molecular clock holds, all lineages in the tree have accumulated substitutions at the same rate; the evolutionary rate is constant (see also Chapter 10). However, the evolutionary rate is dependent on many factors, including the metabolic rate in a species, the generation time, bottleneck events, and selective pressure. Therefore, an absolute molecular clock does not exist. There is always some difference in evolutionary rate along the branches of a tree; this is especially true for viruses that have high replication rates, change hosts – and thus selective pressure environment – and frequently go through bottleneck events. However, statistical tests can be performed, as is explained in Chapter 10, that provide an idea of how different the evolutionary rates along the branches in a tree are from a uniform rate. In many situations, these differences are so small that a molecular clock can be safely assumed to calculate coalescence times (or divergence times when starting from ancestral nodes) in a tree. To apply a molecular clock to a tree, the ancestor has to be known; that is, the direction of time in a tree has to be known and the tree has to be rooted.

Such a tree – in which the direction of time is known, the molecular clock holds, and the taxa are organisms – represents a ***cladogram***. A cladogram maps the ancestor–descendant relationship between organisms or groups of organisms. A ***phenogram*** simply represents the relationships among a group of taxa. Because of the effects, of a population tree, a species tree, and a gene tree, and because an absolute molecular clock does not exist, a cladogram will never be identical to a phenogram. Although the topology can be identical, the branch lengths may differ slightly. Cladograms can be drawn based on morphological characters of fossils, and the branches can be calculated from independent dating methods such as radiocarbon dating. A cladogram also can be based on a phylogenetic tree; a phenogram is always based on a phylogenetic tree.

1.5   Methods to infer phylogenetic trees

Reconstructing the phylogeny from gene or amino-acid sequence alignments is, unfortunately, not as straightforward as one might hope, and it is rarely possible to verify that one has arrived at the true conclusion. The reconstruction results in an inferred phylogenetic tree, which may or may not differ from the true phylogenetic tree. There are no uniquely correct methods for inferring phylogenies, and many methods are used.

The methods for constructing phylogenetic trees from molecular data can be grouped first according to whether the method uses **discrete character** states or a **distance matrix** of pairwise dissimilarities, and second according to whether the method clusters OTUs stepwise, resulting in only one best tree, or considers all theoretically possible trees.

Character-state methods can use any set of discrete characters, such as morphological characters, physiological properties, restriction maps, or sequence data. When comparing sequences, each position in the aligned sequences is a "character," and the nucleotides and amino acids at that position are the "states." All characters are analyzed separately and usually independently from each other. Character-state methods retain the original character status of the taxa and, therefore, can be used to reconstruct the character state of ancestral nodes.

In contrast, distance-matrix methods start by calculating some measure of the dissimilarity of each pair of OTUs to produce a pairwise distance matrix, and then estimate the phylogenetic relationships of the OTUs from that matrix. These methods seem particularly well suited for analyzing sequence data. Although it is possible to calculate distances directly from pairwise aligned sequences, more consistent results are obtained when all sequences are aligned. Distance-matrix methods allow for scoring multiple hits. When two sequences are divergent, it is likely that at a certain position, two or more consecutive mutations have occurred. These multiple events result in two sequences being more distantly related than can be deduced from the percentage difference in sequence. The more divergent the sequences, the bigger the impact of multiple events. Mathematical models allow for correcting the percentage difference between sequences. This is called the **genetic** or **evolutionary distance**, which is always bigger than the distance calculated by direct comparison of sequences (see Chapter 4). Distance methods discard the original character state of the taxa; as a result, the information to reconstruct character states of ancestral nodes is lost. The major advantage of distance methods is that they are much less computer intensive, which is important when many taxa have to be compared.

Exhaustive-search methods are tree-evaluation methods that examine the theoretically possible tree topologies for a given number of taxa, using certain criteria to choose the best one. In particular, maximum-likelihood methods (discussed later in this chapter and in Chapter 6) share the main advantage of producing a large number of different trees and estimate for each tree the conditional probability that it represents the true phylogeny, given the data (i.e., the aligned sequences) and a specific evolutionary model (see also Chapters 6 and 7). This allows the investigator to compare the support for the best tree with the support for the second

Table 1.3 Number of possible rooted and unrooted trees for up to 10 OTUs

Number of OTUs	Number of rooted trees	Number of unrooted trees
2	1	1
3	3	1
4	15	3
5	105	15
6	954	105
7	10,395	954
8	135,135	10,395
9	2,027,025	135,135
10	34,459,425	2,027,025

best, and to estimate the confidence in the tree obtained. Unfortunately, the number of possible trees, and thus the computing time, grows quickly as the number of taxa increases; the number of bifurcated rooted trees for n OTUs is given by $(2n - 3)!/(2^{n-2}(n - 2))!$ (Table 1.3). This means that for a data set of more than 10 OTUs, only a subset of possible trees can be examined. Thus, various strategies are used to search the "tree space," but there is no algorithm that guarantees that the best possible tree was examined.

The ***stepwise-clustering*** methods avoid this problem by examining local subtrees first. They are tree-construction methods because they follow specific algorithms to construct a single tree. Typically, the two most closely related OTUs are combined to form a cluster. The cluster is then treated like a single OTU representing the ancestor of the OTUs it replaces; therefore, the complexity of the data set is reduced by one OTU. This process is repeated, clustering the next closest related OTUs, until all OTUs are combined. The various stepwise-clustering algorithms differ in their methods of determining the relationship of OTUs and in combining OTUs into clusters. They are usually fast and can accommodate large numbers of OTUs. Because they produce only one tree, the confidence estimators of the exhaustive search methods are not available, although various other statistical methods have been developed to estimate the confidence in the correctness of a tree obtained. The majority of distance-matrix methods use stepwise clustering to compute the "best" tree, whereas most character-state methods adopt the exhaustive-search approach.

Table 1.4 lists the currently most used phylogenetic tree construction and tree-analysis methods, classified according to the strategy used: character state or distance matrix, exhaustive search or stepwise clustering. All methods use particular evolutionary assumptions, which do not necessarily apply to the data set. Therefore, it is important to realize which assumptions were made when evaluating the best tree given by each method. The methods themselves and their assumptions are extensively explained in the following chapters.

Table 1.4 Most used phylogenetic analysis methods and their strategies

	Exhaustive search	Stepwise clustering
Character State	Maximum parsimony (MP)	
	Maximum likelihood (ML)	
Distance Matrix	Fitch-Margoliash	UPGMA
		Neighbor-joining (NJ)

Maximum parsimony (*MP*) aims to find the tree topology for a set of aligned sequences that can be explained with the smallest number of character changes (i.e., mutations). The MP algorithm starts by considering a tree with a particular topology. It then infers the minimum number of character changes required to explain all nodes of the tree at every sequence position. Another topology is then evaluated. When all reasonable topologies have been evaluated, the tree that requires the minimum number of changes is chosen as the best tree (see Chapter 7).

Maximum likelihood (*ML*) is similar to the MP method in that it examines every reasonable tree topology and evaluates the support for each by examining every sequence position. In principle, the ML algorithm calculates the probability of expecting each possible nucleotide (amino acid) in the ancestral (internal) nodes and infers the likelihood of the tree structure from these probabilities. The likelihood of all reasonable tree topologies is searched in this way, and the most likely tree is chosen as the best tree. The actual process is complex, especially because different tree topologies require different mathematical treatments, so it is computationally demanding (see Chapters 6 and 7).

UPGMA is the acronym for *unweighted pair group method with arithmetic means*. This is probably the oldest and simplest method used for reconstructing phylogenetic trees from distance data. Clustering is done by searching for the smallest value in the pairwise distance matrix. The newly formed cluster replaces the OTUs it represents in the distance matrix. The distances between the newly formed cluster and each remaining OTU are then calculated. This process is repeated until all OTUs are clustered. In UPGMA, the distance of the newly formed cluster is the average of the distances of the original OTUs. This process of averaging assumes that the evolutionary rate from the node of the two clustered OTUs to each of the original OTUs is identical. The whole process of clustering thus assumes that the evolutionary rate is the same in all branches, meaning that no one strain has accumulated mutations faster than any other strain. This assumption is almost never true. Therefore, UPGMA tends to give the wrong tree when evolutionary rates are different along the branches (see also Chapter 5).

The *neighbor-joining* (*NJ*) *method* constructs the tree by sequentially finding pairs of neighbors, which are the pairs of OTUs connected by a single interior node. The clustering method used by this algorithm is quite different from the one described previously, because it does not attempt to cluster the most closely related OTUs, but rather minimizes the length of all internal branches and thus the length of the entire tree. So it can be regarded as parsimony applied to distance data. The NJ algorithm starts by assuming a bush-like tree that has no internal branches. In the first step, it introduces the first internal branch and calculates the length of the resulting tree. The algorithm sequentially connects every possible OTU pair and finally joins the OTU pair that yields the shortest tree. The length of a branch joining a pair of neighbors, X and Y, to their adjacent node is based on the average distance between all OTUs and X for the branch to X, and all OTUs and Y for the branch to Y, subtracting the average distances of all remaining OTU pairs. This process is then repeated, always joining two OTUs (neighbors) by introducing the shortest possible internal branch (see also Chapter 5).

The ***Fitch-Margoliash method*** is a distance-matrix method that evaluates all possible trees for the shortest overall branch length, using a specific algorithm that considers the pairwise distances.

There have been some reports of comparisons of different sets of algorithms using different sets of data. However, it is difficult to decide which method or methods are best, perhaps because different data sets seem to favor different algorithms. The reason is that different tree-making algorithms are based on different assumptions. If these assumptions are met by the data, the algorithm will perform well. The use of statistical methods helps to estimate the reliability of certain clusters (i.e., tree topologies) and/or branch lengths. However, they are also dependent on the phylogeny method used and suffer from the same bias. There is no evidence that any one method is superior to others, so it is advisable to employ more than one method with each set of data. The ML method intrinsically estimates the standard error on the branch length and therefore already gives some statistical support for each branch length and for the entire tree. The most used tree evaluation method is the ***bootstrapping resampling method***, which is explained in detail in the next chapters.

1.6   Is evolution always tree-like?

The algorithms discussed in the previous section usually generate ***strictly bifurcating trees*** (i.e., trees where any internal node is always connected to only three other branches; see Figure 1.5). This is the standard way of representing evolutionary relationships among organisms through a phylogenetic tree, and it presumes that the underlying evolutionary processes are therefore ***bifurcating*** (i.e., during the course of evolution, any ancestral sequence [internal nodes in the tree] can give rise to only two separate lineages [leaves]). However, there are phenomena in nature, such as the explosive evolutionary radiation of HIV or HCV, that might be best modeled by a ***multifurcating tree***, such as the one shown in Figure 1.7A, or by a ***nonstrictly bifurcating tree*** that allows for some multifurcations, such as the one

(A)   (B)

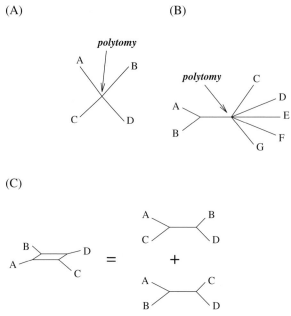

(C)

Figure 1.7   Nonbifurcating trees and networks; arrows indicate polytomy. (A) Star-like (or multifurcating) tree. (B) Tree with an internal politomy. (C) Networks representation: the network on
the left is one way of displaying simultaneously the two conflicting tree topologies on the
right.

in Figure 1.7B. Multifurcations on a phylogenetic tree are also known as ***polytomies***,
and can be distinguished as ***hard polytomies*** and ***soft polytomies***. Hard polytomies
are meant to represent explosive radiation in which a single common ancestor gave
rise to multiple distinct lineages at the same time. Hard polytomies are difficult
to prove and it is even questionable as to whether they actually do occur (cf. Li,
1997, and Page and Holmes, 1998, for a detailed discussion). Soft polytomies, in
contrast, represent unresolved tree topologies. They reflect the uncertainty about
which branching pattern precisely describes the data. Finally, there are situations –
for example, in the case of recombination – in which the data seem to support
equally well two or more different tree topologies. In such cases, the sequences under investigation may be better represented by a network, such as the one depicted
in Figure 1.7C. These topics are covered in Chapters 6, 12, and 14.

REFERENCES

Allison, A. C. (1956). Sickle cells and evolution. *Scientific American.*
Carroll, S. B. (1995). Homeotic genes and the evolution of arthropods and chordates. *Nature,*
    376, 479–485.

Darwin, C. (1859). *On the Origin of Species by Means of Natural Selection.* London: Murray.

Debyser, Z., E. Van Wijngaerden, K. Van Laethem, K. Beuselinck, M. Reynders, E. De Clercq, J. Desmyter, and A.-M. Vandamme (1998). Failure to quantify viral load with two of the three commercial methods in a pregnant woman harbouring an HIV-1 subtype G strain. *AIDS Research and Human Retroviruses*, 14, 453–459.

Domingo, E., L. Menendezarias, and J. J. Holland (1997). RNA virus fitness. *Review of Medical Virology*, 7, 87–96.

Dorit, R. L., H. Akashi, and W. Gilbert (1995). Absence of polymorphism at the ZFY locus on the human Y chromosome. *Science*, 268, 1183–1185.

Eigen, M. and C. Biebricher (1988). Sequence space and quasispecies distribution. In: *RNA Genetics*, vol. 3, eds. E. Domingo, J. J. Holland, and P. Ahlquist, pp. 211–245. CRC Press, Boca Raton, FL.

Ingman, M., H. Kaessmann, S. Paabo, and U. Gyllensten (2000). Mitochondrial genome variation and the origin of modern humans. *Nature*, 408, 708–713.

Kimura, M. (1983). *The Neutral Theory of Molecular Evolution.* Cambridge: Cambridge University Press.

Li, W.-H. (1997). *Molecular Evolution.* Sunderland: Sinauer Associates.

Linnaeus, C. (1758). *Systema Naturae*, 10th ed. Stockholm.

Page, R. D. M. and E. C. Holmes (1998). *Molecular Evolution: A Phylogenetic Approach.* Oxford: Blackwell Science.

Patterson, C., ed. (1987). *Molecules and Morphology in Evolution: Conflict or Compromise?* Cambridge: Cambridge University Press.

Thomson, R., J. K. Pritchard, P. Shen, P. J. Oefner, and M. W. Feldman (2000). Recent common ancestry of human Y chromosomes: Evidence from DNA sequence data. *Proceedings of the National Academy of Sciences of the USA*, 97(13), 7360–7365.

Vigilant, L., M. Stoneking, H. Harpending, K. Hawkes, and A. C. Wilson (1991). African populations and the evolution of human mitochondrial DNA. *Science*, 253(5027), 1503–1507.

2

Sequence databases

THEORY

Guy Bottu and Marc Van Ranst

Phylogenetic analyses are often based on sequence data accumulated by many investigators. Faced with a rapid increase in the number of available sequences, it is not possible to rely on the printed literature; thus, scientists had to turn to digitalized databases. Databases are an essential feature in bioinformatics: they serve as information storage and retrieval locations; however, modern databases come loaded with powerful query tools and are cross-referenced to other databases. This section provides an overview of the available public sequence databases and how to consult them. A list of the databases' uniform resource locators (URLs) discussed in this section is in Box 2.1.

2.1 General nucleic acid sequence databases

There are parallel efforts (probably due to geopolitical reasons) in Europe, the United States, and Japan to maintain public databases with all the nucleic-acid sequences that have been published, as follows:

EMBL (European Molecular Biology Laboratory) database, maintained at EMBL-EBI (European Bioinformatics Institute, Hinxton, UK), three monthly major releases, daily updates.
GenBank, maintained at NCBI (National Center for Biotechnology Information, Bethesda, Maryland, USA), bimonthly major release, daily updates.
DDBJ (DNA Data Bank of Japan), maintained at NIG/CIB (Mishima, Japan), three monthly major releases, daily updates.

In the 1980s, the data bank curators scanned the printed literature for novel sequences, but today the sequences are submitted directly by the authors through World Wide Web (WWW) submission tools. There is an agreement between the

---

Box 2.1 URLs of the major sequence, protein, and structural databases

---

Blocks and Blocks+ : http://blocks.fhcrc.org/
DBCAT "databank of databanks" : http://www.infobiogen.fr/services/dbcat/
DDBJ : http://www.ddbj.nig.ac.jp/
DOMO : http://www.infobiogen.fr/srs6bin/cgi-bin/wgetz?-page+LibInfo+-
   id+3W7Bv1ILMcp+-lib+DOMO
EMBL : http://www.ebi.ac.uk/embl/
Entrez : http://www.ncbi.nlm.nih.gov/Entrez/
FSSP : http://www2.ebi.ac.uk/dali/fssp/
GenBank : http://www.ncbi.nlm.nih.gov/Genbank/
NRL_3D : http://www-nbrf.georgetown.edu/pirwww/dbinfo/nrl3d.html
HIV database : http://hiv-web.lanl.gov/
HPVDB : http://hpv-web.lanl.gov/
HSSP : http://www.sander.ebi.ac.uk/hssp/
IMGT/LIGM : http://imgt.cines.fr:8014/
IMGT/HLA : http://www.ebi.ac.uk/imgt/hla/
InterPro : http://www.ebi.ac.uk/interpro/
MEDLINE : http://www.ncbi.nlm.nih.gov/PubMed/
OWL : http://www.bioinf.man.ac.uk/dbbrowser/OWL/
PDB : http://www.rcsb.org/pdb/
PIR : http://www-nbrf.georgetown.edu/
Pfam : http://www.sanger.ac.uk/Pfam/
PRF/SEQDB : http://www.prf.or.jp/en/os.htm
PRINTS : http://www.bioinf.man.ac.uk/dbbrowser/PRINTS/PRINTS.html
ProDom : http://protein.toulouse.inra.fr/prodom.html
PROSITE : http://www.expasy.ch/prosite/prosite.html
SRS "mother" server : http://srs.ebi.ac.uk/
SWISS-PROT and TrEMBL at EBI : http://www.ebi.ac.uk/swissprot/
SWISS-PROT and TrEMBL at SIB : http://www.expasy.ch/sprot/

---

curators of the three data banks to cross-submit the submitted sequences to each other.

The amount of information in these public databases is staggering. In May 2001, more than 12 billion basepairs were in GenBank. These databases are growing at an enormous rate (Figure 2.1).

The data banks contain ribonucleic acid (RNA) and deoxyribonucleic acid (DNA) sequences but, by convention, a sequence is always written with Ts and not Us; often, sequence-analysis software does not make a distinction between U and T. Modified bases are replaced by their "parent" base ACGT, but the modification is mentioned in the documentation that accompanies the sequence in the data bank.

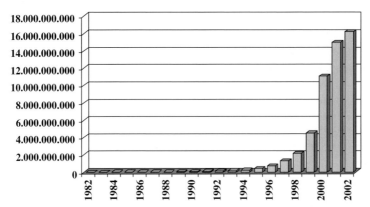

Figure 2.1   Growth of the number of nucleotides in the public databases from January 1982 to March 2002.

Each sequence in a data bank is identified by the following:

entry name, locus name, or identifier (ID): A sequence has one ID, which is unique in the data bank. The ID was originally designed to be more or less mnemonic, but this is not true anymore. Note that the ID can change from release to release.

accession number (AC): An AC is unique in the data bank, but a sequence can have more than one AC (i.e., when several sequences are merged into one, the new sequence inherits the old ACs). The AC remains constant between releases, and there is an agreement between the three data banks to give the same AC to the same sequence. The AC is thus very useful for retrieving a sequence.

version number: The version number is derived from the AC and a number that is incremented each time the sequence is modified. Therefore, the version number is useful if it is necessary to find the exact sequence that was used for a certain study. Note, however, that the version number was introduced in the beginning of 1999 and does not cover previous years.

2.2   General protein sequence databases

A similar parallel effort exists for proteins as follows:

SWISS-PROT, maintained at the University of Geneva/Swiss Institute of Bioinformatics (Geneva, Switzerland) by Professor Amos Bairoch's group in collaboration with the EMBL-EBI (Hinxton, UK). SWISS-PROT is updated weekly, and can be obtained or used freely by academic users, but (in theory) is not free for commercial users.

PIR (Protein Information Resource), maintained at the NBRF (National Biomedical Research Foundation, Washington D.C., USA) in collaboration with the University of Tokyo/JIPID (Japanese International Protein Information Database; Tokyo, Japan) and the MIPS (Martinsried Institute for Protein Sequences; Martinsried, Germany), three-monthly major release.

PRF/SEQDB (Protein Resource Foundation), maintained at the PRF (Osaka, Japan), bimonthly major release.

PDB (Protein Data Bank), maintained by the University of New Jersey, the San Diego Supercomputer Center at the University of California, and the National Institute of Standards and Technology. It is the largest data bank of three-dimensional (3-D) biological macromolecular structure data.

Most protein sequences are obtained by translation of open reading frames identified in the DNA, but some are still the result of genuine protein sequencing or are derived from 3-D structures. The method used to obtain the protein sequence is mentioned in the documentation that accompanies the sequence.

The following data banks are generated automatically by a computer and contain the translations of coding sequences (CDS) from a certain DNA data bank:

TrEMBL: Collection of CDS from EMBL, with the exclusion of sequences already in SWISS-PROT, maintained at the EMBL-EBI (Hinxton, UK) in collaboration with the University of Geneva/Swiss Institute of Bioinformatics (Geneva, Switzerland), three monthly major releases, weekly updates. There are two sections to this data bank:

  SP-TrEMBL: Sequences destined to be entered in a future release of SWISS-PROT; they already have as ID+AC their future SWISS-PROT AC.

  REM-TrEMBL: Sequences not meant to be entered in SWISS-PROT (e.g., minor variants of the same sequence, small fragments, "doubtful" sequences, synthetic sequences, idiotypes).

GenPept: Collection of CDS from `GenBank`, maintained at the NCI-FCRDC (National Cancer Institute; Frederick Cancer Research and Development Center; Frederick, Maryland, USA), two monthly major releases, daily updates.

There are differences in scope and quality between these data banks. SWISS-PROT is a highly curated database that contains excellent documentation. SWISS-PROT systematically merges variants and fragments into a single entry, but is greatly lagging behind the growth of the DNA data banks. PIR contains more sequences, including numerous "really sequenced" oligopeptides, but is not that tightly curated. The "automatic" data banks such as TrEMBL and GenPept are even larger, but contain little documentation and sometimes conceptual translations that are not actually found in nature.

2.3   Nonredundant sequence databases

None of the existing general databases is really complete. To make up for this, efforts have been undertaken to make composite databases. Redundancy is avoided by comparison of sequences, with the elimination of exact duplicates and sequences that differ only trivially. The strict redundancy criteria render these composite databases relatively small and, therefore, efficient in database searches.

The NCBI maintains on a daily basis an up-to-date nucleic-acid data bank (Gen-Bank + EMBL + DDBJ) and a protein data bank (PDB + SWISS-PROT + PIR + PRF + CDS). These data banks can be consulted online using ENTREZ (discussed later in this subsection).

OWL is a nonredundant protein data bank based on SWISS-PROT, with the addition of sequences originating from NBRF/PIR, GenPept, and the Brookhaven PDB 3-D structural database (NRL3D). OWL is maintained at the University of Leeds (UK) in collaboration with A. Bleasby from the HGMP-RC (Human Genome Mapping Project-Resource Centre; Hinxton, UK). Unfortunately, it is available only as major three monthly releases, not as regular updates.

2.4 Specialized sequence databases

In addition to the general databases, more than 50 specialized databases exist. Some of these are made starting from the general databases, whereas some accept author submissions and can contain sequences that are not yet found in the general databases. They offer one or more of the following advantages:

- The database forms a well-defined set of sequences, which is often difficult to extract from the general databases. Searching the specialized database instead of the general database will provide a list of sequences that is less "polluted" by "background noise" of unwanted sequences and that will consume less computing time.
- The specialized data bank is often nonredundant ("cleaned") and does not contain "doubles."
- The data fields *definition* or *keywords* are sometimes (better) standardized, which allows the user to find the relevant sequences in a single keyword search. In a data bank without standardization, searching must often be repeated with different keywords (discussed later in this subsection).
- The documentation is often more extensive than that found in the general data banks.

Following are examples of specialized databases:

HIV Databases: DNA sequences from HIV and SIV, maintained at LANL (Los Alamos National Laboratory; Los Alamos, New Mexico, USA), yearly major release.

HPVSD: DNA and protein sequences from human and animal papillomaviruses and cellular proteins affected by papilloma viral infection, maintained at LANL (Los Alamos, New Mexico, USA), yearly major release.

IMGT (ImMunoGeneTics): A collection of data banks of immunological interest.

    IMGT/LIGM (Laboratoire d'ImmunoGénétique Moléculaire): Genes of immuno-globulins and T-cell receptors, extracted from EMBL with improved documentation, maintained at the University of Montpellier (France).

IMGT/HLA (human histocompatibility locus A): Genes for the major human histo-compatibility locus, maintained at the Anthony Nolan Research Institute (London, UK).

NRL_3D: Subset from PIR with protein sequences for which the 3-D structure is known and documented in the PDB, maintained at the NBRF (Washington D.C., USA), three-monthly major release.

2.5   Databases with aligned protein sequences

Numerous databases with proteins are grouped into (sub)families that are already prealigned. The alignments have been made using different algorithms, and therefore contain alignments with different lengths and different numbers of proteins. The most important are as follows:

Blocks: Contains *local alignments without gaps*, made from protein sequences in SWISS-PROT + SP-TrEMBL, as grouped in the PROSITE-related entries of InterPro, using the BlockMaker software, maintained at Fred Hutchinson Cancer Research Center (Seattle, Washington, USA).

Blocks +: Nonredundant data bank, Blocks + PRINTS + proteins grouped as families of homolog (ProDom + Pfam + DOMO) aligned with BlockMaker.

DOMO (Protein Domain Database): Database of families of **homologous** domains. Its major fields provide information about the related proteins, their functional families, domain decomposition, multiple-sequence alignment, conserved residues, and evolutionary classification tree.

PRINTS (Protein Motif Fingerprint Database): Contains *local alignments without gaps*, made from protein sequences in SWISS-PROT + SP-TrEMBL, using the ADPS package, starting with handmade "seed" alignment, followed by iterative database searches, maintained at the University of Manchester (UK).

ProDom: Contains *local alignments with gaps*, made from protein sequences in SWISS-PROT + SP-TrEMBL, using PSI-BLAST, partly starting with the "seed" alignments from Pfam-A, maintained at INRA (French Institute for Agronomy Research; Toulouse, France) in collaboration with the Sanger Centre (Hinxton, UK).

Pfam: Contains *local alignments with gaps*, made from protein sequences in SWISS-PROT + SP-TrEMBL, maintained at the Sanger Centre (Hinxton, UK).

HSSP (Homology-derived Secondary Structure of Proteins): Contains *global alignments*, made by searching proteins from PDB (with known 3-D) against SWISS-PROT + SP-TrEMBL using BLAST, followed by selection of related sequences according to Schneider-Sander criterion and alignment with MaxHom, maintained at EMBL-EBI (Hinxton, UK).

FSSP (Fold-derived Structurally Similar Proteins): Contains *global alignments* of proteins from PDB, made with the Dali structural comparison software, maintained at EMBL-EBI (Hinxton, UK).

2.6   Database documentation search

Informatized databases are maintained using specialized software. The most popular software packages for larger systems are Sybase and ORACLE. Data-bank management software allows the data-bank manager to define the data bank's structure,
allows the curator to enter or modify entries, and allows the user to consult the data.
Those who want to install a collection of databases on their computer might face a
problem: not only is database management software costly, but also more than one
program will often be needed because the managers of the different database main-
tenance sites do not all choose the same package. Therefore, databases are often
copied and distributed, not in their "native" format, but under the form of simple
text, so-called flat files. This simple text is coded in ASCII (American Standard
Code for Information Interchange), which has de facto become the most popular
standard for data representation and diffusion.

2.6.1   Text-string searching

The simplest method to search a database is to scan the text for words or, more
generally, for strings of characters. This is equivalent to reading the complete contents of a book to find the desired word. In the case of sequence databases, the user
normally scans only the documentation that accompanies the sequences, not the
sequences themselves. Several tools are available. The database file can be opened
with a text editor or text processor that has a built-in search function. Users of
computers with UNIX (or Linux) should be aware of the useful commands **more**
and **grep**. GCG has a program **stringsearch** and EMBOSS **textsearch**. The major
drawback of this simple type of search is the large consumption of computer time.
Not only does it take a lot of time to search a data bank, but also this time increases
linearly with its size. However, the method has a virtue that can sometimes be useful: any string of characters can be sought (e.g., "isolate SBL-1959"). Note that only
what has been written in the text in the first place can be found. Often data banks
are not very standardized. For example, both "HIV-1" and "HIV1" can be found.
Worse are typographical errors, such as "psuedogene" instead of "pseudogene."
Which strings of characters to include in the search should be determined carefully.
To be reasonably sure that all relevant entries are found, it is often necessary to
repeat the search with different words or minor variants of the same word.

2.6.2   Searching by index

Data-bank management software allows searching the data with the help of alphabetic indexes created from words found in the documentation. This search is much
faster than a simple text scan. Furthermore, when a data bank doubles in size, the
time needed for a text scan also doubles; however, when an index doubles in size,

the time needed to search a keyword by the "cut-in-half" method increases only slightly (on average, it will take one supplementary step).

There are tools that generate indexes from a simple text and provide the user with the access functionality of a database-management software. A simple example is **WAISindex**, which is not a creation of the bioinformatics community but was developed for the Internet database consultation tool WAIS (Wide Area Information Service). Few WAIS servers still exist, but WAISindex is still used for local data banks or is "hidden" behind WWW servers. WAISindex generates alphabetical indexes with keywords by extracting contiguous strings of letters and/or digits from the text, while excluding trivial words such as "and" and "the." Search systems using WAIS usually allow the user to select one or more indexes and to define a query using simple logical combinations (**and or not**) of keywords.

The bioinformaticians have developed "high-performance" tools. The most popular is **SRS** (Sequence Retrieval System), developed by T. Etzold at the EMBL-EBI (Hinxton, UK). The copyright of SRS belongs to the private company LIONS, Ltd.; from Version 6 on, the software is still free for academic users, but not for commercial users.

For each database, SRS generates alphabetical indexes for the different fields found in each data-bank entry. For certain fields, there are numerical indexes. SRS also generates link tables that allow the user to search entries in one database and then find related entries in another database. The link tables are made either on the basis of a common ID/AC in the two databases (e.g., EMBL and GenBank or PROSITE and Blocks) or by using the cross references to other databases, which many databases contain. For example, these link tables allow a protein to be searched for in SWISS-PROT; then by linking to PROSITE, all the motifs identified in that protein can be obtained; and linking again to SWISS-PROT, all the proteins (including false positives!) known to share one or more of these motifs with the original protein can be retrieved.

SRS has a WWW interface that can insert "on-the-fly" hyperlinks in the pages retrieved from the WWW server. By their nature, the hyperlinks establish only one-to-one relationships, but have the advantage of being able to point to other servers, including servers located on other computers and servers using software other than SRS. Alternatively, the SRS link tables also can establish many-to-many relationships, but are limited in scope to the local SRS server. The WWW SRS servers can temporarily store results of the queries the user made during a session, and they can be recalled with the "Query Manager" (SRS 5) or the "RESULTS" page (SRS 6). Numerous freely accessible SRS servers exist. The "mother of all servers" at the EMBL-EBI maintains a list with the public servers and the databases they offer.

When using a search system involving indexed keywords, it is not only limited to what is actually in the text, but also to what the software extracted from the text

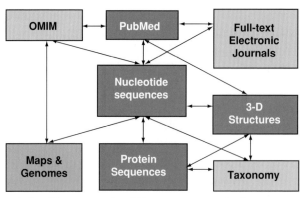

Figure 2.2   Relationships among the different ENTREZ databases.

to build the indexes. An index often contains only simple words; therefore, it serves no purpose to search "reverse transcriptase" if "reverse" and "transcriptase" are separate keywords. It is possible to ask for "reverse and transcriptase" (WAISindex style) or "reverse & transcriptase" (SRS style). However, there are exceptions: the managers of SRS servers have usually configured their system so that the indexes for "Keywords" and "Organism" have strings of characters with blanks included, such as "complete genome" or "hepatitis C virus," and so that the keywords are always in the style "van ranst,m." or "11-feb-2000" for the fields "Authors" and "Date." Formulating the query must be thought out carefully. Again, it may be necessary to repeat the search, exhausting the different possible variations.

2.7   ENTREZ database

ENTREZ, maintained at the NCBI (Bethesda, Maryland, USA), is a relational database using Sybase. Every two months, a major release is distributed on CD-ROM. However, most people consult a daily updated ENTREZ online on the NCBI computers using the WWW or a specific Internet tool called Network ENTREZ (NENTREZ). This TCP/IP-based client–server version offers basically the same functionality as the WWW access, but it has a somewhat nicer look and feel and requires less network traffic. NENTREZ is, of course, freeware obtainable by anonymous ftp from the NCBI; there are client software files available for PC, Macintosh, and UNIX.

ENTREZ provides integrated access to nucleotide and protein sequence data, along with 3-D protein structures, genomic mapping information, and the literature. ENTREZ also contains precomputed similarity searches for each database record, producing a list of related sequences, structures, and literature records. ENTREZ links the following eight databases (Figure 2.2):

Nucleotide: Nonredundant nucleic-acid sequences data bank composed of `GenBank` PDB, EMBL, and DDBJ.

Protein: Nonredundant protein sequences data bank composed of PDB, SWISS-PROT, PIR, PRF, and translations from `GenBank`, EMBL, and DDBJ.

MEDLINE: Database with abstracts from the medical literature, maintained at the National Library of Medicine (Bethesda, Maryland, USA).

Structure: MMDB (Molecular Modelling Database). 3-D structures of nucleic acids, proteins, and complex carbohydrates, derived from the PDB data bank. The NCBI offers a tool for viewing the structures, Cn3D, which is incorporated in NENTREZ. This tool can also be obtained and installed as a helper application of a WWW client.

Genome: Genomic maps for each chromosome or genomic element (e.g., plasmid or phage) that has been sequenced completely or for which an elaborate map is available.

PopSet: Multiple sequence alignments of DNA sequences of different populations or species, submitted as such to `GenBank` and used for phylogeny.

OMIM (Online Mendelian Inheritance in Man): This database is a catalog of human genes and genetic disorders authored and edited by Dr. Victor A. McKusick at the Johns Hopkins University (Baltimore, Maryland, USA).

Taxonomy: The NCBI taxonomy database contains the names of all organisms represented in the genetic databases with at least one nucleotide or protein sequence.

ENTREZ is one of the best search engines. It goes beyond simply retrieving sequences by providing a precomputed list of "neighbor" relationships with every sequence, precalculated using the Basic Local Alignment Search Tool (BLAST) algorithm.

Figure 2.3 provides an example of an ENTREZ nucleotide search for the "coxsackie adenovirus receptor." By clicking the first hyperlink "BC003684," the cDNA sequence of the human coxsackie virus and adenovirus receptor is retrieved (Figure 2.4). Subsequently, clicking the OMIM hyperlink (Figure 2.5) and selecting *Homo sapiens* in the `source organism` field shows that the gene for the coxsackie virus and adenovirus receptor is located on chromosome 21q11.

2.8   Sequence similarity searching: BLAST

After identifying a novel gene or protein sequence, a database search is often conducted to find homologous or similar sequences. In 1990, Altschul and colleagues developed BLAST: Basic Local Alignment Search Tool. The BLAST algorithm breaks the query sequence into short fragments, or "words," and looks for an identical or close match between those words and words from the database sequences. When such a match or "hit" is encountered, the hit is extended in both directions to generate a local alignment segment. The quality of each alignment is quantified in a score, and the ***high-scoring segment pairs*** (***HSPs***) are reported in a table, with the matches ascribed the greatest statistical significance at the top of this table (Figure 2.6).

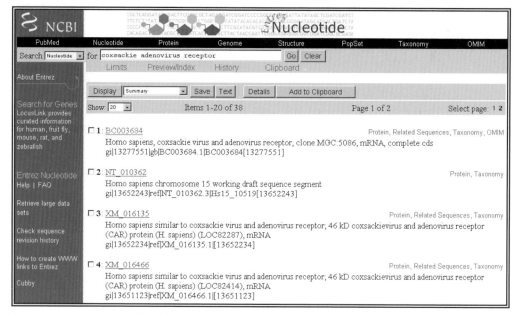

Figure 2.3   ENTREZ nucleotide search for "coxsackie adenovirus receptor."

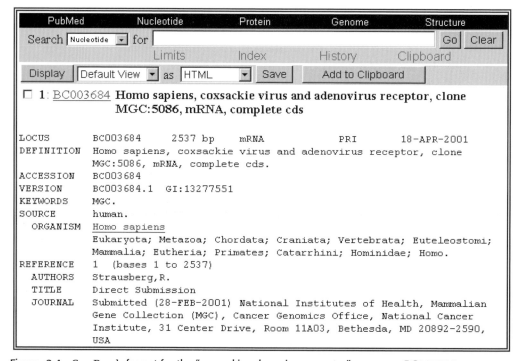

Figure 2.4   GenBank format for the "coxsackie adenovirus receptor" sequence BC003684.

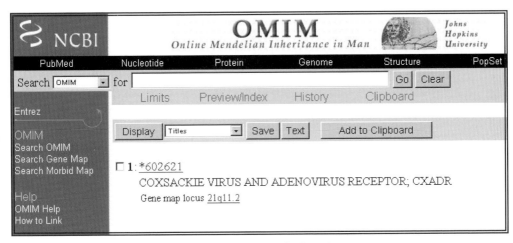

Figure 2.5   OMIM file for the human coxsackie virus and adenovirus receptor.

The BLAST family of programs includes BLASTN, which compares a nucleotide query sequence with a nucleotide sequence database; BLASTP, which compares a protein query sequence with a protein sequence database; BLASTX, which compares a nucleotide query sequence translated in all six open reading frames with a protein sequence database; TBLASTN, which compares a protein query sequence with a nucleotide sequence database dynamically translated in all six open reading frames; and TBLASTX, which compares a six-frame translation of a

```
                                                                   Score    E
Sequences producing significant alignments:                       (bits)  Value

gi|9626612|ref|NP_040902.1|   (NC_001458) putative L1 ORF [Hu...     555   e-157
gi|9626068|ref|NP_040309.1|   (NC_001356) major capsid protei...    536   e-151
gi|73328|pir||P1WL  L1 protein - human papillomavirus type 1a       535   e-151
gi|9628501|ref|NP_043380.1|   (NC_001684) major capsid protei...    532   e-150
gi|9627173|ref|NP_041388.1|   (NC_001535) major capsid protei...    530   e-149
gi|9627164|ref|NP_041380.1|   (NC_001534) 5' end of the codin...    529   e-149
gi|9627741|ref|NP_056819.1|   (NC_001619) late protein [Canin...    529   e-149
gi|4574728|gb|AAD24188.1|AF131950_8   (AF131950) putative maj...     529   e-149
gi|549342|sp|P27232|VL1_HPV35   MAJOR CAPSID PROTEIN L1 >gi|4...     528   e-149
gi|138647|sp|P27233|VL1_HPV42   MAJOR CAPSID PROTEIN L1 >gi|7...     526   e-148
gi|1764168|gb|AAB39894.1|   (U72630) major capsid protein [Co...     525   e-148
gi|9627347|ref|NP_041818.1|   (NC_001588) late protein [Human...    525   e-148
gi|9628715|ref|NP_043578.1|   (NC_001711) Method: conceptual ...    524   e-147
gi|9627970|ref|NP_042523.1|   (NC_001635) ORF putative L1 [Hu...     523   e-147
gi|9627348|ref|NP_041819.1|   (NC_001588) late protein [Human...    523   e-147
```

Figure 2.6   Example of a BLAST output from the NCBI server.

nucleotide query sequence with the six-frame translations of a nucleotide sequence database.

The original BLAST programs were not able to incorporate **gaps** in the alignment. Gapped BLAST programs, which can generate gapped alignments, were introduced in 1997 (Altschul et al., 1997). Another improvement increased the chances of finding subtle but biologically significant similarities between distantly related sequences. The **Position-Specific Iterated BLAST** (PSI-BLAST) program builds a position-specific scoring matrix, or profile, from the multiple alignment of the highest-scoring matches in an initial protein BLAST search. Positions that are highly conserved yield large positive or negative values when a corresponding match or mismatch is found in a subsequent BLAST search; weakly conserved positions yield scores close to zero. This procedure can be repeated or iterated to further refine the profile and discover more distantly related matches in the next BLAST round (Altschul et al., 1997).

PRACTICE

Marco Salemi

2.9 File formats

In the next chapters different computer programs performing phylogenetic analyses will be discussed. Each program will recognize (as input) and be capable of producing (as output) files in a limited range of formats. Therefore, much of the effort involved in carrying out a comprehensive analysis is devoted to preparing and/or converting files. The most common file formats are GenBank, a nucleotide-sequence database format from NCBI; EMBL and SWISS-PROT, nucleotide and amino-acid database formats, respectively; FASTA, a simple format with just one line of text per sequence (see Box 2.2); PHYLIP, of great importance because of the widespread use of the package (see Box 2.2); ClustalW, which is produced by default in the Clustal program; NEXUS, used by the phylogeny package PAUP* (see Chapter 7). GenBank, EMBL, and SWISS-PROT formats are normally used for just one sequence at a time, and most of each entry will be devoted to information about the sequence (see the previous section). The easiest file format is FASTA, which uses just one line of text per sequence, starting with an angle bracket (">") to delimit each sequence (see Box 2.2).

Input files in the appropriate format could be prepared by hand with a common text editor, but this is tedious and impractical, especially for large data sets. Some freeware phylogeny packages, however, offer extensive capabilities for writing and converting input files in numerous formats. Two, in particular, are described: ClustalX and DAMBE (Xia, 2000). The installer application of ClustalX can be downloaded from **ftp://ftp-igbmc.u-strasbg.fr/pub/ClustalX/** (see also Chapter 3, Section 3.12), and it requires the use of WinZip (Windows version) or Stuffit Expander (Mac version) for unpacking. ClustalX is mainly used to carry out multiple sequence alignments (see Chapter 3), but it is also a useful tool for creating and converting input files. DAMBE is an integrated software package for analyzing molecular sequence data (see Chapter 13), with a user-friendly Windows 95/98/2000/NT interface. It also has extensive capabilities to write and convert sequence files in most of the common formats. The application can be executed on Mac computers using VirtualPC, a program that simulates the Windows environment. The package in self-extracting format can be downloaded from **http://web.hku.hk/~xxia/software/software.htm**. Before reading the next section, install both programs on the local computer by following the instructions on the Web site.

Box 2.2 Common formats of sequence data input files

The FASTA format

The seven globin sequences from Figure 3.1 (see Chapter 3) shown in FASTA format. *Taxon* names are preceded by ">" and followed by a line space before the actual sequence, which can be nucleotides or amino acids. In this case, the file represents a complete multiple alignment because hyphens were included in each sequence at gap positions. If the hyphens are removed, this is simply a raw sequence file, suitable for entry to many analysis packages and sites on the Internet.

```
>Humbeta
--------VHLTPEEKSAVTALWGKVN--VDEVGGEALGRLLVVYPWTQRFFESFGDLST
PDAVMGNPKVKAHGKKVLGAFSDGLAHLDN-----LKGTFATLSELHCDKLHVDPENFRL
LGNVLVCVLAHHFGKEFTPPVQAAYQKVVAGVANALAHKYH------
>Horbeta
--------VQLSGEEKAAVLALWDKVN--EEEVGGEALGRLLVVYPWTQRFFDSFGDLSN
PGAVMGNPKVKAHGKKVLHSFGEGVHHLDN-----LKGTFAALSELHCDKLHVDPENFRL
LGNVLVVVLARHFGKDFTPELQASYQKVVAGVANALAHKYH------
>Humalpha
---------VLSPADKTNVKAAWGKVGAHAGEYGAEALERMFLSFPTTKTYFPHF-DLS-
----HGSAQVKGHGKKVADALTNAVAHVDD-----MPNALSALSDLHAHKLRVDPVNFKL
LSHCLLVTLAAHLPAEFTPAVHASLDKFLASVSTVLTSKYR------
>Horalpha
---------VLSAADKTNVKAAWSKVGGHAGEYGAEALERMFLGFPTTKTYFPHF-DLS-
----HGSAQVKAHGKKVGDALTLAVGHLDD-----LPGALSNLSDLHAHKLRVDPVNFKL
LSHCLLSTLAVHLPNDFTPAVHASLDKFLSSVSTVLTSKYR------
>Whalemyo
---------VLSEGEWQLVLHVWAKVEADVAGHGQDILIRLFKSHPETLEKFDRFKHLKT
EAEMKASEDLKKHGVTVLTALGAILKKKGH-----HEAELKPLAQSHATKHKIPIKYLEF
ISEAIIHVLHSRHPGDFGADAQGAMNKALELFRKDIAAKYKELGYQG
>Lamprey
PIVDTGSVAPLSAAEKTKIRSAWAPVYSTYETSGVDILVKFFTSTPAAQEFFPKFKGLTT
ADQLKKSADVRWHAERIINAVNDAVASMDDT — EKMSMKLRDLSGKHAKSFQVDPQYFKV
LAAVIADTVAAG---D------AGFEKLMSMICILLRSAY-------
>Lupin
--------GALTESQAALVKSSWEEFNANIPKHTHRFFILVLEIAPAAKDLFSFLKGTSE
VP--QNNPELQAHAGKVFKLVYEAAIQLQVTGVVVTDATLKNLGSVHVSKGVAD-AHFPV
VKEAILKTIKEVVGAKWSEELNSAWTIAYDELAIVIKKEMNDAA---
```

The MEGA format

The data file begins with "#", followed by the word "mega" on the first line. Comments, which are ignored by the program, can be written on the next lines preceded by "!". Each sequence starts with "#", followed by the name of the sequence and at least one blank space before the actual sequence (DNA or amino acid). In the noninterleaved MEGA format, sequences are separated only by a carriage return character.

```
#mega
!This is an example of noninterleaved mega format
!Three taxa ten nucleotides long
#taxon1 GGCTCCCACC
#taxon2 GGCCATCACC
#taxon3 GGCCCCCACT
```

The PHYLIP format

The file begins by indicating the number of sequences and characters in the data set. Sequences are pasted on the next lines. The sequence name must be exactly ten characters long, including blank spaces. Gaps in the sequence are indicated with "–", whereas "?" can be used to indicate a missing (unknown) character. In the PHYLIP sequential format, sequences are presented one at a time, with a carriage return at the end of each sequence:

```
3 167
Humbeta --------VHLTPEEKSAVTALWGKVN--
VDEVGGEALGRLLVVYPWTQRFFESFGDLSTPDAVMGNPKVKAHGKKVLGAFSDGLAHLDN
-----
LKGTFATLSELHCDKLHVDPENFRLLGNVLVCVLAHHFGKEFTPPVQAAYQKVVAGVANAL
AHKYH------
Horbeta --------VQLSGEEKAAVLALWDKVN--
EEEVGGEALGRLLVVYPWTQRFFDSFGDLSNPGAVMGNPKVKAHGKKVLHSFGEGVHHLDN
-----
LKGTFAALSELHCDKLHVDPENFRLLGNVLVVVLARHFGKDFTPELQASYQKVVAGVANAL
AHKYH------
Humalpha --------V-
LSPADKTNVKAAWGKVGAHAGEYGAEALERMFLSFPTTKTYFPHF-DLS-----
HGSAQVKGHGKKVADALTNAVAHVDD-----MPNALSALSDLHAHKLRVDPVNFKL
LSHCLLVTLAAHLPAEFTPAVHASLDKFLASVSTVLTSKYR------
```

The sequences can also be written in PHYLIP interleaved format:

```
3 167
Humbeta     --------VHLTPEEKSAVTALWGKVN--VDEVGGEALGRLLVVYPWTQR
Horbeta     --------VQLSGEEKAAVLALWDKVN--EEEVGGEALGRLLVVYPWTQR
Humalpha    --------V-LSPADKTNVKAAWGKVGAHAGEYGAEALERMFLSFPTTKT

FFESFGDLSTPDAVMGNPKVKAHGKKVLGAFSDGLAHLDN-----LKGTF
FFDSFGDLSNPGAVMGNPKVKAHGKKVLHSFGEGVHHLDN-----LKGTF
YFPHF-DLS-----HGSAQVKGHGKKVADALTNAVAHVDD-----MPNAL

ATLSELHCDKLHVDPENFRLLGNVLVCVLAHHFGKEFTPPVQAAYQKVVA
AALSELHCDKLHVDPENFRLLGNVLVVVLARHFGKDFTPELQASYQKVVA
SALSDLHAHKLRVDPVNFKLLSHCLLVTLAAHLPAEFTPAVHASLDKFLA
```

Box 2.2 (continued)

```
GVANALAHKYH------
GVANALAHKYH------
SVSTVLTSKYR------
```

In this example, each line contains exactly fifty nucleotides of each sequence. The file can be prepared with any text editor.

Translating different formats

Different formats can be translated with the program ForCon, a PC application that can be downloaded for free from **http://www.evolutionsbiologie.uni-konstanz.de/peer-lab/forcon.html**. In practice, the easist thing to do when preparing an input file is to write the sequences in FASTA format (explained previously) and then, if necessary, translate the file in a different format with ForCon.

Another sequence conversion tool, ReadSeq, is available online at **http://iubio.bio.indiana.edu/cgi-bin/readseq.cgi**. Users can paste their sequences in the Web site window and submit them to the server. The server automatically recognizes the input format and can translate the sequences in a different format chosen by the users. ReadSeq-supported formats are provided in the following table.

Name	Interleaved	Content-type	Extension (PC)
IG|Stanford	—	biosequence/ig	.ig
GenBank|GB	—	biosequence/genbank	.gb
NBRF	—	biosequence/nbrf	.nbrf
EMBL	—	biosequence/embl	.embl
GCG	—	biosequence/gcg	.gcg
DNAStrider	—	biosequence/strider	.strider
Pearson|Fasta	—	biosequence/fasta	.fas
Phylip3.2	Yes	biosequence/phylip2	.phylip2
Phylip|Phylip4	Yes	biosequence/phylip	.phy
Plain|Raw	—	biosequence/plain	.seq
PIR|CODATA	—	biosequence/codata	.pir
MSF	Yes	biosequence/msf	.msf
PAUP|NEXUS	Yes	biosequence/nexus	.nex
XML	—	biosequence/xml	.xml

2.10   Three example data sets

In the remaining part of this book, three data sets are mainly used for the practical exercises – two nucleotide alignments and one amino-acid alignment:

1. The first data set consists of 14 HIV-1/SIV sequences of the entire *env* region coding for the two envelope viral proteins gp120 and gp41: 11 HIV-1 (human

immunodeficiency virus) group M strains; two SIV (simian immunodeficiency virus) strains isolated from African chimpanzees (SIVcpz); and one HIV-1 group O strain. The sequences were translated into amino acid and then aligned using the Clustal algorithm (see Chapter 3) implemented in the DAMBE software package. The resulting alignment, translated back to amino acid, is 2,352 nt long after gap removal (see also Chapter 3, Section 3.16). Throughout the book we will refer to such data set as the HIV data set.

2. The second data set consists of 17 mtDNA concatenated sequences of mitochondrial 12S rRNA, 16S rRNA, and tRNA(val) genes from different species, and is described in detail by Hedges (1994). Sequences were aligned using ClustalX (see Chapter 3), and the resulting alignment was 2,903 nt long. After removing ambiguous nucleotide positions and gaps, the final alignment was 1,998 nt. Throughout the book we will refer to this data set as the mtDNA data set.

3. The third data set consists of 422 amino-acid sequences of the coenzyme NAD-dependent glycerol-3-phosphate dehydrogenase from 12 different species, and they were aligned using ClustalX (see Chapters 3 and 8).

Additional details and the unaligned and aligned data sets can be found at **http://www.kuleuven.ac.be/aidslab/phylogenyBook/datasets.htm**. They should be downloaded on the local computer before starting the exercises.

2.10.1 Preparing input files: HIV/SIV example data set

The file hivULN, which can be downloaded from **http://www.kuleuvem.ac.be/ aidslab/phylogenyBook/datasets/hivULN.htm**, contains the unaligned viral sequences of the HIV/SIV data set, in FASTA format, ready to be opened and aligned with ClustalX or DAMBE (see Chapter 3). The following exercise demonstrates how to prepare such a file; if you want to skip this part, go immediately to Chapter 3.

Table 2.1 shows the accession numbers of the 14 HIV/SIV sequences used in the hivULN file. An easy way to retrieve them is to browse the Los Alamos HIV sequence database (**http://hiv-web.lanl.gov**), click the Search DB link and enter the accession number in the Accession field. Try sequence L20571 (full genome of HIV-1 group O, isolate MVP5180). The sequence will appear in the Web browser in GenBank format. Going through the features description, it can be seen that the *env* gene corresponds to the fragment 6260–8890 of the full-length sequence. To download it in FASTA format, go to the top of the file and perform the following:

1. Enter 6260 in the Start field, and 8887 (in this way, the stop codon will not be included) in the end field of SEQUENCE FRAGMENT.
2. Select FASTA in the FORMAT field.
3. Click the Download button.

Table 2.1 HIV/SIV strains in `hivULN` file

Accession	Sequence name	Subtype	Sampling year
L20571	MVP5180	HIV group O	1991
AF103818	SIVcpzUS	SIV chimpanzee	1985
X52154	SIVcpzGAB	SIV chimpanzee	1988
U09127	92UG037	A	1992
U27426	92UG975	G	1992
U27445	92RU131	G	1992
AF067158	93IN905	C	1993
U09126	92BR025	C	1992
U27399	92UG021	D	1992
U43386	92UG024	D	1992
L02317	BSSG3	B	1987
AF025763	SFMHS20	B	1989
U08443	91TH652	B	1991
AF042106	MBCC18R01	B	1993

Depending on the browser, a window should appear asking the name of the document about to be saved. Name the document as preferred – for example, L20571.fas or HIVO.fas – and save it. The extension .fas in the file name is not necessary for `ClustalX` because the program is able to open files without an extension type. However, DAMBE can only open sequence files with extensions describing their format (e.g., `.fas`, for sequences in FASTA format, `.phy` for sequences in PHYLIP format). When all the sequences have been downloaded on the local computer in a similar way, any text editor can be used to copy and paste them into a single file. If preparing this file with Microsoft Word, remember to save the final file in text-only format because phylogeny programs cannot read files saved as Microsoft Word documents. If working in a Windows environment, the text file just saved will automatically take the extension `.txt`. It will be necessary to change it to `.fas` to enable DAMBE to read the file. An easier way to prepare the input files is to import all the sequences downloaded from the database directly into `ClustalX`, as follows:

1. Place the files with the sequences in the `ClustalX` folder (the folder is automatically created during the installation of the package). Double-click the `ClustalX.exe` icon to run `ClustalX`. Select `Load Sequence` from the `File` menu and open the first sequence so it appears in the `ClustalX` main window.
2. Select `Append Sequence` from the `File` menu and open a second sequence. The second sequence will appear below the first one in the `ClustalX` window.
3. Repeat Step 2 to import the remaining sequences into the `ClustalX` window.

Sequences are now ready to be aligned by ClustalX. Unaligned sequences can also be exported in a single file by choosing Save Sequences as...in the File menu. ClustalX allows you to save them in several formats, such as PHYLIP or NEXUS, which can be read by DAMBE and several other packages (remember to add the .fas extension for use with DAMBE).

A similar procedure can be followed to import the sequences in the DAMBE program (see Section 2.8). Choose Open (sequence or gene frequency) from the File menu. DAMBE asks the user to select the format of the sequence file (e.g., extension .fas for FASTA). Browse to the correct file. A window appears asking whether the sequence is a nonprotein, amino-acid, or protein-coding nucleotide sequence. Select protein-coding nucleotides and Universal genetic code, and click the Go! button. After importing the first sequence in the DAMBE window, the others can be added by selecting Open (sequence or gene frequency) from the File menu and answering "No" when the program asks whether to replace the sequences already in memory. In this way, the new sequences will be appended below the ones already present in the main window. As with ClustalX, it is possible to export the unaligned (or aligned) sequences in several formats by choosing Save or Convert Sequence Format from the File menu.

A useful feature of DAMBE is that it can read the qualifiers of a GenBank file. The program is able to discern several CDS features, so that the user can download a single file with all the GenBank-formatted sequences and splice out the relevant CDS.

1. Create a text file (e.g., with WordPad in Windows or TeachText in MacOs) with each accession number of the query sequences on a separate line.
2. In the ENTREZ database (see Section 2.7), click the Batch ENTREZ link. Browse to the text file that contains the accession numbers from Step 1 and retrieve it.
3. Select GenBank format in the Display box, and save the file as all files with the extension .gb.
4. Run DAMBE. Open the .gb file, click the CDS button, and click Proceed.
5. In the new window, click the first accession number and select the *env* region in the upper-right panel. Click the Splice button.
6. Repeat Step 5 for each sequence.
7. Click Done. The usual DAMBE window appears with the unaligned sequences in the selected region.

The sequences can now be exported in a different format or aligned with the Clustal algorithm, as discussed previously.

REFERENCES

Altschul, S. F., W. Gish, W. Miller, E. W. Myers, and D. J. Lipman (1990). Basic Local Alignment Search Tool. *Journal of Molecular Biology*, 215, 403–410.

Altschul, S. F., T. L. Madden, A. A. Schaffer, J. Zhang, Z. Zhang, W. Miller, and D. J. Lipman (1997). Gapped BLAST and PSI-BLAST: A new generation of protein database search programs. *Nucleic Acid Research*, 25, 3389–3402.

Hedges, S. B. (1994). Molecular evidence for the origin of birds. *Proceedings of the National Academy of Sciences of the USA*, 91, 2621–2624.

Xia, X. (2000). *Data Analysis in Molecular Biology and Evolution*. Boston: Kluwer Academic Publishers.

Multiple alignment

THEORY

Des Higgins

3.1 Introduction

The example of a multiple-sequence alignment shown in Figure 3.1 is a set of amino-acid sequences of globins that have been aligned so that **homologous** residues are arranged in columns as much as possible. The sequences have different lengths, which means that gaps (shown as hyphens in the figure) must be used in some positions to achieve the alignment. If the sequences were all the same length, the gaps would not be needed. The generation of alignments is one of the most common tasks in computational sequence analysis because alignments are required for many other analyses, such as structure prediction or to demonstrate sequence *similarity* within a family of sequences. Of course, one of the most common reasons for generating them is that they are an essential prerequisite for most phylogenetic analyses. Rates or patterns of change in sequences cannot be analyzed unless the sequences can be aligned.

The final goal should clearly be recognized when discussing how to carry out an alignment, either manually or using a computer program. Any phylogeny inference based on molecular data begins by comparing the homologous residues (i.e., those that descend from a common ancestral residue) with different deoxyribonucleic acid (DNA) or protein sequences. The best way to do this is to align sequences one on top of another, so that homologous residues from different sequences line up in the same column. If the sequences are evolutionarily related, they began as identical to each other and diverged over time by the accumulation of substitutions, as well as insertions and deletions. Given a set of sequences, it is then necessary to pad them out with gaps, with the expected result that they will correspond to the insertion or deletion positions and leave the columns neatly aligned. Ideally, all the residues in each column will derive from one residue in the ancestral sequence or the deletion

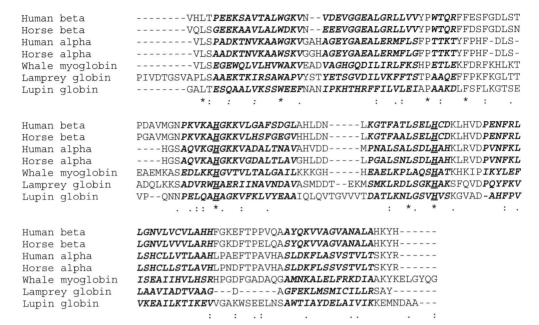

```
Human beta        --------VHLTPEEKSAVTALWGKVN--VDEVGGEALGRLLVVYPWTQRFFESFGDLST
Horse beta        --------VQLSGEEKAAVLALWDKVN--EEEVGGEALGRLLVVYPWTQRFFDSFGDLSN
Human alpha       ---------VLSPADKTNVKAAWGKVGAHAGEYGAEALERMFLSFPTTKTYFPHF-DLS-
Horse alpha       ---------VLSAADKTNVKAAWSKVGGHAGEYGAEALERMFLGFPTTKTYFPHF-DLS-
Whale myoglobin   ---------VLSEGEWQLVLHVWAKVEADVAGHGQDILIRLFKSHPETLEKFDRFKHLKT
Lamprey globin    PIVDTGSVAPLSAAEKTKIRSAWAPVYSTYETSGVDILVKFFTSTPAAQEFFPKFKGLTT
Lupin globin      --------GALTESQAALVKSSWEEFNANIPKHTHRFFILVLEIAPAAKDLFSFLKGTSE
                          *:   :    :   *  .              :  .:    *  :   *    :     .

Human beta        PDAVMGNPKVKAHGKKVLGAFSDGLAHLDN-----LKGTFATLSELHCDKLHVDPENFRL
Horse beta        PGAVMGNPKVKAHGKKVLHSFGEGVHHLDN-----LKGTFAALSELHCDKLHVDPENFRL
Human alpha       ----HGSAQVKGHGKKVADALTNAVAHVDD-----MPNALSALSDLHAHKLRVDPVNFKL
Horse alpha       ----HGSAQVKAHGKKVGDALTLAVGHLDD-----LPGALSNLSDLHAHKLRVDPVNFKL
Whale myoglobin   EAEMKASEDLKKHGVTVLTALGAILKKKGH-----HEAELKPLAQSHATKHKIPIKYLEF
Lamprey globin    ADQLKKSADVRWHAERIINAVNDAVASMDDT--EKMSMKLRDLSGKHAKSFQVDPQYFKV
Lupin globin      VP--QNNPELQAHAGKVFKLVYEAAIQLQVTGVVVTDATLKNLGSVHVSKGVAD-AHFPV
                      .  .::  *.   :    .                   :  *.  *    .         :  .

Human beta        LGNVLVCVLAHHFGKEFTPPVQAAYQKVVAGVANALAHKYH------
Horse beta        LGNVLVVVLARHFGKDFTPELQASYQKVVAGVANALAHKYH------
Human alpha       LSHCLLVTLAAHLPAEFTPAVHASLDKFLASVSTVLTSKYR------
Horse alpha       LSHCLLSTLAVHLPNDFTPAVHASLDKFLSSVSTVLTSKYR------
Whale myoglobin   ISEAIIHVLHSRHPGDFGADAQGAMNKALELFRKDIAAKYKELGYQG
Lamprey globin    LAAVIADTVAAG---D------AGFEKLMSMICILLRSAY-------
Lupin globin      VKEAILKTIKEVVGAKWSEELNSAWTIAYDELAIVIKKEMNDAA---
                      :     :    .:        .         . .          .      :
```

Figure 3.1   Multiple alignment of seven amino-acid sequences. Identical amino-acid positions are marked with asterisks (*) and biochemically conserved positions are marked with colons and periods (less conserved). The lupin sequence is a leghaemoglobin and the lamprey sequence is a cyanohaemoglobin. The whale sequence is from the sperm whale. The approximate positions of the alpha helixes are typed in italics and bold font. The positions of two important histidine residues are underscored, and are responsible for binding the prosthetic haem and oxygen.

of that residue in some sequences corresponds to a hyphen (if a hyphen represents a deletion). The hyphens also may arise from an insertion, or a combination of insertions and deletions, in one or more sequences of the family without a hyphen at that specific position.

3.2   The problem of repeats

It can be difficult to find this ideal alignment for several reasons. First, there may be repeats in one or all members of the sequence family; this problem is shown in the simple diagram in Figure 3.2A. It is not clear which example of a repeat unit should line up with different members of the family. If there are large-scale repeats, such as with duplications of entire protein domains, the problem can be partially solved by excising the domains or repeat units and conducting a ***phylogenetic analysis*** of the repeats. This is only a partial solution because a single domain in one ancestral protein can give rise to two equidistant repeat units in one descendent protein and three in another; therefore, it will not be obvious how the repeat units should

(A)

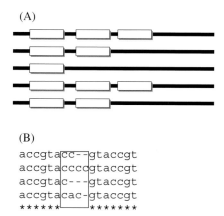

(B)

```
accgtacc--gtaccgt
accgtacccdgtaccgt
accgtac---gtaccgt
accgtacac-gtaccgt
******      *******
```

Figure 3.2   (A) A simple diagram showing the distribution of a repeated domain (marked with a rectangle) in a set of sequences. When there are different numbers of repeats in the sequences, it can be difficult to know exactly how to arrange the alignment. Some domains will be equally similar to several domains in other sequences. (B) A simple example of some nucleotide sequences in which the central boxed alignment is completely ambiguous.

line up with one another. With small-grain repeats, such as those involving single nucleotides or with microsatellite sequences, the problem is worse.

As shown in Figure 3.2B, it is not obvious in the highlighted box if one alignment of sequences is better than any other, except cosmetically. There are some repeated C residues. Fortunately, it often makes no difference phylogenetically which alignment is used and, indeed, will not make any difference if these regions are simply ignored for phylogenetic analysis. It will be difficult for all computer programs to disentangle these repeats and, in most cases, there is no ideal alignment. These small-scale repeats are especially abundant in some nucleotide sequences; for example, they accumulate in some positions as a result of mistakes during DNA replication. Fortunately, they tend to be localized to nonprotein coding or hypervariable regions; these will be almost impossible to align unambiguously and should be treated with caution. In amino-acid sequences, such localized repeats of residues are unusual, whereas large-scale repeats of entire protein domains are common.

3.3   The problem of substitutions

If the sequences in a data set accumulate few substitutions over time, they will remain similar and will be easy to align. The many columns of unchanged residues will make the alignment obvious. In these cases, both manual and computerized alignment will be clear and unambiguous; in reality, sequences can and do change considerably.

For two amino-acid sequences, it becomes increasingly difficult to find a good alignment when the sequences drop below approximately 25% identity (see

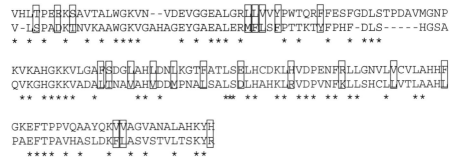

VHLTPEEKSAVTALWGKVN--VDEVGGEALGRLLMVYPWTQRFFESFGDLSTPDAVMGNP
V-LSPADKINVKAAWGKVGAHAGEYGAEALERMFLSFPTTKTYFPHF-DLS-----HGSA
*   *   *   *   * * *   ****   *   * *** *   *   *   *   * ***

KVKAHGKKVLGAFSDGLAHLDNLKGTFATLSELHCDKLHVDPENFRLLGNVLVCVLAHHF
QVKGHGKKVADALTNAVAHVDDMPNALSALSDLHAHKLRVDPVNFKLLSHCLLVTLAAHL
**   *****   *   ** *   *** **   ** *** ** **   *   **

GKEFTPPVQAAYQKVVAGVANALAHKYH
PAEFTPAVHASLDKFLASVSTVLTSKYR
**** * *   *   * *   *   **

Figure 3.3   An alignment of the amino-acid sequences of human alpha globin (below) and human beta globin (above). The boxed pairs of residues are all between pairs of similar amino acids (i.e., biochemically similar side chains). The most conserved pairs are S and T (small polar); F and Y (aromatic); D and E (acidic, negatively charged); any two of H, K, and R (positively charged); and any two of F, I, L, M, or V (hydrophobic). Many of the other unmarked pairs also represent some conservation of biochemical property.

Chapter 8). *Sequence identity*, a commonly used and simple measure of sequence similarity, is the number of identical residues in an alignment divided by the number of aligned positions, excluding any positions with a gap. Of course, a nonsensical alignment with a gap every two or three residues, for example, can give a misleading measure of sequence similarity. Typically, however, the measure is rather robust. Consider any pair of aligned homologous protein sequences with some positions in which one or both residues have changed during evolution. When examining the nature of these changes, two patterns emerge (e.g., Figure 3.3). First, the identities and differences are not evenly distributed along the sequence. Blocks of alignment with between 5 and 20 residues will have more identities and more similar amino acids (i.e., biochemically similar; discussed in the next section) than elsewhere. These blocks typically correspond to the conserved secondary-structure elements of α-helixes and β-strands in the proteins, and change more slowly than the connecting loops of irregular structure. The same pattern is observed when examining gap locations in the sequences. This pattern of blocks is more clearly seen in most multiple alignments of protein sequences (see Figure 3.1). One benefit of this block-like similarity is that there will be regions of alignment that are clear and unambiguous and that which most computer programs can find easily, even between distantly related sequences. However, it also means that there will be regions of alignment that are more difficult to align and, if the sequences are sufficiently dissimilar, impossible to align unambiguously.

The second observed pattern is that most pairs of aligned, but nonidentical, residues are biochemically similar (i.e., their side chains are similar). Similar amino acids such as Leucine and Isoleucine or Serine and Threonine tend to replace each other more often than dissimilar ones (see Figure 3.3). This conservation of biochemical character is especially true in the conserved secondary-structure regions and at important sites such as active sites or ligand-binding domains. By far, the most important biochemical property to be conserved is polarity/hydrophobicity. Size matters but the exact conservation pattern depends on the exact role of the amino acid in the protein, which significantly aids sequence alignment.

To carry out an automatic alignment, it is necessary to quantify biochemical similarity. Previous attempts focused on chemical properties and/or the genetic code; however, for both database searching and multiple alignment, the most current amino-acid scoring schemes are based on empirical studies of aligned proteins. Until the early 1990s, the most powerful method was from the work of Margaret Dayhoff and colleagues (1978), who produced the famous PAM series of weight matrixes (see Chapter 8), which are 20×20 matrixes that provide scores for all possible pairs of aligned amino acids. The higher the score, the more weight is attached to a pair of aligned residues. The PAM matrixes were derived from the original empirical data using a sophisticated evolutionary model, which allowed for the use of different scoring schemes depending on the similarity of the sequences. Although this was a powerful capability, most biologists simply used the default table offered by whatever software they were using. Currently, the PAM matrixes have largely been superseded by the BLOSUM matrixes of Jorja and Steven Henikoff (1992) (see Chapter 8), which are also available as a series, depending on the similarity of the sequences to be aligned. The one most commonly used, the BLOSUM 62 matrix, is shown in Figure 3.4, which illustrates how different pairs of identical residues get different scores. Less weight is given to residues that change easily during evolution, such as Alanine or Serine, and more weight is given to those that change less frequently, such as Tryptophan. The remaining scores are either positive or negative, depending on whether a particular pair of residues is more or less likely to be observed in an alignment.

Given the table in Figure 3.4, the alignment problem is finding the arrangement of amino acids resulting in the highest score, which is the basis of almost all commonly used computer programs. One notable exception is software based on the so-called **hidden Markov models** (**HMMs**), which use probabilities rather than scores; however, these methods use a related concept of amino-acid similarity.

With nucleotide sequences, it is not possible to distinguish conservative from nonconservative substitutions. Furthermore, two random sequences of equal-base compositions will be 25% identical. It may be difficult to make a sensible alignment if the sequences have diverged significantly, especially if the nucleotide compositions are biased or if there are many repeats. This is one reason why, if there is a choice, *it is important to always try to align protein coding sequences at the amino-acid level*.

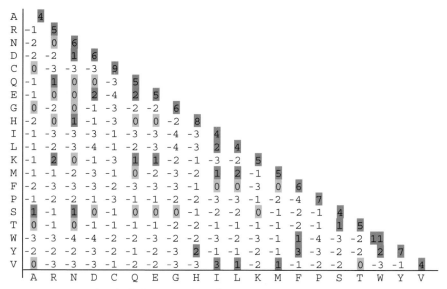

Figure 3.4   BLOSUM 62 matrix. Similarity scores for all 20 × 20 possible pairs of amino acids, including identical pairs (the diagonal elements). Negative numbers represent pairs that are not commonly observed in real alignments; positive numbers represent commonly found pairs (shaded). These scores allow us to quantify the similarity of sequences in an alignment, when combined with gap scores (i.e., gap penalties).

3.4   The problem of gaps

Insertions and deletions also accumulate as sequences diverge from each other. In proteins, these are concentrated between the main secondary-structure elements, just as for substitutions. Typically, no distinction is made between insertions and deletions, which are sometimes collectively referred to as *indels* (see Chapter 1). In alignments, they appear as gaps inserted in sequences to maximize an alignment score.

If there were no indels, the best alignment of two sequences could be found by simply sliding one over the other. To choose the alignment with the highest score, the weight-matrix scores are counted for each pair of aligned residues. With the presence of indels, a proper alignment algorithm is needed to find the arrangement of gaps that maximizes the score. This is usually accomplished with a technique called *dynamic programming* (Needleman and Wunsch, 1970; Gotoh, 1982), which is an efficient method for finding the best alignment and its score. However, there is a complication: if gaps of any size can be placed at any position, it is possible to generate alignments with more gaps than residues. To prevent excessive use of gaps, indels are usually penalized (i.e., a penalty is subtracted from the alignment score) using so-called gap penalties (GPs). The most common formula for

```
VHLTPEEKSAVTALWGKVNVDEVGGEAL...
V-------------------NEEEVGGEAL...
```

Figure 3.5 A simple example to illustrate the problem of end gaps. The second sequence is missing a section from its N-terminus, and the end V aligns with the N-terminal V of the first sequence. This is clearly a nonsensical alignment, but it gets exactly the same score as the one with the V moved across to join the rest of the second sequence – unless end gaps are free (there is no GP for end gaps).

calculating GPs follows:

$$GP = g + hl$$

where l is the length of the gap, g is a gap-opening penalty (charged once per gap), and h is a gap-extension penalty (charged once per hyphen in a gap). These penalties are often referred to as *affine gap penalties*. The formula is quite flexible in that it allows the number and the lengths of gaps to be controlled separately by setting different values for g and h. However, there is no particular mathematical, statistical, or biological justification for this formula. It is widely used because it often works well and it is easy to write computer programs to implement it. In practice, the values of g and h are chosen arbitrarily, and there is no reason to believe that gaps evolve as simply as the formula suggests. Significantly, the alignment with the highest alignment score may or may not be the correct alignment biologically. Finally, it is common to make end gaps free, which allows for the fact that, for various biological and experimental reasons, many sequences are missing sections from the ends. Not making end gaps free risks getting nonsensical alignments, such as the one shown in Figure 3.5.

3.5 Testing multiple-alignment methods

The most common way to test an alignment method is by comparing test cases. Exaggerated claims have been made about how effective some methods are or why others should never be used. The user-friendliness of the software and its availability are important secondary considerations, but it is the quality of the alignments that matter most. In the past, many new alignment programs have been shown to perform better than older programs, using one or two carefully chosen test cases. The superiority of the new method is then verbally described by authors who have selected parameter values (i.e., GPs and weight matrixes) that favorably show their program. Of course, authors do not choose test cases for which their programs perform badly and, to be fair, some programs perform well and are potentially useful.

One solution is to use test cases generated by experts with extensive reference to secondary and tertiary structural information, rather than by automatic programs. For structural ribonucleic acid (RNA) sequences, the huge alignments of ribosomal RNA sequences can be used as tests. These comparisons are difficult and highly specific. For proteins, an excellent collection of test cases is called BaliBase (Thompson et al., 1999b), which is 141 alignments from five different types of alignment situations: (1) equidistant (small sets of phylogenetically equidistant sequences); (2) orphan (as for Type 1 but with one distant member of the family); (3) two families (two sets of related sequences, distantly related to each other); (4) long insertions (one or more sequences have a long insertion); and (5) long deletions. These test cases are made from sets of sequences in which there is extensive tertiary-structure information. The sequences and structures were carefully compared and multiple alignments were produced by manual editing.

The use of these test cases involves several assumptions. First, contradictory scenarios can be invented, but it is generally assumed that an alignment maximizing structural similarity is somehow "correct." A more serious assumption is that the reference alignment has been made correctly. Although parts of alignments will be difficult to judge, even with tertiary information, this can be largely circumvented by selecting those regions of core alignment from each test case that are clearly and unambiguously aligned. These core blocks can then be used to test programs. These tests are referred to in the next section in which various multiple-alignment methods are described.

3.6   Multiple-alignment algorithms

This section discusses the main algorithms developed for multiple sequence alignment. However, we will first examine a simple and fast method to investigate the similarity between two sequences, the dot matrix method, which can be used as an exploratory tool to check whether two sequences can be aligned unambiguously.

3.6.1   Dot-matrix sequence comparison

Probably the easiest and certainly the most common way to look for regions with high similarity scores among nucleotide or amino-acid sequences is to obtain a *dot-matrix representation*, or *dot plot*. In a dot plot, two sequences are compared using a sliding-window approach. The sequences are shown on the x and y axes of a two-dimensional (2-D) graph (Figure 3.6). Each sequence residue is compared pairwise and identity is scored as a dot in the 2-D dot plot. For long sequences (i.e., more than 500 residues) or divergent ones, better results can be obtained by choosing a window size of N residues (usually 5 to 20 amino acids or nucleotides; see the Practice section) and a mismatch limit. For example, the user can look for

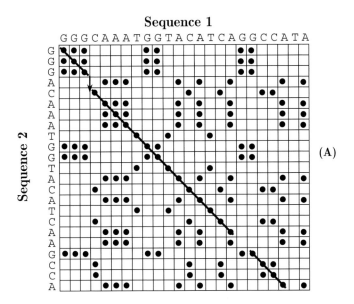

Figure 3.6   (A) DNA dot plot representation with stringency 1 and window size 1 (see text). The nucleotides at each site of Sequence 1 and Sequence 2 are reported on the x (5′ → 3′ from left to right) and y axes (5′ → 3′ from top to bottom), respectively. Each nucleotide site of a sequence is compared with another sequence, and a dot is placed in the graph in case of identical residues. The path drawn by the diagonal line represents regions where the two sequences are identical; the vertical arrow indicates a deletion in Sequence 1 or an insertion in Sequence 2. (B) Sequence alignment deduced folowing the path indicated in the dot plot representation.

DNA or amino-acid stretches of length 10 (window size = 10) with the mismatch limit equal to 3, which means that a dot will be placed in the graph only if at least 7 of 10 paired residues in the two 10-nucleotide stretches compared have identical residues (see Section 3.10). In Figure 3.6A, Sequence 1 and Sequence 2 are compared by choosing a window size of 1 (and a mismatch limit of 1). A high similarity region shows as a diagonal in the dot plot. This method is an excellent way to quickly infer homology between two sequences. Only sequence stretches that are part of a diagonal in a dot-plot analysis are similar enough to be used in an unambiguous alignment and are informative for phylogenetic analysis. In practice, it is assumed that highly similar regions between two sequences are homologous (i.e., they evolved from a common ancestral sequence).

3.6.2 Dynamic programming

For two sequences, dynamic programming can find the best alignment by giving scores for all possible pairs of aligned residues and GPs. This is a relatively rapid procedure on modern computers, requiring time and memory proportional to the product of the sequence lengths (N); this is usually referred to as an N^2 algorithm or as having complexity $O(N^2)$. In practice, unless the two sequences are enormous, dynamic programming can be achieved in a matter of seconds. The approach can easily be generalized to more than two sequences. In this case, first the alignment maximizing the similarity of the sequences in each column (using an amino-acid weight matrix) is found, allowing a minimal number and lengths of gaps. This task is most commonly expressed as finding the alignment that gives the best score for the following formula, called the weighted sum of pairs (WSP) objective function:

$$\Sigma\Sigma\, W_{ij}\, D_{ij}$$

For any multiple alignment, a score between each pair of sequences (D_{ij}) is calculated. Then, the WSP function is simply the sum of all of these scores, 1 for each possible pair of sequences. There is an extra term W_{ij} for each pair, which is a weight; by default, it will always be equal to 1, but it gives extra weight to some pairs. This can be extremely useful because it gives more weight to pairs that might be more reliable than others. Alternatively, it can be used to give less weight to sequences with many close relatives because these are overrepresented in the data set. Dynamic programming can then be used to find a multiple alignment that gives the best possible score for the WSP function. However, the time and memory required for this grow exponentially with the number of sequences as the complexity is $O(N^M)$, where M is the number of sequences and N is the sequence length. This quickly becomes impossible to compute for more than four sequences of even modest lengths.

A solution to the computational complexity came from the MSA program of Lipman et al. (1989). It used a so-called branch-and-bound technique to eliminate many unnecessary calculations and make it possible to compute the WSP function for five to eight sequences. The precise number of sequences depended on length and similarity; the more similar the sequences, the faster the calculation and the number that could be aligned. Although this is an important program, its use is limited by the small number of sequences that can be managed. In tests with BaliBase, this program performed extremely well but it cannot handle all test cases (i.e., those with more than eight sequences). The FastMSA program, a highly optimized version of MSA, is faster and uses less memory, but is still limited to small data sets.

3.6.3 Genetic algorithms

The SAGA program (Notredame and Higgins, 1996) is based on the WSP objective function but uses a genetic algorithm rather than dynamic programming to find the best alignment. This stochastic optimization technique grows a population of alignments and evolves over time using a process of selection and crossing to find the best alignment. In terms of the WSP function, SAGA can find the optimal alignment compared to the performance of MSA. The advantage of SAGA, however, is its ability to deliver good alignments for more than eight sequences; the disadvantage is that it is still relatively slow, perhaps taking many hours of computing to deliver a good alignment for 20 or 30 sequences. The program also must be run several times because results can differ between runs. Despite these reservations, SAGA has proven useful because it allows the testing of alternative WSP scoring functions.

3.6.4 Other algorithms

The DCA (Stoye et al., 1997) and PRRP (Gotoh, 1996) programs also compute alignments according to the WSP scoring function. The former program finds sections of alignment that, when joined together head to tail, give the best alignment. Each section is found using the MSA program, so, ultimately, it is limited in the number of sequences it can handle, even if more than MSA. PRRP uses an iterative scheme to gradually work toward the optimal alignment. At each cycle of iteration, the sequences are split into two groups (randomly). Within each group, the sequences are kept in fixed alignment, and are then aligned to each other using dynamic programming. The cycle is repeated until the score converges. The alignments produced by PRRP are excellent, as judged by test cases using `BaliBase` and other programs (Notredame et al., 1998; Thompson et al., 1999a). The program is slow with more than 20 sequences, but it is faster than SAGA.

The `DIALIGN` method of Morgenstern (1999) is based on finding sections of local multiple alignment in a set of sequences; that is, sections similar across all sequences, but relatively short. Other methods described herein are based on finding global alignments, and they perform the best when the sequences are similar in length and have a similar domain structure, with some sequence similarity along the full length of the alignment. `DIALIGN` is useful if the similarity is localized to isolated domains. In practice, the algorithm performed well in two of the `BaliBase` test sets: those with long insertions and deletions; otherwise, `ClustalW` (see Section 3.7) easily outperforms `DIALIGN`.

3.7 Progressive alignment

It is clear that multiple alignments are useful for phylogenetic analysis. Conversely, if the phylogenetic relationships in a set of sequences were known, this information

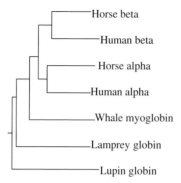

Figure 3.7   A rooted tree showing the possible phylogenetic relationships between the seven globin sequences in Figure 3.1. Branch lengths are drawn to scale.

could be used to help generate an alignment. Indeed, this mutual relationship was used as the basis of an early multiple-alignment method (Sankoff, 1985), which simultaneously generated the tree and alignment, even if the particular method was too complex for routine use. A simple shortcut is to create an approximate tree of the sequences and use it to make a multiple alignment; this approach was first suggested by Hogeweg and Hesper (1984). The method is heuristic in a mathematical sense insofar as it makes no guarantees to produce an alignment with the best score according to a formula. Nonetheless, this method is extremely important because it is the way the majority of automatic alignments are generated. As judged by test cases, it performs well, although not quite as well as those methods that use the WSP objective function. This lack of sensitivity is only visible with the most difficult test cases; for similar sequences, progressive alignment is sufficient. However, the sheer speed of the method and its great simplicity make it extremely attractive for routine work.

A possible phylogenetic tree showing the relatedness of the sequences in Figure 3.1 is shown in Figure 3.7. A similar tree can be generated quickly by making all possible pairwise alignments between all the sequences and calculating a distance (i.e., the proportion of residues that differ between the two sequences) in each case. Such distances are used to make the tree with one of the widely available distance methods, such as the **Neighbor-Joining (NJ)** method of Saitou and Nei (1987) (see Chapter 5). NJ trees, for example, can be calculated quickly for as many as a few hundred sequences. Next, the alignment is gradually built up by following the branching order in the tree. The two closest sequences are aligned first, using dynamic programming with GPs and a weight matrix. For further alignment, the two sequences are treated as one, such that any gaps created between the two cannot be moved. Again, the two closest remaining sequences or prealigned groups of sequences are aligned to each other. Two unaligned sequences or two subalignments can be aligned or a sequence can be added to a subalignment, depending on which is closer. The process is repeated until all the sequences are aligned. Once the initial tree is generated, the multiple alignment can be accomplished with only N−1 separate alignments for N sequences. This process is fast enough to allow the alignment of many hundreds of sequences.

3.7.1 Clustal

The most commonly used software for this type of alignment is `ClustalW` (Thompson et al., 1994) and `ClustalX` (Thompson et al., 1997). These programs are freely available as source code and/or executable images for most computer platforms. They also can be run using servers on the Internet at a number of locations. These programs are identical in terms of alignment method, but offer either a simple text-based interface (`ClustalW`) suitable for high-throughput tasks or a graphical interface (`ClustalX`). In the discussion that follows, only `ClustalW` is referred to, but it applies equally to `ClustalX`. `ClustalW` can take a set of input sequences and automatically perform the entire progressive alignment procedure. The sequences are aligned in pairs to generate a distance matrix that can be used to make a simple initial tree of the sequences. This guide tree is stored in a file and generated using the NJ method, which produces an unrooted tree (see Chapter 1) that is used to guide the multiple alignment. The multiple alignment is finally carried out using the progressive approach, described previously.

`ClustalW` has specific features that help it make more accurate alignments. First, sequences are downweighted according to how closely related they are to other sequences (as judged by the guide tree). This is useful because it prevents large groups of similar sequences from dominating an alignment. Second, the weight matrix used for protein alignments varies, depending on how closely related are the next two sequences or sets of sequences. One weight matrix can be used for closely related sequences that gives high scores to identities and low scores otherwise. For distantly related sequences, the reverse is true; it is necessary to give high scores to conservative amino-acid matches and lower scores to identities. `ClustalW` uses a series of four matrixes chosen from either the BLOSUM or PAM series (Henikoff and Henikoff, 1992; Dayhoff et al., 1978, respectively). During alignment, the program attempts to vary GPs in a sequence- and position-specific manner, which helps align sequences of different lengths and different similarities. Position-specific GPs are used to concentrate gaps in the loops between the secondary-structure elements, which can be set either manually using a mask or automatically. In the latter case, GPs are lowered in runs of hydrophilic residues (likely loops) or at positions where there are already many gaps. They are also lowered near some residues, such as Glycine, which are empirically known to be common near gaps (Pascarella and Argos, 1992). GPs are raised adjacent to existing gaps and near certain residues. These and other parameters can be set by the user before each alignment.

3.7.2 T-Coffee

Progressive alignment is fast and simple but it does have one obvious drawback: a local minimum problem. Any alignment errors (i.e., misaligned residues or whole domains) that occur during the early alignment steps cannot be corrected later as more data are added. This may be due to an incorrect topology in the guide tree, but it is more likely due to simple errors in the early alignments. The latter occurs when the alignment with the best score is not the best one biologically or if all sequences in a data set are considered. An effective way to overcome this problem is to use the recently developed ***T-Coffee method*** (Notredame et al., 2000).

   T-Coffee is slower than `ClustalW` but provides more accurate alignments when tested with `BaliBase`, as well as the PRRP or DIALIGN programs (Gotoh, 1996; Morgenstern, 1999). Indeed, this increase in accuracy is greatest for the most difficult test cases and is found across all of the `BaliBase` test sets. The method is based on finding the multiple alignment that is most consistent with a set of pairwise alignments between the sequences. The pairwise alignments can derive from a combination of sources, such as different alignment programs or types of data including structure superpositions and sequence alignments. These are processed to find the aligned pairs of residues in the initial data set that are most consistent across different alignments. This information then is used to compile data on which residues are most likely to align in which sequences. The final stage is to create the multiple alignment using normal progressive alignment, which is fast, simple, and requires no other parameters such as GPs or weight matrixes. The disadvantages of T-Coffee over `ClustalW` are the extra computer time required for alignment and the lack of functionality in the software. The former is of concern for only 50 sequences; the latter will change over time as new software is developed.

3.8 Hidden Markov models

An interesting approach to alignment is the HMMs, which are based on probabilities of residue substitution and gap insertion and deletion (Krogh et al., 1994). HMMs have been shown to be extremely useful in a wide variety of situations in computational molecular biology, such as locating introns and exons or predicting promoters in DNA sequences. HMMs are also useful for summarizing the diversity of information in an existing alignment of sequences and predicting whether new sequences belong to the family (e.g., Eddy, 1998; Bateman et al., 2000). Some packages simultaneously generate such an HMM and find the alignment from unaligned sequences. These methods used to be inaccurate (e.g., see results in Notredame et al., 1998); however, recent progress has been made and the SAM method of Hughey and Krogh (1996) is now roughly comparable to `ClustalW` in accuracy, although not as easy or as fast to use.

3.9   Nucleotide sequences versus amino-acid sequences

Nucleotide sequences may be coding or noncoding. In the former case, they may code for structural or catalytic RNA species but more commonly for proteins. In the case of protein-coding genes, the alignment can be accomplished based on the nucleotide or the amino-acid sequences. This choice may be biased by the type of analysis to be carried out after alignment (see Chapter 8); for example, silent changes in closely related sequences may be counted. In this case, an amino-acid alignment will not be useful for later analysis because there will be few differences between the sequences. By contrast, if the sequences are distantly related, analysis can be by either amino-acid or nucleotide differences. Regardless of the end analysis desired, amino-acid sequence alignments are easier to carry out and less ambiguous than nucleotide alignments. This is also true of sequence database searching; in addition to the reasons already discussed, most alignment programs do not recognize a codon as a unit of sequence, separating them during alignment. This is not true for certain two-sequence alignment and database search programs (e.g., Searchwise and Pairwise [Birney et al., 1996]), but is the case for most multiple-alignment programs. A typical approach is to carry out the alignment at the amino-acid level and use it to generate a corresponding nucleotide sequence alignment, which can then be analyzed as usual. Numerous computer programs are available; for example, PROTAL2DNA by Catherine Letondal (**http://bioweb.pasteur.fr/seqanal/ interfaces/protal2dna.html**) or DAMBE (discussed later in this section).

If the sequences are not protein coding then the only choice is to carry out a nucleotide alignment. If the sequence code is for structural RNA (e.g., small subunit ribosomal RNA [SSU rRNA]), these will be constrained by function to conserve primary and secondary structure over at least some of their lengths. Typically, there are regions of clear nucleotide identity interspersed by regions that are free to change rapidly; software performance depends on the circumstances. With rRNA, the most commonly used programs manage to align the large core sections that are conserved across broad phylogenetic distances. These core alignment blocks, however, will be interspersed by the highly variable expansion segments; most programs have difficulty with them. Consideration of the secondary structures, perhaps using a dedicated RNA editor, can help but it may still be difficult to find an unambiguous alignment. There may be no clear alignment in these regions that can be chosen over any others. Excluding these regions from further analysis should be seriously considered. If the alignment is arbitrary, further analysis will be unnecessary. Fortunately, there are usually enough clearly conserved blocks of alignment to make a phylogenetic analysis possible.

If the nucleotide sequences are noncoding (e.g., SINES or introns), then alignment may be difficult once the sequences diverge beyond a certain level. Sequences that are very unconstrained can accumulate indels and substitutions in all positions;

these rapidly become unalignable. There is no algorithmic solution and the situation will be especially difficult if the sequences have small repeats. Even if a somewhat "optimal" alignment is obtained using an alignment score or *parsimony*, there may be no reason to believe it is biologically reasonable. Such scores are based on arbitrary assumptions about the details of the evolutionary process in the sequences. Even if the details of the assumptions are justifiable, the alignment may be so hidden as to be otherwise unobtainable. If one alignment cannot be justified over an alternative one, be careful (or just pragmatic).

PRACTICE

Des Higgins and Marco Salemi

3.10 Searching for homologous sequences with `BioEdit`

`BioEdit` is a DNA and amino-acid sequence editor running under Windows with several functions available, including alignment, BLAST searches, plasmid drawing, and restriction mapping. The program can be downloaded for free from **http://www.mbio.ncsu.edu/BioEdit/bioedit.html**. The Web site also includes manuals in `pdf`. In the discussion that follows, it is assumed that `BioEdit` is properly installed in the user's computer and will be used to obtain dot plots. A MacOs compatible application, `DottyPlot`, which performs dot plot analysis, is available at **http://bioinformatics.weizmann.ac.il/software/mac/**. This program is not discussed here; however, with its user-friendly interface, `DottyPlot` could easily perform the exercise with `BioEdit`.

Table 2.1 provides the accession numbers for the HIV/SIV data set that will be used as an example throughout the book. Before obtaining a multiple alignment, it is necessary to check whether the sequences are similar enough to be aligned unambiguously. As discussed in Section 3.6.1, this can be done easily with a dot-plot representation. The HIV Subtypes A, B, C, D, and G in the table are in HIV-1 Group M; the two other major groups are HIV-1 Group N and HIV-1 Group O. It is known from the literature that strains belonging to the same phylogenetic group share enough similarity to be aligned without major problems (Sharp et al., 1994; Louwagie et al., 1995). However, as shown in Table 2.1, the SIVcpz strains isolated from African chimpanzees (AF103818 and X52154) and the HIV Group O strain (MVP5180) belong to two distinct lineages and are rather divergent from HIV-1 Group M subtypes (Gao et al., 1999; Hahn et al., 2000). HIV-2, another major type of HIV found mainly in West African populations (Hahn et al., 2000), also exists in several subtypes and is phylogenetically related to simian immunodeficiency viruses isolated from West African sooty mangabeys (SIVsmm). The following exercise investigates whether the *env* region of HIV-1, SIVcpz, HIV-2, and SIVsmm strains can be aligned unambiguously at the nucleotide level.

`BioEdit` requires an input file in the program folder containing the sequence to compare in FASTA format (see Box 2.2), and it can perform dot-plot comparisons of sequences up to two thousand nucleotides long.

1. From the HIV Los Alamos database, download the *env* sequences of the following isolates (see Section 2.9.1): ANT70 (HIV-1, Group O), cpzGAB (SIV isolated from a chimpanzee), bHXB2R (HIV-1 Subtype B), cETH2220 (HIV-1 Subtype C), H2A_ALI (HIV-2 Subtype A), H2B_D205 (HIV-2 Subtype B), and smmSTM

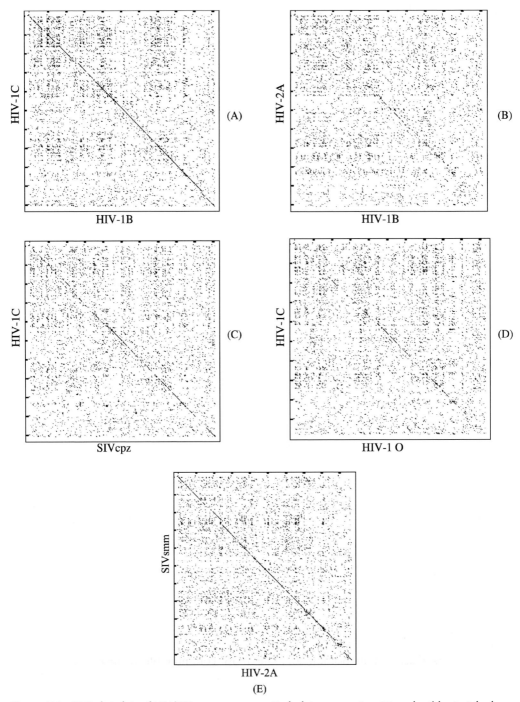

Figure 3.8 DNA dot plots of HIV/SIV *env* sequences. Each dot represents a 10-nucleotide stretch along the two sequences with at least 70% similarity (mismatch limit = 3; see text for more details). (A) HIV-1B (bottom) versus HIV-1C (left). (B) HIV-1B (bottom) versus HIV-2A (left). (C) SIVcpz (bottom) versus HIV-1C (left). (D) HIV-1 O (bottom) versus HIV-1C (left). (E) HIV-2A (bottom) versus SIVsmm (left).

    (SIV isolated from a West African sooty mangabey). Save the unaligned sequences in a single file called DOTPLOT.fas in FASTA format.

2. Double-click on the BioEdit icon and select Open from the File menu. Browse to the DOTPLOT.fas file.

3. Select sequences bHXB2R and cETH222 by right-clicking on the sequence name in the left panel of the window (hold the CTRL key to select more than one sequence). Choose Dot plot at the end of the Sequence menu. Leave the window size at 10 and set the mismatch limit at 3. Click OK twice when the Plot matrix window appears. The diagonal should look like the one shown in Figure 3.8A.

4. Minimize the window and do the same for bHXB2R (HIV-1 Subtype B) and H2A_ALI (HIV-2 Subtype A). Compare the new "diagonal" (Figure 3.8B) to the previous one. Other pairwise comparisons can be explored in the same way.

The dot plot appears on the computer screen in a few seconds (or a few minutes, depending on the computer speed and sequence length). The plot can be saved as a bitmap file by choosing Save as from the File menu and imported in PowerPoint (or a similar program) for further editing or printing.

Figure 3.8 suggests that HIV-1 Subtypes B and C can be aligned unambiguously in the *env* region, and that both SIVcpz and HIV Group O (Figures 3.8C, 3.8D) may be suitable **outgroups**. The comparison of HIV-2A and SIVsmm (see Figure 3.8E) leads to similar conclusions. On the contrary, Figure 3.8B shows that almost no similarity exists in *env* between HIV-1 and HIV-2. Therefore, a nucleotide alignment of HIV-1/HIV-2 sequences, at least in this region, could be unreliable for inferring *homology* (see Section 1.3 and the following discussion).

Dot plots can be used also to identify the sequence overlap between two sequences of unequal length. Multiple alignment works best with sequences that are cut down to the same **homologous regions**. Because the algorithm penalizes the end gaps in the same way as alignment gaps, errors are easily introduced at the beginning and end of the alignment.

3.11   File formats for Clustal

As discussed in Section 2.7, the most common file formats are GenBank, EMBL, SWISS-PROT, FASTA, PHYLIP, and ClustalW. GenBank, EMBL, and SWISS-PROT formats are typically used for just one sequence at a time and most of each entry is devoted to information about the sequence. These formats are not usually used for multiple alignment. Nonetheless, Clustal can read these formats plus hyphens ("-") as gap characters so that an entire multiple alignment can be written in such formats if needed. PHYLIP format can be written by Clustal and is used only for multiple alignments. Clustal format can be written and read by Clustal so that alignments can be recycled and reused in various ways. This not only allows alignments to be read in so that new sequences can be added, but also to be taken in later with a new alignment written in a different format. Clustal can be used as an alignment converter (see Sections 2.7 and 2.8); can read multiple alignments in the following formats: NBRF/PIR, EMBL/SWISS-PROT, FASTA, GDE, Clustal, GCG/MSF, and RSF; and can write alignment files in the following formats: NBRF/PIR, GDE, Clustal, GCG/MSF, and PHYLIP. An alignment can be read in and then output again in a new format without being changed.

3.12 Access to ClustalW and ClustalX

ClustalW and ClustalX are both available for free and can be downloaded from the EMBL/EBI file server (**ftp://ftp.ebi.ac.uk/pub/software/**) or from the ICGEB in Strasbourg, France (**ftp://ftp-igbmc.u-strasbg.fr/pub/ClustalW/** and **ftp://ftp-igbmc.u-strasbg.fr/pub/ClustalX/**). These sites are also accessible through the phylogeny programs Web site maintained by Joe Felsenstein at the University of Seattle, Washington (**http://evolution.genetics.washington.edu/PHYLIP/software.html**). Clustal is available for PCs running MSDOS or Windows, Macintosh computers (although the programs do not conform to the usual Mac look and feel), VAX VMS, and UNIX, including Linux. In each case, ClustalX (X stands for X windows) provides a graphical user interface, users can use a mouse, and there is a colorful display of alignments. ClustalW has an older, text-based interface and is less attractive for occasional use. Nonetheless, it does have extensive command-line facilities, making it extremely useful for embedding in scripts for high-throughput use. Clustal is also available in numerous commercial packages; the GCG package includes a similar program (PILEUP), which does not contain as many features as Clustal but does produce satisfactory alignments.

Access to Clustal is also possible directly from several servers across the Internet, which is especially attractive for occasional users but is not without drawbacks. First, users do not have access to the full set of features that Clustal provides. Second, it can be complicated, sending and retrieving many sequences or alignments. Third, it can take a long time to carry out large alignments with little indication about progress. Nonetheless, recommended excellent Clustal servers in use are those at the EBI (**http://www.ebi.ac.uk/ClustalW/**) and the BCM search launchers, especially **http://www.hgsc.bcm.tmc.edu/SearchLauncher/**.

3.13 Aligning the HIV/SIV sequences with ClustalX

Place the file hivULN (**http://www.kuleuvem.ac.be/aidslab/phylogenyBook/datasets/hivULN.phy**) (see Section 2.9.1) in the ClustalX folder and open it with

ClustalX. Select Do complete Alignment from the Alignment menu. ClustalX creates an OUTPUT Guide Tree File and an OUTPUT Alignment File. The default format for the Alignment File is the Clustal format, but it is possible to choose a different format in the Output Format Options from the Alignment menu; for example, the PHYLIP format, which can be read by most of the phylogeny packages. The Guide Tree File contains the guide tree generated by ClustalX to carry out the multiple alignment (see Section 3.7.1). Note: The guide tree is a rough phylogenetic tree that should not be used to draw conclusions about the phylogeny of the *taxa* under investigation!

ClustalX also allows the alignment parameters to change (from Alignment Parameters in the Alignment menu). Unfortunately, there are no general rules about choosing the best set of parameters, such as the gap-open or the gap-extension penalty. If an alignment shows, for example, too many large gaps, the user can try to increase the gap-opening penalty and redo the alignment. ClustalX shows an alignment quality curve at the bottom of the aligned sequences that can be used to evaluate a given alignment. By selecting Calculate Low-Scoring Segment and Show Low-Scoring Segments from the Quality menu, it is also possible to visualize particularly unreliable parts of the alignments, which may be better to exclude from subsequent analyses. As noted in Section 3.9, the user should remember that alignment scores are based on arbitrary assumptions about details of the evolutionary process and are not always biologically plausible.

Probably the most important problem when editing or refining an alignment is how to handle gaps. Gaps need to be inserted in an alignment because phylogenetic inference requires *positional homology*; that is, the nucleotides observed at a particular position in the taxa under investigation are supposed to trace their ancestry to a single position that occurred in a common ancestor (Swofford et al., 1996). When sequences assumed to be homologous – as the *env* sequences in the HIV/SIV example – have different lengths, inserting gaps should allow recovery of that positional homology. However, especially when large gaps are present in some of the taxa, it is not always clear whether to put a gap in a particular column rather than the preceding or following one to recover the true positional homology. In addition, most of the methods used to infer evolutionary distances and phylogenetic trees do not explicitly treat gaps. For these reasons, gap columns are usually excluded from the final alignment before phylogenetic analysis, which can be done by manually editing the aligned sequences with one of the programs discussed in Section 3.10. Some phylogeny packages, like PAUP* (see Chapter 7), DAMBE (see the next section and Chapter 13), and TREECON (see Chapter 9), can exclude gap columns from an alignment before further analyses. However, excluding gapped columns is not always the best strategy. If in a given data set of aligned sequences only a few of them contain large gaps, removing all the columns with gaps is a waste of phylogenetic information. In such cases, it may be wiser either to remove the few sequences with large gaps from the alignment, or to replace each gap in those few sequences with a question mark (?) character which can be done with a text editor. Most phylogenetic programs, like PHYLIP, PAUP*, or TREE-PUZZLE, interpret "?" as a missing character. In this way, we avoid excluding from the analysis sites containing homologous nucleotides for all but few sequences. Conversely, it is safer to exclude regions where most of the sequences have gaps because the alignment is likely to be wrong there and information about positional homology is probably lost.

3.14   Aligning nucleotide sequences in a coding region with DAMBE

The HIV/SIV alignment generated by ClustalX shows several regions with gaps, some of which have been inserted within a codon triplet. If gap columns are excluded from this alignment, the information about codon position is lost. This would make it impossible, for example, to calculate evolutionary distances based on *synonymous* and/or *nonsynonymous substitutions* (see Chapter 11) or to separately analyze the 1st, 2nd, and 3rd codon positions. As discussed in Section 3.9, a common approach to generating a nucleotide-sequence alignment in a coding region is to carry out the alignment at the amino-acid level and then translate it back to nucleotides. In this way, the problem of inserting gaps within a coding triplet is avoided. The task can be easily accomplished with DAMBE. First, it is necessary to translate the DNA sequence into the corresponding amino-acid sequence.

1. Place the hivULN.fas file in the DAMBE folder, run DAMBE.exe, and open the file by choosing Open (sequence or gene frequency) from the File menu. The window Sequence Info appears.
2. Select Protein-coding Nuc. Seq. and Universal. Click Go!. DAMBE warns that the sequences are not of equal length; click OK.
3. Select Work on Amino Acid Sequence from the Sequence menu. Nucleotide sequences are translated into the corresponding amino-acid sequences.

Stop codons cannot be present in the original sequences for DAMBE to successfully carry out Step 3. Moreover, DAMBE starts the translation from the first nucleotide – considered the first codon position – of each sequence. Ensure that the coding sequences are in the correct frame, with no stop codons.

The amino-acid sequences can be aligned by selecting Align Sequences using ClustalW from the Sequences menu. A new window appears in which

```
L20571    MTVTM--             L20571    MTVTM
AF103818  --MKVME            AF103818  MKVME
X52154    --MKVME            X52154    MKVME
U09127    --MRVMG            U09127    MRVMG
U27426    --MKVRG            U27426    MKVRG
U27445    --MRVKG            U27445    MRVKG
AF067158  --MRVRG            AF067158  MRVRG
U09126    --MRVEG            U09126    MRVEG
U27399    --MRVRE            U27399    MRVRE
U43386    --MRVRE            U43386    MRVRE
L02317    --MRVKE            L02317    MRVKE
AF025763  --MRVKG            AF025763  MRVKG
U08443    --MRAKE            U08443    MRAKE
AF042106  --MTVKG            AF042106  MTVKG
```
            (A)   (B)

Figure 3.9   HIV/SIV amino-acid alignment obtained with DAMBE. (A) Before editing. (B) After editing.

it is possible to modify gap-open and gap-extension penalties. Retain the default values and click Go !. After the alignment has been carried out, select No in the window that asks whether the user wishes to see the guide tree for the multiple alignment.

The beginning of the alignment should appear as in Figure 3.9A. It is clear that the algorithm erred by placing two gaps in the first two columns of the alignment below the first two amino-acid residues of sequence L20571. According to positional homology (see the previous section), the expected alignment should look like Figure 3.9B. Therefore, it is necessary to edit the alignment (for example, using the alignment editor discussed in Section 3.10) before translating it back to nucleotides. After editing:

1.   Reopen the edited amino-acid alignment in DAMBE.
2.   Select `Align nuc.seq.to aligned AA seq.in buffer` from the `Sequence` menu, and open the file `hivULN.fas` containing the original unaligned nucleotide sequences.

DAMBE carries out the back translation, and the aligned nucleotide sequences appear in the main window, ready to be saved in the appropriate format for subsequent analyses. Gap columns can be excluded from the alignment before saving by selecting `Site-wise deletion of unresolved nucleotides` from `Get Rid of Segment` in the `Sequence` menu. The final alignment should look like the one provided in file `hivALN.phy`, which will be used in following chapters.

3.15   Adding sequences to preexisting alignments

`Clustal` has an entire subsection devoted to the task of adding one or more sequences to an existing alignment. The user can input two files, one of which is usually an existing alignment. The task is then to align one or more sequences with this program. Doing it this way is better than simply adding all sequences together in one file and performing the complete multiple alignment from scratch because (1) the alignment may be large and time-consuming; (2) the alignment may contain manually edited information; or (3) the user may use structural information to mark the alignment using a structural mask, which gives locations of helixes and strands that can be used to guide gaps into loops. One way or another, the second file can contain (1) a second set of prealigned sequences or a single sequence; or (2) a set of new unaligned sequences. In the first case, the program simply aligns the two together in one step, just as is done during progressive alignment. The gap penalties and so forth are exactly as used during multiple alignment and can be set by the user. In the second case, the sequences can be aligned to the existing alignment one at a time, guiding each so as to align it with its closest relatives in the first alignment.

3.16   Editing and viewing multiple alignments

Changing an alignment is not unusual and is done to correct obvious alignment errors, remove sections of dubious quality, or cosmetically improve an alignment. Standard word-processing software is sufficient if the end result is to be used in a figure in a publication, for example. If the alignment is to be used for further analysis, however, file format is important. It is better to use a dedicated alignment editor and/or viewer so that the user can select sets of sequences or alignment positions and choose from several file formats. The SEAVIEW package (Galtier et al., 1996) is available for UNIX, including Linux and Windows (**http://pbil.univ-lyon1.fr/software/seaview.html**), and has powerful graphics for alignment editing; it is also closely integrated with the PHYLOWIN (Galtier et al., 1996) and Clustal packages. The user can select either sets of sequences to be aligned or realigned by ClustalW or regions of the alignment to be refined. The DCSE editor by Peter DeRijk has specific facilities for editing rRNA alignments (**http://rrna.uia.ac.be/dcse/**).

The ClustalX package provides extensive facilities for viewing amino-acid alignments using a variety of flexible and dynamic coloring schemes. A color can be selected for a residue not only according to an initial table of colors, but also according to the combination of residues in the column. It is also possible to highlight residues, sequences, or sections of protein alignment that appear misaligned as a result of sequencing errors in the original nucleotide sequence from which the amino-acid sequences were derived. The program also clearly highlights an outlier sequence (i.e., one that does not belong in the data set).

3.17   Databases of alignments

In the future, it will be clear how many different protein and RNA sequence families exist and large alignments will be available through computer networks. Meanwhile, extensive alignments for the largest and most complicated families (e.g., rRNA) already exist, which are maintained and updated regularly, and can be accessed from **http://rrna.uia.ac.be/**. Numerous databases of protein alignments exist, each with advantages and disadvantages, different file formats, and other pertinent details. The Interpro project (**http://www.ebi.ac.uk/interpro/index.html**) aims to link some of the most important protein family databases, and is an invaluable resource for information about a family of interest and for downloading sample alignments. These alignments can be used for both the basis of further phylogenetic work and educational or informational purposes.

REFERENCES

Altschul, S. F., T. L. Madden, A. A. Schaeffer, J. Zhang, A. Zhang, W. Miller, and D. J. Lipman (1997). Gapped BLAST and PSI-BLAST: A new generation of protein database search programs. *Nucleic Acids Research*, 25, 3398–3402.

Bateman, A., E. Birney, R. Durbin, S. R. Eddy, K. L. Howe, and E. L. L. Sonnhammer (2000). The Pfam protein families database. *Nucleic Acids Research*, 28, 263–266.

Birney, E., J. D. Thompson, and T. J. Gibson (1996). PairWise and SearchWise: Finding the optimal alignment in a simultaneous comparison of a protein profile against all DNA translation frames. *Nucleic Acids Research*, 24, 2730–2739.

Dayhoff, M. O., R. M. Schwartz, and B. C. Orcutt (1978). In: *Atlas of Protein Sequence and Structure*, vol. 5, suppl. 3, ed. M. O. Dayhoff, p. 345. NBRF, Washington, DC.

Eddy, S. R. (1998). Profile of hidden Markov models. *Bioinformatics*, 14, 755–763.

Galtier, N., M. Gouy, and C. Gautier (1996). SeaView and Phylo_win, two graphic tools for sequence alignment and molecular phylogeny. *Computer Applications in the Biosciences*, 12, 543–548.

Gao F., E. Bailes, D. L. Robertson, Y. Chen, C. M. Rodenburg, S. F. Michael, L. B. Cummins, L. O. Arthur, M. Peeters, G. M. Shaw, P. M. Sharp, and B. H. Hahn (1999). Origin of HIV-1 in the chimpanzee *Pan troglodytes troglodytes*. *Nature*, 397, 436–441.

Gotoh, O. (1982). An improved algorithm for matching biological sequences. *Journal of Molecular Biology*, 162, 705–708.

Gotoh, O. (1996). Significant improvement in accuracy of multiple-protein sequence alignments by iterative refinements as assessed by reference to structural alignments. *Journal of Molecular Biology*, 264, 823–838.

Hahn, B. H., G. M. Shaw, K. M. De Cock, and P. M. Sharp (2000). AIDS as a zoonosis scientific and public health implications. *Science*, 287, 607–614.

Henikoff, S. and J. G. Henikoff (1992). Amino-acid substitution matrixes from protein blocks. *Proceedings of the National Academy of Sciences USA*, 89, 10915–10919.

Hogeweg, P. and B. Hesper (1984). The alignment of sets of sequences and the construction of phylogenetic trees: An integrated method. *Journal of Molecular Evolution*, 20, 175–186.

Hughey, R. and A. Krogh (1996). Hidden Markov models for sequence analysis: Extension and analysis of the basic method. *Computer Applications in Biosciences*, 12, 95–107.

Krogh, A., M. Brown, I. S. Mian, K. Sjolander, and D. Haussler (1994). Hidden Markov models in computational biology: Applications to protein modeling. *Journal of Molecular Biology*, 235, 1501–1531.

Lipman, D. J., S. F. Altschul, and J. D. Kececioglu (1989). A tool for multiple-sequence alignment. *Proceedings of the National Academy of Sciences USA*, 86, 4412–4415.

Louwagie, J., W. Janssens, J. Mascola, L. Heyndrickx, P. Hegerich, G. van der Groen, F. E. McCutchan, and D. S. J. Burke (1995). Genetic diversity of the envelope glycoprotein from human immunodeficiency virus type 1 isolates of African origin. *Journal of Virology*, 69, 263–271.

Morgenstern, B. (1999). DIALIGN2: Improvement of the segment-to-segment approach to multiple-sequence alignment. *Bioinformatics*, 15, 211–218.

Needleman, S. B. and C. D. Wunsch (1970). A general method applicable to the search for similarities in the amino-acid sequences of two proteins. *Journal of Molecular Biology*, 48, 443–453.

Notredame, C. and D. G. Higgins (1996). SAGA: Sequence alignment by genetic algorithm. *Nucleic Acids Research*, 24, 1515–1524.

Notredame, C., L. Holm, and D. G. Higgins (1998). COFFEE: An objective function for multiple-sequence alignments. *Bioinformatics*, 14, 407–422.

Notredame, C., D. G. Higgins, and J. Heringa (2000). T-Coffee: A novel method for fast and accurate multiple-sequence alignment. *Journal of Molecular Biology*, 302, 205–217.

Pascarella, S. and P. Argos (1992). Analysis of insertions and deletions in protein structures. *Journal of Molecular Biology*, 224, 461–471.

Saitou, N. and M. Nei (1987). The neighbor-joining method: A new method for reconstructing phylogenetic trees. *Molecular Biology and Evolution*, 4, 406–425.

Sankoff, D. (1985). Simultaneous solution of the RNA folding, alignment, and protosequence problems. *SIAM Journal on Applied Mathematics*, 45, 810–825.

Sharp, P. M., D. L. Robertson, D. L. Gao, and B. H. Hahn (1994). Origins and diversity of human immunodeficiency viruses. *AIDS*, 8, S27–S42.

Smith, T. F. and M. S. Waterman (1981). Identification of common molecular subsequences. *Journal of Molecular Biology*, 147, 195–197.

Stoye, J., V. Moulton, and A. W. Dress (1997). DCA: An efficient implementation of the divide-and-conquer approach to simultaneous multiple-sequence alignment. *Computer Applications in the Biosciences*, 13, 625–626.

Swofford, D. L., G. J. Olsen, P. J. Waddell, and D. M. Hillis (1996). Phylogenetic inference In: *Molecular Systematics*, 2nd ed., eds. D. M. Hillis, C. Moritz, and B. K. Mable pp. 407–514, Sinauer Associates Inc., Sunderland, MA.

Thompson, J. D., D. G. Higgins, and T. J. Gibson (1994). CLUSTALW: Improving the sensitivity of progressive multiple-sequence alignment through sequence weighting, position-specific gap penalties, and weight matrix choice. *Nucleic Acids Research*, 22, 4673–4680.

Thompson, J. D., T. J. Gibson, F. Plewniak, F. Jeanmougin, and D. G. Higgins (1997). The CLUSTAL_X windows interface: Flexible strategies for multiple-sequence alignment aided by quality analysis tools. *Nucleic Acids Research*, 25, 4876–4882.

Thompson, J. D., F. Plewniak, and O. Poch (1999a). A comprehensive comparison of multiple-sequence alignment programs. *Nucleic Acids Research*, 27, 2682–2690.

Thompson, J. D., F. Plewniak, and O. Poch (1999b). BaliBase: A benchmark alignment database for the evaluation of multiple-alignment programs. *Bioinformatics*, 15, 87–88.

Nucleotide substitution models

THEORY

Korbinian Strimmer and Arndt von Haeseler

4.1 Introduction

The first step in the analysis of aligned nucleotide or amino-acid sequences usually consists of the computation of **genetic distances** (or **evolutionary distances**) between the deoxyribonucleic acid (DNA) sequences. In this chapter, two questions that naturally arise in this context are discussed: first, what is a reasonable definition of a genetic distance; and, second, how can it be estimated using statistical models of the substitution process.

It is well known that a variety of evolutionary forces act on DNA sequences (see Chapter 1); as a result, sequences change in the course of time. Therefore, any two sequences derived from a common ancestor that evolve independently will eventually diverge from each other (Figure 4.1). A measure of this divergence is called a genetic distance. Not surprisingly, this quantity plays an important role in many aspects of sequence analysis. First, by definition, it provides a measure of the dissimilarity between sequences. Second, if a **molecular clock** is assumed (see Chapter 10), then the genetic distance is linearly proportional to the time elapsed. Third, for sequences related by an evolutionary tree, the branch lengths represent the distance between the nodes (i.e., sequences) in the tree. Therefore, if the exact amount of sequence divergence between all pairs of sequences from a set of N sequences is known, the genetic distance provides a basis to infer the evolutionary tree relating the sequences. In particular, if sequences actually evolved according to a tree and if the correct genetic distances between all pairs of sequences are available, then it is computationally simple to reconstruct this tree (see Chapter 5).

The substitution of nucleotides or amino acids in a sequence is usually considered a random event. Consequently, an important prerequisite for computing

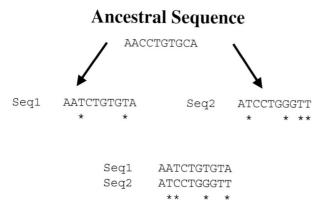

Figure 4.1   Two sequences derived from the same common ancestral sequence mutate and diverge. Mutations are indicated by an asterisk.

genetic distances is the prior specification of a ***model of substitution*** that provides a statistical description of this stochastic process. Once a mathematical model of substitution is assumed, then straightforward procedures exist to infer genetic distances from the data.

In this chapter, the mathematical framework to model the process of nucleotide substitution is described. The most widely used classes of models are discussed and an overview of how genetic distances are estimated using these models is provided, focusing especially on those designed for the analysis of nucleotide sequences.

4.2   Observed and expected distances

The simplest approach to measure the divergence between two strands of aligned DNA sequences is to count the number of sites where they differ. The proportion of different ***homologous*** sites is called ***observed distance***, sometimes also called ***p-distance***, and it is expressed as the number of nucleotide differences per site.

The p-distance is an intuitive measure. Unfortunately, it suffers from a severe shortcoming: if the rate of substitution is high, it may considerably underestimate the number of substitutions that actually occurred. This is the effect of ***homoplasy*** (see Chapter 1). Assume that two or more mutations take place at the same site in the sequence (i.e., multiple hits); for example, suppose an A is being replaced by a C, and then by a G. As result, although two replacements have occurred, only one difference is observed (A to G). Moreover, in case of a back mutation (e.g., A to C to A) or a parallel mutation (e.g., both sequences mutated into an A), not even a single replacement would be detected. As a consequence, the observed distance p underestimates the true genetic distance d; that is, the actual number of substitutions per site that occurred. Figure 4.2 illustrates the general relationship

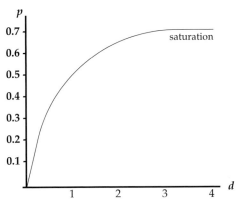

Figure 4.2   Relationships between expected genetic distance d and observed p-distance. As the genetic distance between two sequences increases, the curve reaches a plateau where the p-distance is no longer related with d (i.e., substitution saturation; see Section 4.12).

between d and p. The precise shape of this curve depends on the details of the substitution model used.

Because the genetic distance cannot be observed directly, statistical techniques are necessary to infer this quantity from the data. For example, using the relationship between d and p given in Figure 4.2, it is possible to map an observed distance p to the corresponding genetic distance d. This transformation is generally nonlinear. Conversely, d can also be inferred directly from the sequences using *maximum-likelihood* methods. At high genetic distances, the p-distance reaches a plateau, which is called **saturation** (see Section 4.11), because so many mutations have occurred that virtually every position has mutated. In the example above, sequence dissimilarities are always lower than 75%, and the observed distance is lower than 0.75.

In the following sections, the substitution process as a stochastic process is described. Later, the "mathematical" mechanics of nucleotide substitution and how *maximum-likelihood estimators* (*MLEs*) are derived are outlined.

4.3   Number of mutations in a given time interval *(optional)

Poisson processes are well suited to model processes such as radioactive decay, telephone calls, spread of epidemics, population growth, and so on, where at any time an event can take place at a more or less constant rate. The Poisson process also can be used to estimate the number of mutations $X(t)$ that occurred during a time interval t; that is, per unit of time a mutation occurs with intensity or rate μ, and the events that occur are expressed as an integer number.

Let $P_n(t)$ denote the probability that exactly n mutations occurred during the time t:

$$P_n(t) = P(X(t) = n) \tag{4.1}$$

If t is changed, this probability will change.

Consider a time interval δt. It is reasonable to assume that the occurrence of a new mutation in this interval is independent of the number of mutations that happened so far. When δt is small compared to the rate μ, $\mu \delta t$ equals the probability that exactly one mutation occurs during δt. The probability of no mutation during δt is $1 - \mu \delta t$. In other words, it is assumed that at the time $t + \delta t$, the number of mutations either remains unchanged or increases by one, compared to the number of mutations at time t. More formally,

$$P_0(t + \delta t) = P_0(t) \cdot (1 - \mu \delta t) \tag{4.2}$$

That is, the probability of no mutation up to time $t + \delta t$ is equal to the probability of no mutation up to time t multiplied by the probability that no mutation took place during the interval $(t, t + \delta t)$. If exactly n mutations are observed during this period, then two possible scenarios have to be considered. In the first scenario, $n - 1$ mutations occurred up to time t and exactly one mutation occurred during δt, with the probability of observing n mutations given by $P_{n-1}(t) \cdot \mu \delta t$. In the second scenario, n mutations already occurred at time t and no further mutation takes place during δt, with the probability of observing n mutations given by $P_n(t) \cdot (1 - \mu \delta t)$. Thus, the total probability of observing n mutations at time $t + \delta t$ is given by the sum of the probabilities of the two possible scenarios, as follows:

$$P_n(t + \delta t) = P_{n-1}(t) \cdot \mu \delta t + P_n(t) \cdot (1 - \mu \delta t) \tag{4.3}$$

Equations 4.2 and 4.3 can be rewritten as:

$$[P_0(t + \delta t) - P_0(t)]/\delta t = -\mu P_0(t) \tag{4.4a}$$

$$[P_n(t + \delta t) - P_n(t)]/\delta t = \mu[P_{n-1}(t) - P_n(t)] \tag{4.4b}$$

When δt tends to zero, the left part of Equations 4.4a and 4.4b can be rewritten (ignoring certain regularity conditions) as the first derivative of $P(t)$ with respect to t, as follows:

$$P_0'(t) = -\mu \cdot P_0(t) \tag{4.5a}$$

$$P_n'(t) = \mu \cdot [P_{n-1}(t) - P_n(t)] \tag{4.5b}$$

These are typical differential equations that can be solved to compute the probability $P_0(t)$ that no mutation has occurred at time t. In fact, the solution is the function $P_0(t)$ such that its derivative equals $P_0(t)$ itself multiplied by the rate μ. An obvious solution is the exponential function:

$$P_0(t) = \exp(-\mu t) \tag{4.6}$$

That is, with probability $\exp(-\mu t)$, no mutation occurred in the time interval $(0, t)$. Alternatively, it could be said that the probability that the first mutation occurred at time $x \geq t$ is given by:

$$F(x) = 1 - \exp(-\mu t) \tag{4.7}$$

This is exactly the density function of the **exponential distribution** with parameter μ. In other words, the time to the first mutation is exponentially distributed: the longer the time, the higher the probability that a mutation occurs. Incidentally, the times between any two mutations also are exponentially distributed with parameter μ. This is the result of the underlying assumption that the mutation process "does not know" how many mutations already occurred.

To compute the probability that a single mutation occurred at time t: $P_1(t)$, recall Equation 4.5b:

$$P_1'(t) = \mu \cdot [P_0(t) - P_1(t)] \tag{4.8}$$

From elementary calculus, remember the well-known rule of products to compute the derivative of a function $f(t)$, when $f(t)$ is of the form $f(t) = h(t)g(t)$:

$$f'(t) = g'(t)h(t) + g(t)h'(t) \tag{4.9}$$

Comparing Equations 4.9 and 4.8, the idea is that $P_1(t)$ can be written as the product of two functions; that is, $P_1(t) = h(t)g(t)$, where $h(t) = P_0(t) = \exp(-\mu t)$ and $g(t) = \mu t$. Thus, $P_1(t) = (\mu t)\exp(-\mu t)$. To compute the derivative, Equation 4.8 is reproduced. Induction leads to Equation 4.10:

$$P_n(t) = [(\mu t)^n \exp(-\mu t)]/n! \tag{4.10}$$

This formula describes the **Poisson distribution**; that is, the number of mutations up to time t is **Poisson-distributed** with parameter μt. On average, μt mutations with variance μt are expected. The parameters μ (nucleotide substitutions per site per unit time) and t (the time) are confounded, meaning that they cannot be estimated separately but only through their product μt (number of mutations per site up to time t).

$$Q = \begin{pmatrix} -(a\pi_C + b\pi_G + c\pi_T) & a\pi_C & b\pi_G & c\pi_T \\ g\pi_A & -(g\pi_A + d\pi_G + e\pi_T) & d\pi_G & e\pi_T \\ h\pi_A & i\pi_C & -(h\pi_A + i\pi_C + f\pi_T) & f\pi_T \\ j\pi_A & k\pi_C & l\pi_G & -(j\pi_A + k\pi_C + l\pi_G) \end{pmatrix}$$

$$\begin{array}{cccc} \mathbf{A} & \mathbf{C} & \mathbf{G} & \mathbf{T} \end{array}$$

Figure 4.3 Instantaneous rate Q matrix. Each entry in the matrix represents the instantaneous substitution rate from nucleotide i to nucleotide j (rows and columns follow the order **A, C, G, T**). a, b, c, d, e, f, g, h, i, j, k, l are rate parameters describing the relative rate of each nucleotide substitution to any other. π_A, π_C, π_T, π_G are frequency parameters corresponding to the nucleotide frequencies (Yang, 1994a). Diagonal elements are chosen so that the sum of each row is equal to zero. The expected substitution rate is then given by Eq. 4.24.

4.4 Nucleotide substitutions as a homogeneous Markov process

The **nucleotide substitution process** of DNA sequences outlined in the previous section (i.e., the Poisson process) can be generalized to a so-called **Markov process** that uses a **Q matrix**, which specifies the relative rates of change of each nucleotide along the sequence (see Section 4.5 for the mathematical details). The most general form of the Q matrix is shown in Figure 4.3. Rows follow the order A, C, G, and T so that, for example, the second term of the first row is the instantaneous rate of change from base A to base C. This rate is given by the frequency of base C, times a, a relative-rate parameter, describing (in this case) how often the substitution A to C occurs during evolution with respect to other possible substitutions. In other words, each nondiagonal entry in the matrix represents the flow from nucleotide i to j, while the diagonal elements are chosen to make the sum of each row equal to zero because they represent the total flow that leaves nucleotide i.

Nucleotide substitution models like those summarized by the Q matrix in Figure 4.3 belong to a general class of models known as **time-homogeneous time-continuous stationary Markov models**. When applied to modeling nucleotide substitutions, they all share the following set of underlying assumptions:

- At any given site in a sequence, the rate of change from base i to base j is independent of the base that occupied that site prior to i (**Markov property**).
- Substitution rates do not change over time (**homogeneity**).
- The relative frequencies of A, C, G, and T (π_A, π_C, π_G, π_T) are at equilibrium (**stationarity**).

These assumptions are not necessarily biologically plausible; they are the consequence of modeling substitutions as a stochastic process. Within this general framework, several submodels can still be developed. This book examines only the so-called time-reversible models; that is, those that assume for any two nucleotides that the rate of change from i to j is always the same as from j to i ($a = g, b = h,$ $c = j, d = i, e = k, f = l$ in the Q matrix). As soon as the Q matrix – and thus the evolutionary model – is specified, it is possible to calculate the probabilities of change from any base to any other during the evolutionary time t, $\mathbf{P}(t)$, by computing the matrix exponential:

$$\mathbf{P}(t) = \exp(\mathbf{Q}t) \tag{4.11a}$$

This equation has the same form as Equation 4.6, $P_0(t) = \exp(-\mu t)$, which calculates the probability that no mutation occurred in the time interval $(0, t)$. Here, $\mathbf{P}(t)$ is the probability matrix giving the probability for any $i \rightarrow j$ nucleotide substitution up to time t. As in Equation 4.6, such probabilities are exponentially distributed (i.e., the more time, the more likely that a new mutation occurs). Instead of a single parameter μ, the mean **nucleotide substitution rate**, Equation 4.11a has a Q matrix in the exponential specifying the relative nucleotide substitution rates for each possible $i \rightarrow j$ nucleotide substitution according to a particular evolutionary model (e.g., Jukes and Cantor, Tamura-Nei, or the general time-reversible model; see Section 4.6). When the probabilities $\mathbf{P}(t)$ are known, this equation also can be used to compute the expected genetic distance between two sequences according to the evolutionary models specified by the Q matrix.

Consider two homologous sequences, both sampled at time t_1, having the same nucleotide i at a given site. Two main scenarios are possible. Either their common ancestor at time t_0 possessed the same nucleotide i at that site, or it had any one of the other three bases. In the first scenario, the probability that the ancestral base i remained the same during the interval (t_0, t_1) is given by $P_{ii}(t_1)$ for any of the two sequences, and by $[P_{ii}(t_1)]^2$ for both. Analogously, the probability that an ancestral base i mutated into j in both sequences is given by $[P_{ij}(t_1)]^2$; however, because j could have been any one of the remaining three bases, the total probability of the second scenario is $3[P_{ij}(t_1)]^2$. Therefore, the probability of two sequences sharing the same nucleotide at any given site, $I(t)$, is computed as the sum of the probabilities of all these possible scenarios (assuming that $\pi_A = \pi_C = \pi_G = \pi_T$):

$$I(t) = [P_{ii}(t_1)]^2 + 3[P_{ij}(t_1)]^2 \tag{4.11b}$$

and the probability of the two sequences being different, $p = 1 - I(t)$, as

$$p = 1 - [P_{ii}(t_1)]^2 - 3[P_{ij}(t_1)]^2 \tag{4.11c}$$

It can be demonstrated that an unbiased estimator of p is the **observed proportion** of different sites between two sequences being compared; that is, the p-distance (see Section 4.2). Note that Eq. 4.11c is a special case of Eq. 4.29.

$$
Q = \begin{pmatrix}
-\tfrac{3}{4} & \tfrac{1}{4} & \tfrac{1}{4} & \tfrac{1}{4} \\
\tfrac{1}{4} & -\tfrac{3}{4} & \tfrac{1}{4} & \tfrac{1}{4} \\
\tfrac{1}{4} & \tfrac{1}{4} & -\tfrac{3}{4} & \tfrac{1}{4} \\
\tfrac{1}{4} & \tfrac{1}{4} & \tfrac{1}{4} & -\tfrac{3}{4}
\end{pmatrix}
$$

Figure 4.4   Instantaneous rate Q matrix for the Jukes and Cantor model (JC69).

The next section describes how to calculate $P(t)$ and the expected genetic distance in the case of the simple **Jukes and Cantor model** of evolution (Jukes and Cantor, 1969); for more complex models, only the main results are discussed.

4.4.1   The Jukes and Cantor (JC69) model

The simplest possible **nucleotide substitution model**, introduced by Jukes and Cantor in 1969 (**JC69**), assumes that the equilibrium frequencies of the four nucleotides are 25% each and that during evolution, any nucleotide has the same probability to be replaced by any other. These assumptions correspond to a Q matrix with $\pi_A = \pi_C = \pi_G = \pi_T = 1/4$, and $a = b = c = d = e = f = 1$ (Figure 4.4). The matrix fully specifies the rates of change between pairs of nucleotides in the **JC69 model**. To obtain an analytical expression for p, it is necessary to know how to compute $P_{ii}(t)$ (the probability of a nucleotide to remain the same during the evolutionary time t) and $P_{ij}(t)$ (the probability of replacement). This can be accomplished by solving the exponential $\mathbf{P}(t) = \exp(\mathbf{Q}t)$ (Equation 4.11a), with $\mathbf{Q}$ as the instantaneous rate matrix for the JC69 model. The detailed solution requires the use of matrix algebra (see Section 4.5 for the relevant mathematics), but the result is quite straightforward:

$$P_{ii}(t) = 1/4 + 3/4 \exp(-t) \tag{4.12a}$$

$$P_{ij}(t) = 1/4 - 1/4 \exp(-t) \tag{4.12b}$$

By substituting $P_{ii}(t)$ and $P_{ij}(t)$ from Equations 4.12a and 4.12b in Equation 4.29, the following formula for two sequences with a total divergence of t time units is derived:

$$p = 3/4[1 - \exp(-t)] \tag{4.13}$$

which can be rewritten as follows by solving for t:

$$t = -\ln(1 - 4/3\,p) \tag{4.14}$$

The interpretation of the previous equation is simple: Under the JC69 model, with rate matrix Q as in Figure 4.4, the expected substitution rate per site is $\mu = 3/4$ (see Eq. 4.24), so that the number of expected substitutions in time t is $d =$

$\mu t = 3/4t$. Conversely, p is the observed distance or p-distance; that is, the observed proportion of different nucleotides between the two sequences (see Section 4.4). By substituting t with $4/3d$ in Equation 4.14 and rearranging it somewhat, the *Jukes and Cantor correction formula* for the genetic distance d between two sequences finally is obtained:

$$d = -3/4\, ln(1 - 4/3\, p) \tag{4.15a}$$

It also can be demonstrated that the variance $V(d)$ will be given by:

$$V(d) = 9\, p(1 - p)/(3 - 4p)^2 n \tag{4.15b}$$

(Kimura and Ohta, 1972). Similarly, more complex nucleotide substitution models can be implemented depending on which parameters of the Q matrix are estimated (see Section 4.6). The practical part of this chapter shows how to calculate pairwise genetic distances for the example data sets according to different models. Chapter 10 discusses a statistical test that can help select the best-fitting nucleotide substitution model for a given data set.

4.5 Derivation of Markov process *(optional)

This section shows how the stochastic process for nucleotide substitution can be derived from basic principles such as detailed balance and the *Chapman-Kolmogorov equations*. To model the substitution process on the DNA level, it is commonly assumed that a replacement of one nucleotide by another occurs randomly and independently, and that nucleotide frequencies π_i in the data do not change over time or from sequence to sequence in an alignment. Under these assumptions, a *time-homogeneous stationary Markov process* can model the mutation process. In this model, each site in the DNA sequence is treated as a random variable with a discrete number n of possible states. For nucleotides, there are four states ($n = 4$) that correspond to the four nucleotide bases A, C, G, and T. The Markov process specifies the transition probabilities from one state to another; that is, it gives the probability of the replacement of nucleotide i by nucleotide j after a certain period of time t. These probabilities are collected in the transition probability matrix $\mathbf{P}(t)$. Its components $P_{ij}(t)$ satisfy the following conditions:

$$\sum_{j=1}^{n} P_{ij}(t) = 1 \tag{4.16}$$

and

$$P_{ij}(t) > 0 \text{ for } t > 0 \tag{4.17}$$

Moreover, it also fulfills the requirement that

$$P(t + s) = P(t) \cdot P(s) \qquad (4.18)$$

known as the Chapman-Kolmogorov equation, and the initial condition

$$P_{ij}(0) = 1, \text{ for } i = j \qquad (4.19a)$$

$$P_{ij}(0) = 0, \text{ for } i \neq j \qquad (4.19b)$$

For simplicity, it is also often assumed that the substitution process is reversible; that is,

$$\pi_i P_{ij}(t) = \pi_j P_{ji}(t) \qquad (4.20)$$

This additional condition on the substitution process, known as detailed balance, implies that the substitution process has no preferred direction. For small t, the transition probability matrix $\mathbf{P}(t)$ can be linearly approximated (*Taylor expansion*) by

$$\mathbf{P}(t) \approx \mathbf{P}(0) + t\mathbf{Q} \qquad (4.21)$$

where $\mathbf{Q}$ is called the rate matrix. It provides an infinitesimal description of the substitution process. To not violate Equation 4.16, the rate matrix $\mathbf{Q}$ satisfies

$$\sum_{j=1}^{n} Q_{ij} = 0 \qquad (4.22)$$

which can be achieved by defining

$$Q_{ii} = -\sum_{i \neq j}^{n} Q_{ij} \qquad (4.23)$$

Note that $Q_{ij} > 0$ can be interpreted as the flow from nucleotide i to j. $Q_{ii} < 0$ is then the total flow that leaves nucleotide i; hence, it is less than zero. In contrast to $\mathbf{P}$, the rate matrix $\mathbf{Q}$ does not comprise probabilities; rather, it describes the amount of change of the substitution probabilities per unit time. The rate matrix is the first derivative of $\mathbf{P}(t)$, which is constant for all t in a time-homogeneous Markov process. As a consequence of Equation 4.23, the total number of substitutions per unit time (i.e., the total rate μ) can be written as

$$\mu = -\sum_{i=1}^{n} Q_{ii}\pi_i \qquad (4.24)$$

so that the number of substitutions during time t equals $d = \mu t$. Note that in this equation, μ and t are confounded. As a result, the rate matrix can be arbitrarily scaled; that is, all entries can be multiplied by the same factor without changing the overall substitution pattern. Only the unit in which time t is measured will be affected. For a reversible process $\mathbf{P}$, the rate matrix $\mathbf{Q}$ can be decomposed into rate parameters R_{ij} and nucleotide frequencies π_i:

$$Q_{ij} = R_{ij}\, \pi_j, \text{ for } i \neq j \tag{4.25a}$$

$$Q_{ij} = -\sum_{i \neq m}^{n} Q_{im}, \text{ otherwise} \tag{4.25b}$$

The matrix $\mathbf{R} = R_{ij}$ is symmetric, $R_{ij} = R_{ji}$, and has vanishing diagonal entries, $R_{ii} = 0$.

From the Chapman-Kolmogorov Equation 4.18, the forward and backward differential equations are obtained:

$$\frac{d}{dt}\mathbf{P}(t) = \mathbf{P}(t)\mathbf{Q} = \mathbf{Q}\mathbf{P}(t) \tag{4.26}$$

They can be solved under the initial condition (Equations 4.19a and 4.19b) to give

$$\mathbf{P}(t) = \exp(t\mathbf{Q}) = \sum_{m=0}^{\infty} \frac{(t\mathbf{Q})^m}{m!} \tag{4.27}$$

For a reversible-rate matrix $\mathbf{Q}$ (see Equation 4.20), this quantity can be computed by *spectral decomposition* (Bailey, 1964):

$$P_{ij}(t) = \sum_{m=1}^{n} \exp(\lambda_m t)U_{mi}U_{jm}^{-1} \tag{4.28}$$

where the λ_i are the eigenvalues of $\mathbf{Q}$, $\mathbf{U} = (U_{ij})$ is the matrix with the corresponding eigenvectors, and $\mathbf{U}^{-1}$ is the inverse of $\mathbf{U}$.

Choosing a nucleotide substitution model in the framework of a reversible-rate matrix amounts to specifying explicit values for the matrix $\mathbf{R}$ and the frequencies π_i. Assuming n different states, the model has $n - 1$ independent frequency parameters π_i (as $\Sigma \pi_i = 1$) and $[n(n-1)/2] - 1$ independent rate parameters (because the scaling of the rate matrix is irrelevant, and $R_{ij} = R_{ji}$ and $R_{ii} = 0$). Thus, in the case of nucleotides ($n = 4$), the substitution process is governed by three independent frequency parameters (π_i) and five independent rate parameters (R_{ij}).

4.5.1 Inferring the expected distances

Once the rate matrix $\mathbf{Q}$ or, equivalently, the parameters π_i and R_{ij} are fixed, the substitution model provides the basis to statistically infer the genetic distance d between two DNA sequences. Two different techniques exist, both of which are widely used. The first approach relies on computing the exact relationship between d and p for the given model (see Figure 4.2). The probability that a substitution is observed after time t is

$$p = 1 - \sum_{i=1}^{n} \pi_i P_{ii}(t) \tag{4.29}$$

Using Equation 4.28, the following is obtained:

$$p = 1 - \sum_{i=1}^{n} \pi_i P_{ii}(t) \left(-\frac{d}{\sum_{i=1}^{n} \pi_i Q_{ii}} \right) \tag{4.30}$$

This equation can be used to construct a method of moments estimator of the **expected distance** by solving for d and estimating p (observed proportion of different sites) from the data. This formula is a generalization of Equation 4.13.

Another way to infer the expected distance between two sequences is to use a maximum-likelihood approach. This requires the introduction of a **likelihood function** $L(d)$ (see Chapter 6 for more details). The likelihood is the probability to observe the two sequences given the distance d. It is defined as

$$L(d) = \prod_{s=1}^{l} \pi_{x_{A(s)}} P_{x_{A(s)} x_{B(s)}} \left(-\frac{d}{\sum_{i=1}^{n} \pi_i Q_{ii}} \right) \tag{4.31}$$

where $x_{A(s)}$ is the state at site $s = 1, \ldots l$ in sequence A and $P_{x_{A(s)} x_{B(s)}}$ is the transition probability. A value for d that maximizes $L(d)$ is called a maximum-likelihood estimate (MLE) of the genetic distance. To find this estimate, numerical optimization routines are employed because analytical results are generally not available. Estimates of error of the inferred genetic distance can be computed for both the methods-of-moments estimator (Equation 4.30) and the likelihood estimator (Equation 4.31) using standard statistical techniques. The so-called delta method can be employed to compute the variance of an estimate obtained from Equation 4.30, and **Fisher's information criterion** is helpful to estimate the asymptotic variance of MLEs.

4.6 Nucleotide substitution models

If all of the eight free parameters of a reversible nucleotide rate matrix $\mathbf{Q}$ are specified, the **general time-reversible** (**GTR**) model is derived (Figure 4.5). However, it is

$$Q = \begin{pmatrix} -(a\pi_C + b\pi_G + c\pi_T) & a\pi_C & b\pi_G & c\pi_T \\ a\pi_A & -(a\pi_A + d\pi_G + e\pi_T) & d\pi_G & e\pi_T \\ b\pi_A & d\pi_C & -(b\pi_A + d\pi_C + f\pi_T) & f\pi_T \\ c\pi_A & e\pi_C & f\pi_G & -(c\pi_A + e\pi_C + f\pi_G) \end{pmatrix}$$

$$\begin{matrix} \mathbf{A} & \mathbf{C} & \mathbf{G} & \mathbf{T} \end{matrix}$$

Figure 4.5 Q matrix of the general time-reversible (GTR) model of nucleotide substitutions.

often desirable to reduce the number of free parameters, particularly when parameters are unknown and, therefore, need to be estimated from the data. This can be achieved by introducing constraints reflecting some approximate symmetries in the underlying substitution process. For example, nucleotide exchanges fall into two major groups (Figure 4.6). Substitutions in which a purine is exchanged for a pyrimidine or vice versa (A ↔ C, A ↔ T, C ↔ G, G ↔ T) are called *transversions* (Tv); all other substitutions are *transitions* (Ts). Additionally, it is possible to distinguish between substitutions among purine and pyrimidines: *purine transitions* (A ↔ G) Ts$_R$ and *pyrimidine transitions* (C ↔ T) Ts$_Y$. When these constraints are imposed, only two independent rate parameters (out of five) remain: the ratio κ of the Ts and Tv rates and the ratio γ of the two types of transition rates. This defines the *Tamura-Nei* (*TN93*) *model* (Tamura and Nei, 1993), which can be written as follows:

$$R_{ij}^{TN} = \kappa \left(\frac{2\gamma}{\gamma + 1} \right) \quad \text{for Ts}_Y \tag{4.32a}$$

$$R_{ij}^{TN} = \kappa \left(\frac{2}{\gamma + 1} \right) \quad \text{for Ts}_R \tag{4.32b}$$

$$R_{ij}^{TN} = 1 \quad\quad \text{for Tv} \tag{4.32c}$$

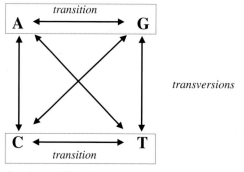

Figure 4.6 The six possible substitution patterns for nucleotide data.

If $\gamma = 1$ and, therefore, the purine and pyrimidine transitions have the same rate, this model reduces to the **HKY85 model** (Hasegawa, Kishino, and Yano, 1985):

$$R_{ij}^{\text{HKY}} = \kappa \text{ for Ts} \tag{4.33a}$$

$$R_{ij}^{\text{HKY}} = 1 \text{ for Tv} \tag{4.33b}$$

If the base frequencies are uniform ($\pi_i = 1/4$), the HKY85 model further reduces to the **Kimura 2-parameters (K80) model** (Kimura, 1980). For $\kappa = 1$, the HKY85 model is called the **F81 model** (Felsenstein, 1981), and the K80 model degenerates to the Jukes and Cantor (JC69) model. The **F84 model** (Thorne et al., 1992; Felsenstein, 1993) is also a special case of the TN93 model; it is similar to the HKY85 model but uses a slightly different parameterization. A single parameter τ generates the κ and γ parameters of the TN93 model (see Equations 4.32a, b, and c) in the following fashion:

First the quantity

$$\rho = \frac{\pi_R \pi_Y [\pi_R \pi_Y \tau - (\pi_A \pi_G + \pi_C \pi_T)]}{(\pi_A \pi_G \pi_Y + \pi_C \pi_T \pi_R)} \tag{4.34}$$

is computed, which determines both

$$\kappa = 1 + \frac{1}{2}\rho \left(\frac{1}{\pi_R} + \frac{1}{\pi_Y} \right) \tag{4.35}$$

and

$$\gamma = \frac{\pi_Y + \rho}{\pi_Y} \frac{\pi_R}{\pi_R + \rho} \tag{4.36}$$

of the TN93 model, where π_A, π_C, and so on are the base frequencies; π_R and π_Y are the frequency of purines and pyrimidines. The hierarchy of the substitution models discussed previously is shown in Figure 4.7 (see Figure 10.1 for more models).

4.6.1 Rate heterogeneity over sites

It is a well-known phenomenon that the rate of nucleotide substitution can vary substantially for different positions in a sequence. For example, in protein-coding genes, third codon positions usually fix mutations faster than first positions, which, in turn, fix mutations faster than second positions. Such a pattern of evolution is commonly explained by the presence of different evolutionary forces for the sites in question. This problem was not addressed in the previous sections and rate homogeneity over sites was silently assumed. However, rate heterogeneity can play a crucial part in the inference of genetic distances. To account for the site-dependent rate variation, a plausible model for distribution of rates over sites is required. The

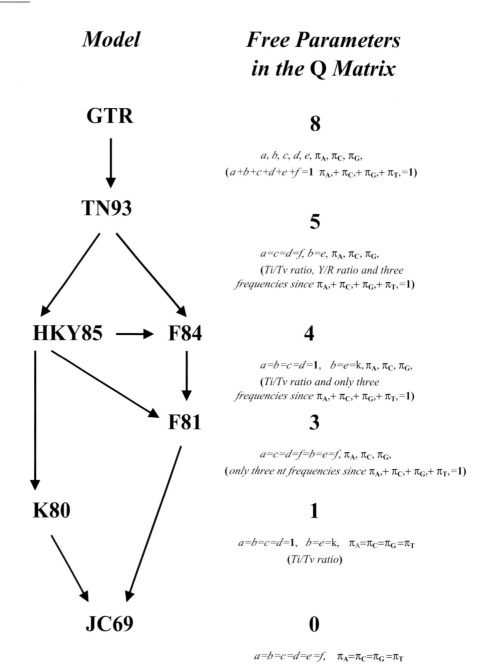

Figure 4.7 Hierarchy of nucleotide substitution models. The meaning of the free parameters to be estimated for each model is given in parentheses. Ti/Tv indicates the transition/transversion ratio; Y/R indicates the pyrimidine transitions/purine transversion ratio.

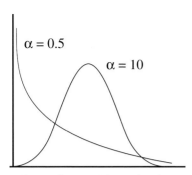

Figure 4.8   Different shapes of the Γ-distribution depending on the α shape parameter.

common "empirical" approach is to use a Γ-***distribution*** with expectation 1.0 and variance $1/\alpha$.

$$\text{Pdf}(r) = \alpha^\alpha r^{\alpha-1}/\exp(\alpha r)\Gamma(\alpha) \tag{4.37}$$

By adjusting the shape parameter α, the Γ-distribution allows varying degrees of rate heterogeneity (Figure 4.8). For $\alpha > 1$, the distribution is bell-shaped and models weak rate heterogeneity over sites. The relative rates drawn from this distribution are all close to 1.0. For $\alpha < 1$, the Γ-distribution takes on its characteristic L-shape, which describes situations of strong rate heterogeneity (i.e., some positions have large substitution rates, but most other sites are practically invariable). Rather than using the ***continuous*** Γ-***distribution***, it is computationally more efficient to assume a ***discrete*** Γ-***distribution*** with a finite number c of equally probable rates $q_1, q_2, \ldots, q_c$. Usually, four to eight discrete categories are sufficient to obtain a good approximation of the continuous function (Yang, 1994b) (see Section 9.2).

PRACTICE: THE PHYLIP AND TREE-PUZZLE SOFTWARE PACKAGES

Marco Salemi

4.7 Software packages

Numerous software packages exist for computing genetic distances from DNA sequences. An exhaustive list is maintained by Joe Felsenstein at **http://evolution. genetics.washington.edu/PHYLIP/software.html**. Among others, the programs PAUP* (see Chapter 7), PAL (Drummond and Strimmer, 2001), MEGA2 (Kumar et al., 1993; see Chapter 11), TREECON (Van de Peer, 1994; see Chapter 9), DAMBE (Xia, 2000; see Chapter 13), and PAML (Yang, 2000; see Chapter 10) provide the possibility to infer *maximum-likelihood distances*.

PHYLIP (Phylogeny Inference Package), one of the first freeware phylogeny softwares to be developed (Felsenstein, 1993), is a package that consists of several programs for calculating genetic distances and inferring *phylogenetic trees* according to different algorithms. Already-compiled executable files are available for Windows3.x/95/98, pre-386 and 386 DOS, Macintosh (non-PowerMac), and PowerMac. A complete description of the package, including instructions for installation on different machines, can be found at **http://evolution. genetics.washington.edu/ PHYLIP.html**. The PHYLIP software modules discussed throughout this book are briefly summarized in Box 4.1.

TREE-PUZZLE (Strimmer and von Haeseler, 1996) was originally developed to reconstruct phylogenetic trees from molecular sequence data by maximum likelihood with a fast tree-search algorithm called *quartet-puzzling* (see Chapter 6). The program also computes pairwise maximum-likelihood distances according to a number of models of nucleotide substitution. Compatible of TREE-PUZZLE for UNIX, MacOS, and Windows95/98/NT can be downloaded for free from the TREE-PUZZLE Web site at **http://www.TREE-PUZZLE.de/**. The program was mainly developed to estimate phylogenetic trees with the quartet-puzzling algorithm, which is described in Chapter 6. In what follows, it is shown how to compute genetic distances according to different *evolutionary models* with TREE-PUZZLE.

As soon as the proper versions of these programs are installed, the PHYLIP and the TREE-PUZZLE folders should be visible on the local computer. These folders contain several files, including executable applications, documentation, and source codes. In the Mac version of PHYLIP 3.5 the executables are in the PHYLIP 3.573c exe folder. The PHYLIP V3.573c directory contains extensive documentation and

---

Box 4.1 The PHYLIP software package

PHYLIP contains several executables for analyzing molecular and morphological data and drawing phylogenetic trees. However, only some of the software modules included in the package are discussed in this book; that is, those dealing with DNA and amino-acid sequences analysis, which are summarized herein. Information about the other modules can be found in the PHYLIP documentation available with the program at **http://evolution.genetics.washington.edu/PHYLIP/software.html**.

PHYLIP Executable	Input data	Type of analysis	Book chapters
DNAdist.exe	aligned DNA sequences	calculates genetic distances according to different nucleotide substitution models	5
ProtDist.exe	aligned protein sequences	calculates genetic distances according to different amino-acid substitution matrixes	8
Neighbor.exe	genetic distances	calculates NJ or UPGMA trees	5, 8
Fitch.exe	genetic distances	calculates Fitch-Margoliash trees	5, 8
Kitch.exe	genetic distances	calculates Fitch-Margoliash trees with the assumption of a molecular clock	5
DNAML.exe	aligned DNA sequences	calculates ML trees	13
ProtPars.exe	aligned protein sequences	calculates parsimony trees	8
SeqBoot.exe	aligned DNA or protein sequences	generates bootstrap replicates of aligned DNA or protein sequences	5
Consense.exe	phylogenetic trees (usually obtained from bootstrap replicates)	generates a consensus tree	5

the source codes. In TREE-PUZZLE Version 5.0, the executable treepuzzle. exe can be found in the TREE-PUZZLE folder. Any of the software modules within PHYLIP and TREE-PUZZLE works in the same basic way: it needs a file containing the input data (e.g., aligned DNA sequences in PHYLIP format [see Box 2.2]) to be placed in the same directory where the program resides. It then produces one or more output files (usually called outfile and treefile) in text format, containing the results of some kind of analysis. By default, any application reads the data from a file named infile (no extension type!) if such a file is present in the same directory; otherwise, the user is asked to enter the name of the input file. Other details about PHYLIP are provided in Box 4.1.

4.8   Jukes and Cantor (JC69) genetic distances

Begin by calculating d for the HIV data set: the aligned sequences can be downloaded from **http://kuleuven.ac.be/aidslab/phylogenyBook/datasets.htm**. Figure 4.9 is a matrix with pairwise p-distances (i.e., number of different sites between two sequences divided by the sequence length) for the HIV/SIV data. The matrix is written in lower triangular format with the number on the first row indicating the number of sequences being compared. The genetic distance d between sequence L20571 and AF103818 according to the JC69 model can be obtained by substituting the **observed distance** $p = 0.3937$ in Equation 4.15a (see Section 4.4.1); it results in 0.5582. Obviously, using p instead of d would grossly underestimate the genetic divergence between the two sequences.

Place the file hivALN in the same directory in which the PHYLIP software module DNAdist is saved. Rename the file infile and start DNAdist. By double-clicking on its icon, a new window appears with the following menu:

```
Nucleic acid sequence Distance Matrix program, version 3.5c

Settings for this run:
    D   Distance (Kimura, Jin/Nei, ML, J-C)?  Kimura 2-parameter
    T   Transition/transversion ratio?        2.0
    C   One category of substitution rates?   Yes
    L   Form of distance matrix?              Square
    M   Analyze multiple data sets?           No
    I   Input sequences interleaved?          Yes
    0   Terminal type (IBM PC, VT52, ANSI)    ANSI
    1   Print out the data at start of run    No
    2   Print indications of progress of run  Yes

Are these settings correct? (type Y or letter for
    one to change)
```

Type a D, followed by the enter key, and again until the model selected is Jukes-Cantor. In the new menu, the option T Transition/transversion

```
14
L20571
AF103818    0.3937
X52154      0.3920 0.3010
U09127      0.3941 0.3104 0.3210
U27426      0.3912 0.3333 0.3244 0.1573
U27445      0.3848 0.3274 0.3240 0.1475 0.0927
AF067158    0.3967 0.3401 0.3299 0.1594 0.1713 0.1671
U09126      0.3958 0.3248 0.3193 0.1543 0.1688 0.1573 0.0816
U27399      0.3895 0.3227 0.3236 0.1641 0.1743 0.1675 0.1794 0.1692
U43386      0.3924 0.3219 0.3236 0.1611 0.1756 0.1684 0.1713 0.1624 0.0855
L02317      0.3912 0.3168 0.3236 0.1590 0.1675 0.1616 0.1611 0.1556 0.1267 0.1386
AF025763    0.3890 0.3121 0.3180 0.1565 0.1684 0.1582 0.1684 0.1565 0.1335 0.1395 0.0710
U08443      0.3907 0.3155 0.3274 0.1599 0.1667 0.1560 0.1662 0.1573 0.1310 0.1365 0.0804 0.0825
AF042106    0.3929 0.3172 0.3231 0.1624 0.1713 0.1641 0.1743 0.1654 0.1344 0.1407 0.0731 0.0829 0.0838
```

Figure 4.9   Pairwise *p*-distance matrix for the HIV/SIV example data set.

```
14
L20571
AF103818    0.5582
X52154      0.5547 0.3848
U09127      0.5591 0.4006 0.4190
U27426      0.5529 0.4408 0.4249 0.1766
U27445      0.5397 0.4302 0.4242 0.1643 0.0989
AF067158    0.5645 0.4532 0.4347 0.1792 0.1945 0.1890
U09126      0.5627 0.4257 0.4160 0.1728 0.1912 0.1766 0.0864
U27399      0.5493 0.4219 0.4234 0.1852 0.1984 0.1896 0.2051 0.1918
U43386      0.5556 0.4205 0.4234 0.1814 0.2001 0.1907 0.1945 0.1830 0.0907
L02317      0.5529 0.4116 0.4234 0.1787 0.1896 0.1820 0.1814 0.1744 0.1388 0.1532
AF025763    0.5485 0.4035 0.4138 0.1755 0.1907 0.1776 0.1907 0.1755 0.1470 0.1543 0.0746
U08443      0.5520 0.4094 0.4302 0.1798 0.1885 0.1749 0.1879 0.1766 0.1439 0.1506 0.0850 0.0874
AF042106    0.5565 0.4123 0.4227 0.1830 0.1945 0.1852 0.1984 0.1868 0.1481 0.1559 0.0769 0.0879 0.0888
```

Figure 4.10   Pairwise genetic distances for the HIV/SIV example data set according to the Jukes and Cantor (JC69) model.

ratio? is no longer present because under the JC69 model, all nucleotide substitutions are equally likely (see Section 4.4.1). Type a Y, followed by the enter key to carry out the computation of the genetic distances. The result is stored in a file called outfile, which can be opened with any text editor (see Figure 4.10). The format of the output matrix, square or lower triangular, can be chosen before starting the computation by selecting option L. Of course, each pairwise distance in Figure 4.10 can be obtained by replacing p in Equation 4.19 with the observed distance given in Figure 4.9. That is exactly what the program DNAdist with the current settings has done: first it calculates the p-distances, then it uses the JC69 formula to convert them in genetic distances.

4.9   Kimura 2-parameters (K80) and F84 genetic distances

The K80 model relaxes one of the main assumptions of the JC69 model allowing for different instantaneous substitution rates between transitions and transversions ($a = c = d = f = 1$ and $b = e = \kappa$ in the Q matrix) (Kimura, 1980). The K80 correction formula for the expected genetic distance between two DNA sequences is obtained by solving the exponential $\mathbf{P}(t) = \exp(\mathbf{Q}t)$ for $\mathbf{P}(t)$, similar to what was done in Section 4.4 for the JC69 model:

$$d = \log(1/(1 - 2P - Q)) + \log(1/(1 - 2Q)) \tag{4.38a}$$

where P and Q are the proportion of the transitional and transversional differences between the two sequences, respectively. The variance of the K80 distances is calculated by

$$V(d) = 1/n[(A^2 P + B^2 Q - (AP + BQ)^2] \tag{4.38b}$$

with $A = 1/(1 - 2P - Q)$ and $B = 1/2[(1/1 - 2P - Q) + (1/1 - 2Q)]$.

K80 distances can be estimated with DNAdist by choosing Kimura 2-parameter from the D option. The user also can type an empirical Ts/Tv ratio by selecting option T from the main menu. The default value for Ti/Tv in DNAdist is 2.0. Considering that there are twice as many possible transversions as transitions, a Ti/Tv = 2.0 is equivelent to assuming that during evolution, transitional changes occur four times faster than transversional ones. When an empirical Ti/Tv value for the set of organisms under investigation is not known from the literature, it is good practice to estimate it directly from the data. A general strategy to estimate the Ti/Tv ratio of aligned DNA sequences is discussed in Chapter 5.

The K80 genetic distance (Ti/Tv = 2.0) between the HIV Group O strain L20571 and the SIV chimpanzee strain AF103818 results in 0.6346, which is 1.6 times larger than the p-distance, but only 1.1 times larger than the JC69 distance.

The K80 model still relies on restricted assumptions such as equal frequency of the four bases at equilibrium. The HKY85 (Hasegawa, Kishino, and Yano, 1985) and F84 (Felsenstein, 1984; Kishino and Hasegawa, 1989) models relax that assumption allowing for unequal frequencies; their Q matrices are slightly different, but both models essentially share the same set of assumptions: a bias in the rate of transitional with respect to the rate of transversional substitutions, and unequal base frequencies (which are usually estimated from the data set). F84 distances can be computed with PHYLIP by selecting Maximum likelihood under option D. A new option, F, also appears in the main menu:

```
F Use empirical base frequencies? Yes
```

By default, DNAdist empirically estimates the frequencies for each sequence and uses the average value over all sequences to compute pairwise distances. When No is selected in option F, the program asks the user to input the base frequencies in order A, C, G, T/U separated by blank spaces.

HKY85, as well as distances according to other models described in this chapter, can be estimated by TREE-PUZZLE.

4.10 More complex models

The TN93 model (Tamura and Nei, 1993) is an extension of the F84 model, allowing different nucleotide substitution rates for purine (A $\leftrightarrow$ G) and pyrimidine (C $\leftrightarrow$ T) transition ($b \neq e$ in the correspondent Q matrix). TN93 genetic distances can be computed with TREE-PUZZLE by selecting from the menu Pairwise distances only (no tree) in option K, and TN (Tamura-Nei 1993) in option M. The user can input from the menu empirical transition/transversion bias and pyrimidine/purine transition bias; otherwise, the program will estimate those parameters from the data set (see Chapter 6). Genetic distances also can be obtained according to simpler models. For example, by selecting HKY in option

M, TREE-PUZZLE computes HKY85 distances. The JC69 model is a further simplification of the HKY85 model in which equilibrium nucleotide frequencies are
equal and there is no nucleotide substitution bias (discussed previously). JC69
distances can be calculated with TREE-PUZZLE by selecting the HKY85 model
and setting nucleotide frequencies equal to 0.25 each (option F in the menu) and
the transition-transversion ratio equal to 0.5 (option T). Because there are twice
as many transversions as transitions (see Figure 4.6), the transition–transversion
ratio needs to be set to 0.5, not to 1, in order to reduce the HKY85 model to the
JC69 model!

The distance matrix in square format is written to the outdist file and can be
opened with any text editor. The program also outputs an outfile with several
statistics about the data set (the file is mostly self-explanatory; see Section 4.11).

4.10.1 Modeling rate heterogeneity over sites

The JC69 model assumes that each site in a sequence changes over time at a uniform
rate. More complex models allow particular substitutions (e.g., transitions) to occur
at a different rate than others (e.g., transversions), but any particular substitution
rate between nucleotide i and nucleotide j is the same across different sites. Section
4.6.1 pointed out that such an assumption is not realistic, and it is especially violated
in coding regions where different codon positions usually evolve at different rates.
Replacements at second codon position are always nonsynonymous; that is, they
change the encoded amino acid (see Chapter 1). Because of the degeneracy of the
genetic code, about 65% of the possible replacements at the 3rd codon position are
synonymous (i.e., no change in the encoded amino acid). Finally, only 4% of the
possible replacements at the 1st codon position are synonymous. Because mutations
in a protein sequence are most likely to reduce the ability of that protein to perform
its biological function, they are rapidly removed from the population by purifying
selection (see Chapter 1). Consequently, over time, fixations will accumulate more
rapidly at the 3rd codon positions rather than at the 2nd or 1st. It has been shown,
for example, that in each coding region of the human T-cell lymphotropic viruses
(HTLVs), a group of human oncogenic retroviruses, 3rd codon positions mutate
about eight times faster than 1st positions and sixteen times faster than 2nd positions
(Salemi et al., 2000). It is possible to model rate heterogeneity over sites by selecting
the option

```
C One category of substitution rates? Yes
```

in the main menu of DNAdist (which toggles this option to No) and to choose up
to nine different categories of substitution rates. The program then asks the user to
input the relative substitution rate for each category as a nonnegative real number.

For example, consider how to estimate the genetic distances for the hivALN data set with the JC69 model, assuming that mutations at the 3rd position accumulate 10 times faster than at the 1st and 20 times faster than at the 2nd codon positions. Because only the relative rates are considered, one possibility is to set the rate at the 1st codon position equal to 1, the rate at the 2nd position to 0.5, and the rate at the 3rd position to 10. It is also necessary to assign each site in the aligned data set to one of the three rate categories. This can be done easily by adding to the `infile` – after the initial line indicating the number of sequences and the number of nucleotides – one or more lines like the following:

```
CATEGORIES 1231231231231231231[...]
```

Each number in the line represents a nucleotide position in the aligned data set. For example, the first four numbers, `1231`, refer to the first four positions in the alignment. They assign the first position to rate Category 1, the second position to rate Category 2, the third position to rate Category 3, the fourth position to rate Category 1 again, and so forth. In the hivALN data set, sequences are in frame, starting at the 1st codon position and ending at the third codon position; there are 2,352 positions. Thus, the input file has to be edited in the following way:

```
14 2352

CATEGORIES 123123123 [and so forth up to 2,352 digits]

L20571     ATGACAG [...]

[...]
```

A file already edited can be found at **http://kuleuven.ac.be/aidslab/phylogeny Book/datasets.htm**:

1. Place the input file in the PHYLIP folder and run DNAdist.
2. Select option C and type 3 to choose three different rate categories.
3. At the prompt of the program asking to specify the relative rate for each category, type 1 0.5 10 and press enter.
4. Choose the desired evolutionary model as usual and run the calculation.

If there is no information about the distribution and the extent of the relative substitution rates across sites, rate heterogeneity can alternatively be modeled using a Γ-distribution (Yang, 1994b). As discussed in Section 4.6.1, a single parameter α describes the shape of the Γ-distribution: L-shaped for $\alpha < 1$ (strong rate hetero­geneity) or bell-shaped for $\alpha > 1$ (weak rate heterogeneity). Which value of α is the most appropriate for a given data set, however, is usually not known. The following chapters discuss how to estimate α with different approaches and how to estimate genetic distances with Γ-distributed rates across sites (see especially Sections 6.5

and 13.6.3). However, remember that although different sites across the genome change at different rates (Li, 1997), the use of a discrete Γ-distribution to model rate heterogeneity over sites has little biological justification; it merely reflects our ignorance about the underlying distribution of rates. It is widely used because of its "computational" simplicity and because it allows both low- and high-rate hetero-geneity to be modeled easily over sites by varying the α parameter. Chapters 7 and 10 show how to compare different evolutionary models. In such a way, it is also pos-sible to test whether a nucleotide substitution model implementing Γ-distributed rates across sites fits the data significantly better than a model assuming uniform rates.

4.11 The problem of substitution saturation

It can be demonstrated that two randomly chosen aligned DNA sequences of the same length and base composition equal to 0.25 for each nucleotide would have, on average, 25% identical residues; if gaps are allowed, as much as 50% can be identical, resulting in 50% similarity. This is why the curve in Figure 4.2, showing the relationship between p-distance and genetic distance, reaches a plateau for p between 0.5 and 0.75 (i.e., for similarity scores between 50% and 25%). Beyond that point, it is no longer possible to infer the expected genetic distance from the observed distance, the sequences are said to be saturated or to have reached **substitution saturation**, and the similarities between them could be the result of chance alone rather then homology (i.e., common ancestry). In other words, when saturation is reached, the phylogenetic signal is lost (i.e., the sequences are no longer informative about the underlying evolutionary process that produced them). The problem of saturation is often overlooked in phylogeny reconstruction, when it is rather crucial (see Section 13.5.2). For example, in coding regions, the 3rd codon positions – which usually evolve much faster than 1st and 2nd positions, – are likely to be saturated, especially when distantly related **operational taxonomic units (OTUs)** are compared. One way to avoid this problem is to exclude 3rd positions from the analysis or to use only **nonsynonymous distances** (see Chapter 11); another alternative is to analyze the corresponding amino-acid sequence (see Section 8.5).

The program DAMBE (see Section 2.8 and Chapter 13) implements a simple but powerful method to check for saturation in a data set of aligned nucleotide se-quences. The method takes advantage of the empirical observation that in most data sets, transitional substitutions happen more frequently than transversional substi-tutions. Therefore, by plotting the estimated number of transitions and transver-sions against the genetic distance for the $n(n-1)/2$ pairwise comparisons in an alignment of n **taxa**, transitions and transversions should both increase linearly with the genetic distance, with transitions being higher than transversions. However,

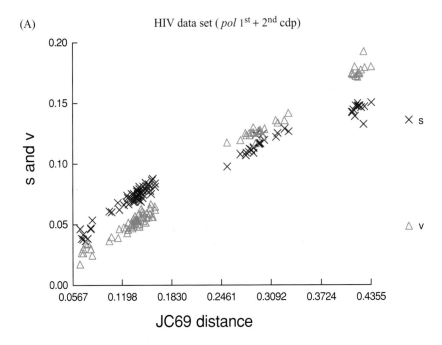

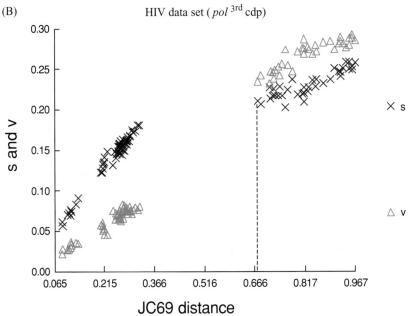

Figure 4.11 Transition and transversion versus divergence plots for the HIV data set (`hivALN.phy` file). The estimated number of transitions and transversions for each pairwise comparison is plotted against the genetic distance calculated with the TN93 model. (A) 1st and 2nd codon positions only. (B) 3rd codon positions only; the dotted line indicates the beginning of saturation.

as the genetic distance (the evolutionary time) gets larger (i.e., more divergent sequences are compared), saturation is reached and transversions outnumber transitions, quickly reaching a plateau. This is because, by chance alone, there are eight possible transversions but only four transitions (see Figure 4.6 and Chapter 1). By choosing `Transition and transversion versus divergence` in the `Graphics` menu of the DAMBE program (see Sections 2.8 and 2.9 for general information about DAMBE), and selecting a nucleotide substitution model in the dialog window, the user can obtain a plot of transitions and transversions versus genetic distance for each pairwise comparison in a data set of aligned nucleotide sequences. Figure 4.11 shows the result for the HIV data set (`hivALN.phy` file). DAMBE allows the first and second codon positions to be analyzed separately (Figure 4.11A) or the 3rd codon position (Figure 4.11B) by selecting the item `Work on codon position 1 and 2` or `Work on codon position 3` from the `Sequences` menu before starting any analysis (the original sequences can be restored by choosing `Restore sequences` from the same menu). It is clear that 3rd codon positions are saturated for pairs of sequences with JC69 estimated genetic distances greater than 0.667 (see Figure 4.11B). When only 1st and 2nd positions are analyzed (Figure 4.11A), transversions still outnumber transitions for genetic distances greater than 0.246. However, in this case, neither transitions or transversions seem to have reached a plateau, indicating that they still retain enough phylogenetic signal.

4.12 Choosing among different evolutionary models

It should be clear after a few exercises that genetic distances inferred according to different evolutionary models can lead to quite different results. Tree-building algorithms such as UPGMA and Neighbor-Joining (see Chapter 5) are based on pairwise distances among taxa; unreliable estimates could lead to the wrong tree topology and, certainly, to wrong branch lengths. It could appear that the more complex the model, the more free parameters it allows and the more accurate the inferred distances should be; however, this is not necessarily true. A model with fewer parameters will produce estimates with smaller variance. One reason is that all evolutionary models discussed throughout this book share the underlying assumption that the number of sites compared between two given sequences is infinite. The violation of such an assumption can lead to serious sampling errors. Although for sequences at least 1000 nucleotides long, the approximation usually holds well; it has been shown that models with more parameters produce a larger error than simpler models (Gojobori et al., 1992; Tajima and Nei, 1984; Zharkikh, 1994). When a simple evolutionary model (e.g., JC69 or F84) fits the data not significantly worse than a more complex model, the former is preferred (see Chapter 10). Finally, complex

(A)

SEQUENCE COMPOSITION (SEQUENCES IN INPUT ORDER)

	5% chi-square test	p-value
L20571	passed	14.60%
AF103818	passed	48.09%
X52154	passed	51.77%
U09127	passed	84.58%
U27426	passed	99.73%
U27445	passed	97.45%
AF067158	passed	77.15%
U09126	passed	59.51%
U27399	passed	94.89%
U43386	passed	92.60%
L02317	passed	97.86%
AF025763	passed	92.40%
U08443	passed	92.86%
AF042106	passed	99.59%

(B)

SEQUENCE COMPOSITION (SEQUENCES IN INPUT ORDER)

	5% chi-square test	p-value
LngfishAu	passed	6.20%
LngfishSA	failed	0.62%
LngfishAf	failed	1.60%
Frog	passed	58.01%
Turtle	passed	44.25%
Sphenodon	passed	59.78%
Lizard	passed	38.67%
Crocodile	failed	2.51%
Bird	failed	0.00%
Human	failed	0.85%
Seal	passed	68.93%
Cow	passed	59.11%
Whale	passed	97.83%
Mouse	failed	1.43%
Rat	passed	39.69%
Platypus	failed	3.46%
Opossum	failed	0.01%

Figure 4.12 Section of the TREE-PUZZLE outfile: Chi-square test comparing the nucleotide composition of each sequence to the frequency distribution assumed in the selected evolutionary model (HKY85, in this case). (A) HIV data set. (B) mtDNA data set.

models can be computationally daunting for analyzing large data sets, even for fast computers.

Another basic assumption of time-homogeneous time-continuous stationary Markov models, the class of nucleotide substitution models discussed in Section 4.4, is that the base composition among the sequences being analyzed is at equilibrium (i.e., each sequence in the data set is supposed to have a similar base composition), which does not change over time. Such an assumption usually holds when closely related species are compared, but it may break down for divergent OTUs. TREE– PUZZLE implements by default a chi-square test comparing whether the nucleotide composition of each sequence is significantly different with respect to the frequency distribution assumed in the model selected for the analysis. The result is written in the `outfile` at the end of any computation (Figure 4.12). Genetic distances among taxa showing significant heterogeneity in nucleotide composition risk being unreliable.

The analysis of the mtDNA data set reveals significantly different base frequences for 8 of the 17 taxa included in the data (see Figure 4.12B). In this case, a better estimate of d can be obtained with the LogDet method, which was developed to deal specifically with this type of problem (Lockart et al., 1994; Steel, 1994). The method estimates the genetic distance d between two aligned sequences by calculating:

$$d = -ln[\det F] \tag{4.39}$$

in which det F is the determinant of a 4×4 matrix in which each entry represents the proportion of sites having any possible nucleotide pair within the two aligned sequences. A mathematical justification for Equation 4.39 goes beyond the scope of this book. An intuitive introduction to the LogDet method can be found in Page and Holmes (1998); a more detailed discussion is provided by Swofford et al. (1996). LogDet distances can be calculated with the PAUP* program; their application to the mtDNA data set is discussed in the practical part of Chapter 7 (see Section 7.8). Chapters 7 and 10 focus on statistical tests for selecting the evolutionary model best fitting the data under investigation.

REFERENCES

Bailey, N. T. J. (1964). *The Elements of Stochastic Processes with Application to the Natural Sciences.* New York: Wiley.

Drummond, A. and K. Strimmer (2001). PAL: An object-oriented programming library for molecular evolution and phylogenetics. *Bioinformatics,* 17, 662–63.

Felsenstein, J. (1981). Evolutionary trees from DNA sequences: A maximum likelihood approach. *Journal of Molecular Evolution,* 17, 368–376.

Felsenstein, J. (1984). Distance methods for inferring phylogenies: A justification. *Evolution,* 38, 16–24.

Felsenstein, J. (1993). *PHYLIP. Phylogenetic Inference Package, Version 3.5c.* Seattle: Department of Genetics, University of Washington.

Hasegawa, M., H. Kishino, and T. Yano (1985). Dating of the human-ape splitting by a molecular clock of mitochondrial DNA. *Journal of Molecular Evolution,* 21, 160–174.

Jukes, T. and C. R. Cantor (1969). Evolution of protein molecules. In: *Mammalian Protein Metabolism,* ed. H. N. Munro, pp. 21–132. New York, Academic Press.

Kimura, M. (1980). A simple method for estimating volutionary rates of base substitutions through comparative studies of nucleotide sequences. *Journal of Molecular Evolution*, 15, 111–120.

Kimura, M. and T. Ohta (1972). On the stochastic model for estimation of mutational distance between homologous proteins. *Journal of Molecular Evolution*, 2, 87–90.

Kishino, H. and M. Hasegawa (1989). Evaluation of the maximum likelihood estimate of the evolutionary tree topologies from DNA sequence data, and the branching order of the Hominoidea. *Journal of Molecular Evolution*, 29, 170–179.

Li, W.-H. (1997). *Molecular Evolution*. Sunderland: Sinauer Associates.

Lockart, P. J., M. D. Steel, M. D. Hendy, and D. Penny (1994). Recovering evolutionary trees under a more realistic model of evolution. *Molecular Biology and Evolution*, 11, 605–612.

Page, R. D. M. and E. C. Holmes (1998). *Molecular Evolution: A Phylogenetic Approach*. Oxford, UK: Blackwell Science, pp. 158–159.

Salemi M., J. Demyter, and A.-M. Vandamme (2000). Tempo and mode of Human and Simian T-Lymphotropic virus (HTLV/STLV) evolution revealed by analyses of full-genome sequences. *Molecular Biology and Evolution*, 17, 374–386.

Steel, M. (1994). Recovering a tree from the Markov leaf colorations it generates under a Markov model. *Applied Mathematics Letters*, 7, 19–23.

Swofford, D. L., G. J. Olsen, P. J. Waddell, and D. M. Hillis (1996). Phylogenetic inference. In: *Molecular Systematics*, 2nd ed., eds. D. M. Hillis, C. Moritz, and B. K. Mable, pp. 407–514. Sunderland, MA, Sinauer Associates.

Tamura, K. and M. Nei (1993). Estimation of the number of nucleotide substitutions in the control region of mitochondrial DNA in humans and chimpanzees. *Molecular Biology and Evolution*, 10, 512–526.

Thorne, J. L., H. Kishino, and J. Felsenstein (1992). Inching toward reality: An improved likelihood model of sequence evolution. *Journal of Molecular Evolution*, 34, 3–16.

Xia, X. (2000). *Data Analysis in Molecular Biology and Evolution*. Boston: Kluwer Academic Publishers.

Yang, Z. (1994a). Estimating the pattern of nucleotide substitution. *Journal of Molecular Evolution*, 39, 105–111.

Yang, Z. (1994b). Maximum-likelihood phylogenetic estimation from DNA sequences with variable rates over sites: Approximate methods. *Journal of Molecular Evolution*, 39, 306–314.

Yang, Z. (2000). *Phylogenetic Analysis by Maximum-Likelihood (PAML) Version 3.0*. London: University College.

Zharkikh, E. (1994). Estimation of evolutionary distances between nucleotide sequences. *Journal of Molecular Evolution*, 39, 315–329.

Phylogeny inference based on distance methods

THEORY

Yves Van de Peer

5.1 Introduction

In addition to **maximum parsimony** (**MP**) and likelihood methods (see Chapters 6 and 7), pairwise **distance methods** form the third large group of methods to infer evolutionary trees from sequence data (Figure 5.1). In principle, distance methods try to fit a tree to a matrix of pairwise **genetic distances** (Felsenstein, 1988). For every two sequences, the distance is a single value based on the fraction of positions in which the two sequences differ, defined as **p-distance** (see Chapter 4). The p-distance is an underestimation of the true genetic distance because some of the aligned nucleotides are the result of multiple events. Indeed, because mutations are fixed in the genes, there has been an increasing chance of multiple substitutions occurring during evolution at the same sequence position. Therefore, in distance-based methods, one tries to estimate the number of substitutions that have actually occurred by applying a specific **evolutionary model** that makes assumptions about the nature of evolutionary changes (see Chapter 4). When all the pairwise distances have been computed for a set of sequences, a tree topology can then be inferred by a variety of methods (Figure 5.2).

Correct estimation of the genetic distance is crucial and, in most cases, more important than the choice of method to infer the tree topology. Using an unrealistic evolutionary model can cause serious artifacts in tree topology, as previously shown in numerous studies (e.g., Olsen, 1987; Lockhart et al., 1994; Van de Peer et al., 1996; see also Chapter 10). However, because the exact historical record of events that occurred in the evolution of sequences is not known, the best method for estimating the genetic distance is not necesserily self-evident (see Chapter 9).

	Character-based methods	Noncharacter-based methods
Methods based on an explicit model of evolution	Maximum-likelihood methods	Pairwise-distance methods
Methods not based on an explicit model of evolution	Maximum-parsimony methods	

Figure 5.1   Pairwise distance methods are non-character-based methods that make use of an explicit substitution model.

Substitution models are discussed in Chapters 4, 9, and 10. Chapters 7 and 10 discuss how to select the best-fitting evolutionary model for a given data set of nucleotide or amino-acid aligned sequences in order to get an accurate estimation of the genetic distances. In the following sections, it is assumed that genetic distances were estimated using an appropriate evolutionary model, and some of the methods used for inferring tree topologies on the basis of these distances are briefly outlined. However, by no means should this be considered a complete discussion of distance methods; additional discussions are in Felsenstein (1982), Swofford et al. (1996), Li (1997), and Page and Holmes (1998).

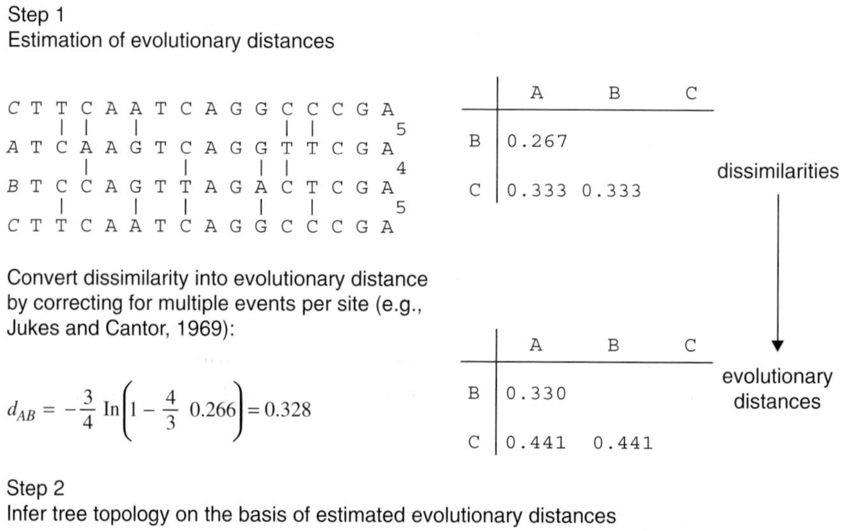

Step 1
Estimation of evolutionary distances

```
C T T C A A T C A G G C C C G A
  | |   |             | |        5
A T C A A G T C A G G T T C G A
        |   |         | |        4
B T C C A G T T A G A C T C G A
  |     |   |     |   |   |      5
C T T C A A T C A G G C C C G A
```

	A	B	C
B	0.267		
C	0.333	0.333	

dissimilarities

Convert dissimilarity into evolutionary distance by correcting for multiple events per site (e.g., Jukes and Cantor, 1969):

$$d_{AB} = -\frac{3}{4} \ln\left(1 - \frac{4}{3} \, 0.266\right) = 0.328$$

	A	B	C
B	0.330		
C	0.441	0.441	

evolutionary distances

Step 2
Infer tree topology on the basis of estimated evolutionary distances

Figure 5.2   Distance methods proceed in two steps. First, the evolutionary distance is computed for every sequence pair. Usually, this information is stored in a matrix of pairwise distances. Second, a tree topology is inferred on the basis of the specific relationships between the distance values.

5.2   Tree-inferring methods based on genetic distances

The main distance-based tree-building methods are cluster analysis and minimum evolution. Both rely on a different set of assumptions, and their success or failure in retrieving the correct phylogenetic tree depends on how well any particular data set meets such assumptions.

5.2.1   Cluster analysis (UPGMA and WPGMA)

Clustering methods are tree-building methods that were originally developed to construct taxonomic phenograms (Sokal and Michener, 1958; Sneath and Sokal, 1973); that is, trees based on overall phenotypic similarity. Later, these methods were applied to phylogenetics to construct **ultrametric trees**. **Ultrametricity** is satisfied when, for any three **taxa**, A, B, and C,

$$d_{AC} \leq \max(d_{AB}, d_{BC}). \tag{5.1}$$

In practice, Equation 5.1 is satisfied when two of the three distances under consideration are equal and as large (or larger) as the third one. Ultrametric trees are **rooted trees** in which all the end nodes are equidistant from the root of the tree, which is only possible by assuming a **molecular clock** (see Chapters 1 and 10). Clustering methods such as the **unweighted-pair group method with arithmetic means** (**UPGMA**) or the **weighted-pair group method with arithmetic means** (**WPGMA**) use a sequential clustering algorithm. A tree is built in a stepwise manner, by grouping sequences or groups of sequences – usually referred to as **operational taxonomic units** (**OTUs**) – that are most similar to each other; that is, for which the genetic distance is the smallest. When two OTUs are grouped, they are treated as a new single OTU (Box 5.1). From the new group of OTUs, the pair for which the similarity is highest is again identified, and so on, until only two OTUs are left. The method applied in Box 5.1 is actually the WPGMA, in which the averaging of the distances is not based on the total number of OTUs in the respective clusters. For example, when OTUs A, B (which have been grouped before), and C are grouped into a new node 'u', then the distance from node 'u' to any other node 'k' (e.g., grouping D and E) is computed as follows:

$$d_{uk} = \frac{d_{(A, B)k} + d_{Ck}}{2} \tag{5.2}$$

Conversely, in UPGMA, the averaging of the distances is based on the number of OTUs in the different clusters; therefore, the distance between 'u' and 'k' is

Box 5.1 Cluster analysis (Sneath and Sokal, 1973)

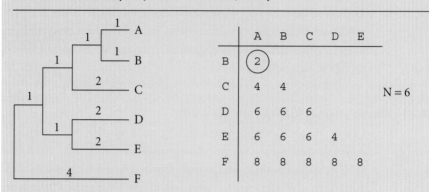

	A	B	C	D	E
B	2				
C	4	4			
D	6	6	6		
E	6	6	6	4	
F	8	8	8	8	8

$N = 6$

Cluster analysis proceeds as follows:

1. Group together (cluster) these OTUs for which the distance is minimal; e.g., A and B together. The depth of the divergence is the distance between A and B divided by 2.

2. Compute the distance from cluster (A, B) to every other OTU.

$$d_{(AB)C} = (d_{AC} + d_{BC})/2 = 4$$
$$d_{(AB)D} = (d_{AD} + d_{BD})/2 = 6$$
$$d_{(AB)E} = (d_{AE} + d_{BE})/2 = 6$$
$$d_{(AB)F} = (d_{AF} + d_{BF})/2 = 8$$

	(AB)	C	D	E
C	4			
D	6	6		
E	6	6	4	
F	8	8	8	8

Repeat Steps 1 and 2 until all OTUs are clustered (repeat until $N = 2$).

$$N = N - 1 = 5$$

1. Group together (cluster) these OTUs for which the distance is minimal; e.g., group D and E together. Alternatively, (AB) could be grouped with C.

2. Compute the distance from cluster (D, E) to every other OTU (cluster).

$d_{(DE)(AB)} = (d_{D(AB)} + d_{E(AB)})/2 = 6$
$d_{(DE)C} = (d_{DC} + d_{EC})/2 = 6$
$d_{(DE)F} = (d_{DF} + d_{EF})/2 = 8$

	(AB)	C	(DE)
C	4		
(DE)	6	6	
F	8	8	8

$N = N - 1 = 4$

1. Group together these OTUs for which the distance is minimal; e.g., group (A, B) and C.

2. Compute the distance from cluster (A, B, C) to every other OTU (cluster).

$d_{(ABC)(DE)} = (d_{(AB)(DE)} + d_{C(DE)})/2 = 6$
$d_{(ABC)F} = (d_{(AB)F} + d_{CF})/2 = 8$

	(ABC)	(DE)
(DE)	6	
F	8	8

$N = N - 1 = 3$

1. Group together these OTUs for which the distance is minimal; e.g., group (A, B, C) and (D, E).

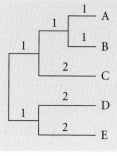

Box 5.1 (continued)

2. Compute the distance from cluster (A, B, C, D, E) to OTU F.

$d_{(ABCDE) F} = (d_{(ABC) F} + d_{(DE) F})/2 = 8$

	(ABC) , (DE)
F	8

$N = N - 1 = 3$

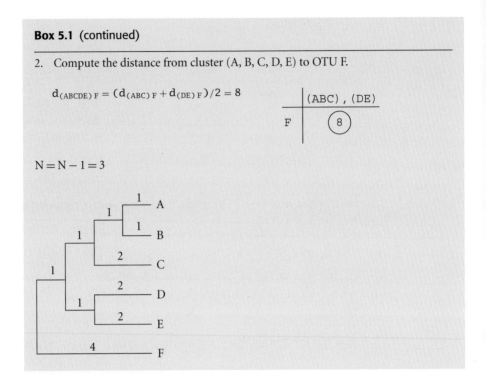

computed as follows:

$$d_{uk} = \frac{\left(N_{AB} d_{(A, B)k} + N_C d_{Ck}\right)}{(N_{AB} + N_C)} \qquad (5.3)$$

where N_{AB} equals the number of OTUs in cluster AB (i.e., 2) and N_C equals the number of OTUs in cluster C (i.e., 1). When the data are **ultrametric**, UPGMA and WPGMA have the same result. However, when the data are not ultrametric, they can differ in their inferences.

Until about fifteen years ago, clustering was often used to infer evolutionary trees based on sequence data, but this is no longer the case. Many computer-simulation studies have shown that clustering methods such as UPGMA are extremely sensitive to unequal rates in different lineages (e.g., Sourdis and Krimbas, 1987; Huelsenbeck and Hillis, 1993). To overcome this problem, some have proposed methods that convert nonultrametric distances into ultrametric distances. Usually referred to as **transformed distance methods**, these methods correct for unequal rates among different lineages by comparing the sequences under study to a reference sequence or an outgroup (Farris, 1977; Klotz et al., 1979; Li, 1981). Once the distances are made ultrametric, a tree is constructed by clustering, as explained previously. Nevertheless, because there are now better and more effective methods to cope

$$d_{AB} + d_{CD} \leq \min(d_{AC} + d_{BD}, d_{AD} + d_{BC})$$

$d_{AB} = a + b$ $d_{AC} = a + e + c$ $d_{AD} = a + e + d$
$d_{CD} = c + d$ $d_{BD} = b + e + d$ $d_{BC} = b + e + a$

$$(a + b + c + d) \leq [(a + b + c + d + 2e)(a + b + c + d + 2e)]$$

Figure 5.3   Four-point condition. Letters on the branches of the unrooted tree represent branch lengths. The function min[ ] returns the minimum among a set of values.

with nonultrametricity and non-clock-like behavior, there is little reason left to use cluster analysis or transformed distance methods to infer distance trees for nucleic-acid or amino-acid sequence data.

5.2.2   Minimum evolution and neighbor-joining

Because of the serious limitations of ordinary clustering methods, algorithms were developed that reconstruct so-called **additive distance trees**. Additive distances satisfy the following condition, known as the **four-point metric condition** (Buneman, 1971): for any four taxa, A, B, C, and D,

$$d_{AB} + d_{CD} \leq \max(d_{AC} + d_{BD}, d_{AD} + d_{BC}) \tag{5.4}$$

Only **additive distances** can be fitted precisely into an **unrooted tree** such that the genetic distance between a pair of OTUs equals the sum of the lengths of the branches connecting them, rather than an average, as in the case of cluster analysis. Why Equation 5.4 needs to be satisfied is explained by the example shown in Figure 5.3. When A, B, C, and D are related by a tree in which the sum of branch lengths connecting two terminal taxa is equal to the genetic distance between them, such as the tree in Figure 5.3, $d_{AB} + d_{CD}$ is always smaller or equal than the minimum between $d_{AC} + d_{BD}$ and $d_{AD} + d_{BC}$ (see Figure 5.3). In fact, the latter case occurs when the four sequences are related by a star-like tree; that is, only when the internal branch length of the tree in Figure 5.3 is $e = 0$ (see Figure 5.3). If Formula 5.4 is not satisfied, A, B, C, and D cannot be represented by an additive distance tree because, to maintain the additivity of the genetic distances, one or more branch lengths of any tree relating them should be negative, which would be biologically meaningless. Real data sets often fail to satisy the four-point condition; this problem is the origin of the discrepancy between **actual distances** (i.e., those estimated from pairwise comparisons among nucleotide or amino-acid sequences) and tree distances (i.e., those actually fitted into a tree) (see Section 5.2.3).

If the genetic distances for a certain data set are ultrametric, then both the ultra-metric tree and the additive tree will be the same if the additive tree is rooted at the same point as the *ultrametric tree*. However, if the genetic distances are not ultrametric due to non-clock-like behavior of the sequences, additive trees will al-most always be a better fit to the distances than ultrametric trees. However, because of the finite amount of data available when working with real sequences, stochas-tic errors usually cause deviation of the estimated genetic distances from perfect tree additivity. Therefore, some systematic error is introduced and, as a result, the estimated tree topology may be incorrect.

Minimum evolution (*ME*) is a distance method for constructing additive trees that was first described by Kidd and Sgaramella-Zonta (1971); Rzhetsky and Nei (1992) described a method with only a minor difference. In ME, the tree that minimizes the lengths of the tree, which is the sum of the lengths of the branches, is regarded as the best estimate of the phylogeny:

$$S = \sum_{i=1}^{2n-3} v_i \qquad (5.5)$$

where n is the number of taxa in the tree and v_i is the ith branch (remember that there are $2n-3$ branches in an unrooted tree of n taxa). For each tree topology, it is possible to estimate the length of each branch from the estimated pairwise distances between all OTUs. In this respect, the method can be compared with the maximum parsimony (MP) approach (see Chapter 6), but in ME, the length of the tree is inferred from the genetic distances rather than from counting individual nucleotide substitutions over the tree (Rzhetsky and Nei, 1992, 1993; Kumar, 1996). The minimum tree is not necessarily the "true" tree. Nei et al. (1998) have shown that, particularly when few nucleotides or amino acids are used, the "true" tree may be larger than the minimum tree found by the optimization principle used in ME and MP. A drawback of the ME method is that, in principle, all different tree topologies have to be investigated to find the minimum tree. However, this is impossible in practice because of the explosive increase in the number of tree topologies as the number of OTUs increases (Felsenstein, 1978); an exhaustive search can no longer be applied when more than ten sequences are being used (see Chapter 1).

A good heuristic method for estimating the ME tree is the *neighbor-joining* (*NJ*) *method*, developed by Saitou and Nei (1987) and modified by Studier and Keppler (1988). Because NJ is conceptually related to clustering, but without assuming a clock-like behavior, it combines computational speed with uniqueness of results.

Box 5.2 The neighbor-joining method (Saitou and Nei, 1987; modified by Studier and Keppler, 1988)

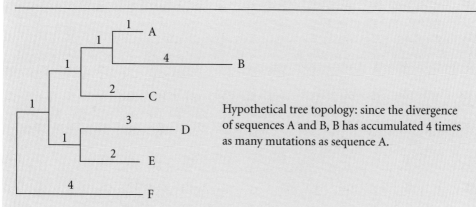

Hypothetical tree topology: since the divergence of sequences A and B, B has accumulated 4 times as many mutations as sequence A.

Assume the following matrix of pairwise evolutionary distances:

	A	B	C	D	E
B	5				
C	4	7			
D	7	10	7		
E	6	9	6	5	
F	8	11	8	9	8

Clustering methods (discussed in Box 5.1) would erroneously group sequences A and C because they assume clock-like behavior. Although sequences A and C look more similar, sequences A and B are more closely related.

Neighbor-joining proceeds as follows:

1. Compute the net divergence r for every end node (N = 6)

$$r_A = 5 + 4 + 7 + 6 + 8 = 30 \qquad r_D = 38$$
$$r_B = 5 + 7 + 10 + 9 + 11 = 42 \qquad r_E = 34$$
$$r_C = 32 \qquad\qquad\qquad\qquad\quad r_F = 44$$

2. Create a rate-corrected distance matrix; the elements are defined by $M_i = d_{ij} - (r_i + r_j)/(N-2)$

$$M_{AB} = d_{AB} - (r_A + r_B)/(N-2) = 5 - (30+42)/4 = -13$$
$$M_{AC} = ...$$
...

Box 5.2 (continued)

	A	B	C	D	E
B	$\boxed{-13}$				
C	-11.5	-11.5			
D	-10	-10	-10.5		
E	-10	-10	-10.5	$\boxed{-13}$	
F	-10.5	-10.5	-11	-11.5	-11.5

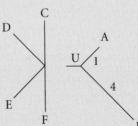

3. Define a new node that groups OTUs i and j for which M_i is minimal.
    For example, sequences A and B are neighbors and form a new node U (but, alternatively, OTUs D and E could have been joined; see below).

4. Compute the branch lengths from node U to A and B:

$S_{AU} = d_{AB}/2 + (r_A - r_B)/2(N-2) = 1$
$S_{BU} = d_{AB} - S_{AU} = 4$

or, alternatively:

$S_{BU} = d_{AB}/2 + (r_B - r_A)/2(N-2) = 4$
$S_{AU} = d_{AB} - S_{BU} = 1$

5. Compute new distances from node U to every other terminal node:

$d_{CU} = (d_{AC} + d_{BC} - d_{AB})/2 = 3$
$d_{DU} = (d_{AD} + d_{BD} - d_{AB})/2 = 6$
$d_{EU} = (d_{AE} + d_{BE} - d_{AB})/2 = 5$
$d_{FU} = (d_{AF} + d_{BF} - d_{AB})/2 = 7$

	U	C	D	E
C	3			
D	6	7		
E	5	6	5	
F	7	8	9	8

6. $N = N - 1$; repeat Steps 1 through 5.

1. Compute the net divergence r for every end node (N = 5):

$r_B = 21$ $r_E = 24$
$r_C = 24$ $r_F = 32$
$r_D = 27$

2. Compute the modified distances:

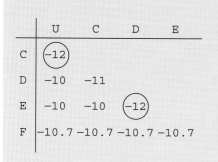

	U	C	D	E
C	-12			
D	-10	-11		
E	-10	-10	-12	
F	-10.7	-10.7	-10.7	-10.7

3. Define a new node: e.g., U and C are neighbors and form a new node V; alternatively, D and E could be joined.

1. Compute the net divergence r for every end node (n = 4):

$r_V = 15$ $r_E = 17$
$r_D = 19$ $r_F = 23$

2. Compute the modified distances:

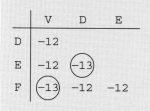

	V	D	E
D	-12		
E	-12	-13	
F	-13	-12	-12

3. Define a new node: e.g., D and E are neighbors and form a new node W; alternatively, F and V could be joined.

4. Compute the branch lenghts from node V to C and U:

$S_{UV} = d_{CU}/2 + (r_U - r_C)/2(N - 2) = 1$
$S_{CV} = d_{CU} - S_{UV} = 2$

5. Compute distances from V to every other terminal node:

$d_{DV} = (d_{DU} + d_{CB} - d_{CU})/2 = 5$
$d_{EV} = (d_{EU} + d_{CB} - d_{CU})/2 = 4$
$d_{FV} = (d_{FU} + d_{CF} - d_{CU})/2 = 6$

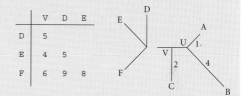

	V	D	E
D	5		
E	4	5	
F	6	9	8

6. N = N − 1; repeat Steps 1 through 5.

4. Compute the branch lengths from node W to E and D:

$S_{DW} = d_{DE}/2 + (r_D - r_E)/2(N - 2) = 3$
$S_{DW} = d_{DE} - S_{DW} = 2$

5. Compute distances from W to every other terminal node:

$d_{VW} = (d_{DV} + d_{EV} - d_{DE})/2 = 2$
$d_{FW} = (d_{Dr} + d_{gr} - d_{DE})/2 = 6$

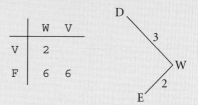

	W	V
V	2	
F	6	6

6. N = N − 1; repeat Steps 1 through 5.

Box 5.2 (continued)

1. Compute the net divergence r for every end node ($n = 3$):

   $r_V = 8$ $r_F = 17$ $r_W = 8$

2. Compute the modified distances:

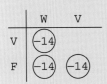

3. Define a new node: e.g., V and F are neighbors and form a new node X;

alternatively, W and V or W and F could be joined.

4. Compute the branch lengths from node X to V and F:

   $S_{VX} = d_{FV}/2 + (r_V - r_F)/2(N-2) = 1$
   $S_{FX} = d_{FV} - S_{VX} = 5$

5. Compute distances from X to every other terminal node:

   $d_{WX} = (d_{FW} + d_{VW} - d_{FV})/2 = 1$

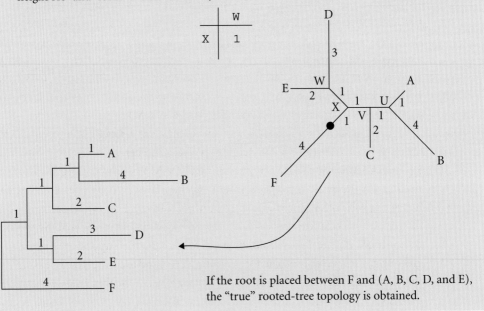

If the root is placed between F and (A, B, C, D, and E), the "true" rooted-tree topology is obtained.

NJ is today the method most commonly used to construct distance trees. Box 5.2 is an example of a tree constructed with the NJ method. The method adopts the ME criterion and combines a pair of sequences by minimizing the S value (see Equation 5.5) in each step of finding a pair of neighboring OTUs. Because the S value is not minimized globally (Saitou and Nei, 1987; Studier and Keppler, 1988), the NJ tree may not be the same as the ME tree if pairwise distances are not additive (Kumar, 1996). However, NJ trees have proven to be the same or similar to the ME tree (Saitou and Imanishi, 1989; Rzhetsky and Nei, 1992, 1993; Russo et al.,

1996; Nei et al., 1998). Several methods have been proposed to find ME trees, starting from an NJ tree but evaluating alternative topologies close to the NJ tree by conducting local rearrangements (e.g., Rzhetsky and Nei, 1992). Nevertheless, it is questionable whether this approach is really worth considering (Saitou and Imanishi, 1989; Kumar, 1996), and it has been suggested that combining NJ and **bootstrap analysis** (Felsenstein, 1985) might be the best way to evaluate trees using distance methods (Nei et al., 1998).

Recently, alternative versions of the NJ algorithm have been proposed, including **BIONJ** (Gascuel, 1997), **generalized neighbor-joining** (Pearson et al., 1999), and **weighted neighbor-joining** or **weighbor** (Bruno et al., 2000). BIONJ and weighbor both consider that long genetic distances present a higher variance than short ones when distances from a newly defined node to all other nodes are estimated (see Box 5.2). This should result in higher accuracy when distantly related sequences are included in the analysis. Furthermore, the weighted neighbor-joining method of Bruno et al. (2000) uses a likelihood-based criterion rather than the ME criterion of Saitou and Nei (1987) to decide which pair of OTUs should be joined. The generalized neighbor-joining method of Pearson et al. (1999) keeps track of multiple, partial, and potentially good solutions during its execution, thus exploring a greater part of the tree space. As a result, the program is able to discover topologically distinct solutions that are close to the ME tree.

Figure 5.4 shows two trees based on **evolutionary distances** inferred from 20 small subunit ribosomal RNA sequences (Van de Peer et al., 2000a). The tree in Figure 5.4A was constructed by clustering (UPGMA) and shows some unexpected results. For example, the sea anemone, *Anemonia sulcata,* clusters with the fungi rather than the other animals, as would have been expected. Furthermore, neither the basidiomycetes nor the ascomycetes form a clear-cut **monophyletic** grouping. In contrast, on the NJ tree all animals form a highly supported monophyletic grouping, and the same is true for basidiomycetes and ascomycetes. The NJ tree also shows why clustering could not resolve the right relationships. Clustering methods are sensitive to unequal rates of evolution in different lineages; as is clearly seen, the branch length of *Anemonia sulcata* differs greatly from that of the other animals. Also, different basidiomycetes have evolved at different rates and, as a result, they are split into two groups in the tree obtained by clustering (see Figure 5.4A).

5.2.3 Other distance methods

It is possible for every tree topology to estimate the length of all branches from the estimated pairwise distances between all OTUs (e.g., Fitch and Margoliash, 1967; Rzhetsky and Nei, 1993). However, when summing the branch lengths between sequences, there is usually some discrepancy between the distance obtained (referred to as the **tree distance** or **patristic distance**) and the distance as estimated directly

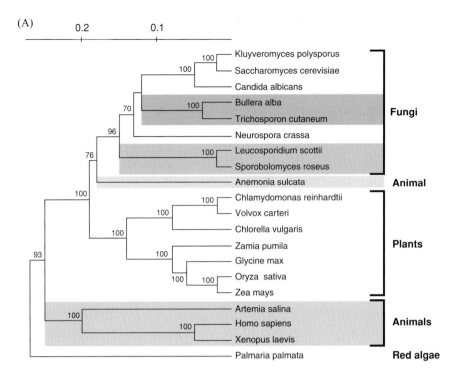

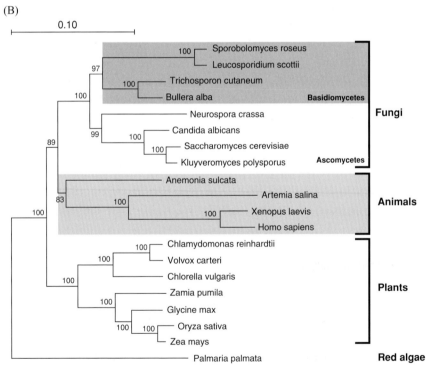

Figure 5.4 Phylogenetic trees based on the comparison of 20 small subunit ribosomal RNA sequences. Animals are indicated by light gray shading; dark gray shading indicates the basidiomycetes. The scales on top measure evolutionary distance in substitutions per nucleotide. The red alga *Palmaria palmata* was used to root the tree. (A) Ultrametric tree obtained by clustering. (B) Neighbor-joining tree.

from the sequences themselves (the observed or actual distances) due to deviation from tree additivity (see Section 5.2.2). Whereas ME methods try to find the tree for which the sum of the lengths of branches is minimal, other distance methods have been developed to construct additive trees depending on goodness of fit measures between the actual distances and the tree distances. The best tree, then, is that tree that minimizes the discrepancy between the two distance measures. When the criterion for evaluation is based on a **least-squares fit**, the goodness of fit F is given by the following:

$$F = \sum_{i,j} w_{ij} \left(D_{ij} - d_{ij} \right)^2 \tag{5.6}$$

where D_{ij} is the observed distance between i and j, d_{ij} is the tree distance between i and j, and w_{ij} is different for different methods. For example, in the Fitch and Margoliash method (1967), w_{ij} equals $1/D_{ij}^2$; in the Cavalli-Sforza and Edwards approach (1967), w_{ij} equals 1. Other values for w_{ij} are also possible (Swofford et al., 1996) and using different values can influence which tree is regarded as the best. To find the tree for which the discrepancy between actual and tree distances is minimal, one has in principle to investigate all different tree topologies. However, as with ME, distance methods that are based on the evaluation of an explicit criterion, such as goodness of fit between observed and tree distances, suffer from the explosive increase in the number of different tree topologies as more OTUs are examined. Therefore, heuristic approaches, such as stepwise addition of sequences and local and global rearrangements, must be applied when trees are constructed on the basis of ten or more sequences (e.g., Felsenstein, 1993).

5.3 Evaluating the reliability of inferred trees

The two techniques used most often to evaluate the reliability of the inferred tree or, more precisely, the reliability of specific clades in the tree are bootstrap analysis (Box 5.3) and **jackknifing** (see Section 5.3.2).

5.3.1 Bootstrap analysis

Bootstrap analysis is a widely used sampling technique for estimating the statistical error in situations in which the underlying sampling distribution is either unknown or difficult to derive analytically (Efron and Gong, 1983). The bootstrap method offers a useful way to approximate the underlying distribution by resampling from the original data set. Felsenstein (1985) first applied this technique to the estimation of confidence intervals for phylogenies inferred from sequence data. First, the sequence data are bootstrapped, which means that a new alignment is obtained from the original by randomly choosing columns from it. Each column in the alignment

Box 5.3 Bootstrap analysis (Felsenstein, 1985)

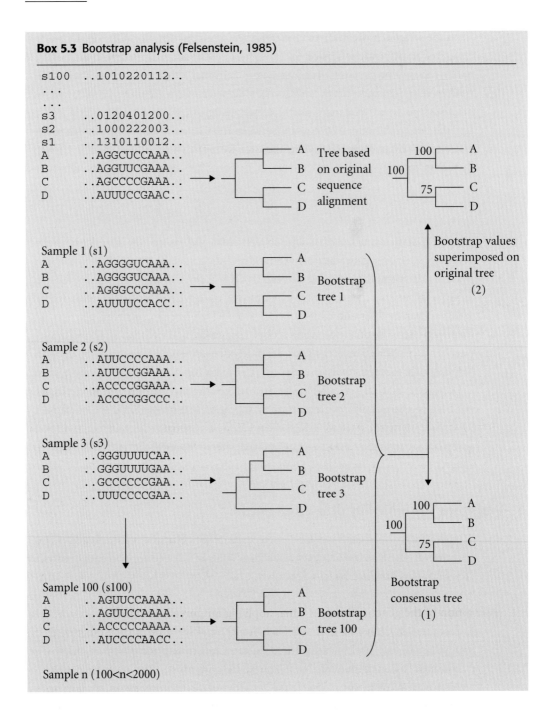

```
s100   ..1010220112..
...
...
s3     ..0120401200..
s2     ..1000222003..
s1     ..1310110012..
A      ..AGGCUCCAAA..
B      ..AGGUUCGAAA..
C      ..AGCCCCGAAA..
D      ..AUUUCCGAAC..
```

Tree based on original sequence alignment

Sample 1 (s1)
```
A      ..AGGGGUCAAA..
B      ..AGGGGUCAAA..
C      ..AGGGCCCAAA..
D      ..AUUUUCCACC..
```
Bootstrap tree 1

Bootstrap values superimposed on original tree (2)

Sample 2 (s2)
```
A      ..AUUCCCCAAA..
B      ..AUUCCGGAAA..
C      ..ACCCCGGAAA..
D      ..ACCCCGGCCC..
```
Bootstrap tree 2

Sample 3 (s3)
```
A      ..GGGUUUUCAA..
B      ..GGGUUUUGAA..
C      ..GCCCCCCGAA..
D      ..UUUCCCCGAA..
```
Bootstrap tree 3

Sample 100 (s100)
```
A      ..AGUUCCAAAA..
B      ..AGUUCCAAAA..
C      ..ACCCCCAAAA..
D      ..AUCCCCAACC..
```
Bootstrap tree 100

Bootstrap consensus tree (1)

Sample n (100<n<2000)

can be selected more than once or not at all until a new set of sequences, a ***bootstrap replicate***, the same length as the original one has been constructed. Therefore, in this resampling process, some characters will not be included at all in a given bootstrap replicate and others will be included once, twice, or more. Second, for each reproduced (i.e., artificial) data set, a tree is constructed, and the proportion of each clade among all the bootstrap replicates is computed. This proportion is taken as the statistical confidence supporting the monophyly of the subset.

Two approaches can be used to show ***bootstrap values*** on ***phylogenetic trees***. The first summarizes the results of bootstrapping in a majority-rule ***consensus tree*** (see Box 5.3, Option 1), as done for example in the PHYLIP software package (Felsenstein, 1993). The second approach superimposes the bootstrap values on the tree obtained from the original sequence alignment (see Box 5.3, Option 2). In this case, all bootstrap trees are compared with the tree based on the original alignment and the number of times a cluster (as defined in the original tree) is also found in the bootstrap trees is recorded. This approach is used in TREECON (see Chapter 9) (Van de Peer and De Wachter, 1994). Although in terms of general statistics the theoretical foundation of the bootstrap has been well established, the statistical properties of the bootstrap estimation applied to sequence data and evolutionary relationships are less well understood; several studies have reported on this problem (Zharkikh and Li, 1992a,b; Felsenstein and Kishino, 1993; Hillis and Bull, 1993). Bootstrapping itself is a neutral process that only reflects the phylogenetic signal (or noise) in the data as detected by the tree-construction method used. If the tree-construction method makes a bad estimate of the phylogeny due to systematic errors (caused by incorrect assumptions in the tree-construction method), or if the sequence data are not representative of the underlying distribution, the resulting confidence intervals obtained by the bootstrap are not meaningful. Furthermore, if the original sequence data are biased, the bootstrap estimates will be too. For example, if two sequences are clustered together because they both share an unusually high GC content, their artificial clustering will be supported by bootstrap analysis at a high confidence level. Another example is the artificial grouping of sequences with an increased *evolutionary rate*. Due to the systematic underestimation of the genetic distances when applying an unrealistic substitution model (as described in Chapter 9), distant species either will be clustered together or drawn toward the root of the tree. When the bootstrap trees are inferred on the basis of the same incorrect evolutionary model, the early divergence of long branches or the artificial clustering of long branches (the so-called ***long-branch attraction***) will be supported at a high bootstrap level. Therefore, when there is evidence of these types of artifacts, bootstrap results should be interpreted with caution.

In conclusion, bootstrap analysis is a simple and effective technique to test the relative stability of groups within a phylogenetic tree. The major advantage of the bootstrap technique is that it can be applied to basically all tree-construction methods, although it must be remembered that applying the bootstrap method multiplies the computer time needed by the number of bootstrap samples requested. Between 200 and 2000 resamplings are usually recommended (Hedges, 1992; Zharkikh and Li, 1992a). Overall, under normal circumstances, considerable confidence can be given to branches or groups supported by more than 70 or 75%; conversely, branches supported by less than 70% should be treated with caution (Zharkikh and Li, 1992a; see also Van de Peer et al., 2000b for a discussion about the effect of species sampling on bootstrap values).

5.3.2 Jackknifing

An alternative resampling technique often used to evaluate the reliability of specific clades in the tree is the so-called **delete-half jackknifing** or jackknife. Jackknife randomly purges half of the sites from the original sequences so that the new sequences will be half as long as the original. This resampling procedure typically will be repeated many times to generate numerous new samples. Each new sample (i.e., new set of sequences) – no matter whether from bootstrapping or jackknifing – will then be subjected to regular phylogenetic reconstruction. The frequencies of subtrees are counted from reconstructed trees. If a subtree appears in all reconstructed trees, then the jackknifing *value* is 100%; that is, the strongest possible support for the subtree. As for bootstrapping, branches supported by a jackknifing *value* less than 70% should be treated with caution.

5.4 Conclusions

Pairwise distance methods are tree-construction methods that proceed in two steps. First, for all pairs of sequences, the genetic distance is estimated (Swofford et al., 1996) from the observed sequence dissimilarity (p-distance) by applying a correction for multiple substitutions. The genetic distance thus reflects the expected mean number of changes per site that have occurred since two sequences diverged from their common ancestor. Second, a phylogenetic tree is constructed by considering the relationship between these distance values. Because distance methods strongly reduce the phylogenetic information of the sequences (to basically one value per sequence pair), they are often regarded as inferior to character-based methods (see Chapters 6 and 7). However, as shown in many studies, this is not necessarily so, provided that the genetic distances were estimated accurately (see Chapters 9 and 10). Moreover, contrary to maximum parsimony, distance methods have the advantage – which they share with maximum-likelihood methods – that an appropriate substitution model can be applied to correct for multiple mutations. Popular distance methods such as the NJ and the Fitch and Margoliash methods have long proven to be quite efficient in finding the "true" tree topologies or those that are close (Saitou and Imanishi, 1989; Huelsenbeck and Hillis, 1993; Charleston et al., 1994; Kuhner and Felsenstein, 1994; Nei et al., 1998). NJ has the advantage of being very fast, which allows the construction of large trees including hundreds of sequences; this significant difference in speed of execution compared to other distance methods has undoubtedly accounted for the popularity of the method (Kuhner and Felsenstein, 1994; Van de Peer and De Wachter, 1994).

Distance methods are implemented in many different software packages, including PHYLIP (Felsenstein, 1993), MEGA2 (Kumar et al., 1993), TREECON (Van de Peer and Dewachter, 1994; see also Chapter 9), PAUP* (Swofford, 2002), DAMBE (Xia, 2000), and many more.

PRACTICE

Marco Salemi

5.5 The `TreeView` program

`TreeView` is a user-friendly application for displaying phylogenetic trees such as those estimated with `PHYLIP` or `TREE-PUZZLE`, and it can be downloaded for free from **http://taxonomy.zoology.gla.ac.uk/rod/treeview.html**. The program runs on both Mac and PC using almost identical interfaces, and it should be installed before starting the next exercise.

5.6 Procedure to estimate distance-based phylogenetic trees with `PHYLIP`

The `PHYLIP` software package implements four different tree-building methods: the NJ and the UPGMA methods, carried out by the program `Neighbor.exe`; the Fitch-Margoliash method, carried out by the program `Fitch.exe`; and the Fitch and Margoliash method with the assumption of a molecular clock, carried out by the program `Kitch.exe`. The UPGMA method and the algorithm implemented in the program `Kitch.exe` both assume the ultrametricity of the sequences in the data set; that is, that each sequence accumulates mutations over time at a more or less constant rate following a sort of molecular clock (see Section 5.2). As discussed previously, when mutations occur at significantly different rates in different lineages, ultrametric methods tend to result in less reliable phylogenetic trees. Conversely, nonultrametric methods, such as NJ or Fitch-Margoliash, do not assume a molecular clock and are better at recovering the correct phylogeny when different lineages exhibit a strong heterogeneity in evolutionary rates.

Chapter 10 discusses how to test the molecular clock hypothesis in a data set of molecular sequences. However, the statistical evaluation of the molecular clock, as well as the evaluation of which *nucleotide substitution model* best fits the data (see Chapter 4), requires knowledge of the tree topology relating the OTUs under investigation. Therefore, the first step in phylogeny studies based on molecular sequence data usually consists of the construction of trees with a simple evolutionary model (e.g., the Kimura 2-parameters or the F84 model) and tree-building algorithms not assuming a molecular clock. The reliability of each clade in the tree is then tested with bootstrap analysis or jackknifing (see Section 5.3). In this way, it is possible to infer one or more trees that – although they may not represent the true phylogeny – are likely to be not too far from it. These "not-too-wrong" trees can be used to test a variety of evolutionary hypotheses, including the nucleotide substitution model and the molecular clock, using ***maximum-likelihood*** methods (see Chapters 6, 7, and 10). Moreover, when a "reasonable" tree topology is known, the free parameters of any model (e.g., the transition/transversion ratio or the shape parameter α of the Γ-distribution; see Chapter 4) can be estimated from the data set by maximum-likelihood methods. These topics are discussed in Chapters 6, 7, and 10.

5.7   Inferring an NJ tree for the mtDNA data set

To infer an NJ tree with the program `Neighbor.exe` of the PHYLIP package for the mtDNA alignment (`mtDNA.phy` file), the program needs an input file with pairwise evolutionary distances. Therefore, before starting the NJ program, first calculate the distance matrix with the program `DNAdist.exe` (as explained in Chapter 4), employing the F84 model and an empirical transition/transversion ratio of 2. The matrix in the `outfile` is already in the proper format, and it can be used directly as an input file for `Neighbor.exe`. Rename the `outfile` file as `infile` and run `Neighbor.exe`; the following menu appears:

```
Neighbor-Joining/UPGMA method version 3.5

Settings for this run:
  N         Neighbor-joining or UPGMA tree?  Neighbor-joining
  O                          Outgroup root?  No, use as outgroup
                                                     species 1
  L         Lower-triangular data matrix?  No
  R         Upper-triangular data matrix?  No
  S                         Subreplicates?  No
  J       Randomize input order of species?  No, use input order
  M             Analyze multiple data sets?  No
  0      Terminal type (IBM PC, VT52, ANSI)?  ANSI
  1      Print out the data at start of run?  No
  2   Print indications of progress of run?  Yes
  3                          Print out tree?  Yes
  4         Write out trees onto tree file?  Yes

Are these settings correct? (type Y or the letter for one to
    change)
```

Option N allows choosing between NJ and UPGMA as the tree-building algorithm. Option O asks for an outgroup. Because NJ trees do not assume a molecular clock, the choice of an outgroup merely influences the way the tree is drawn. Thus,

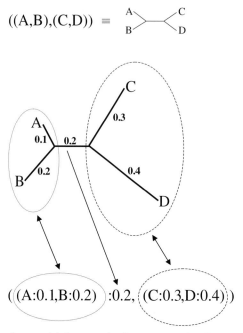

Figure 5.5   The Newick format. The linear text is the description of the indicated tree in Newick format.

Option O can be left unchanged. Options L and R allow the use of a pairwise distance matrix written in lower-triangular or upper-triangular format as input. The rest of the menu is self-explanatory: enter Y to start the computation. Outputs are written into outfile and treefile, respectively. The outfile contains a description of the tree topology and a table with the branch lengths, and it can be viewed using any text editor. The treefile contains a description of the tree in the so-called Newick format (Figure 5.5). Run TreeView.exe and select Open from the File menu. Choose All Files in Files of type and open the treefile just created in the exe subfolder of the PHYLIP folder; the tree in Figure 5.6A will appear. The **cladogram** represents the phylogenetic relationships among the taxa in the data set. In this type of graph, branch lengths are not drawn proportionally to evolutionary distances; therefore, only the topology of the tree matters. Remember that an NJ tree is unrooted, and the fact that the "seal" sequence appears to be the most external does not mean that "seal" is the eldest lineage! Choose Unrooted from the Tree menu to view the tree as an unrooted **phylogram** (Figure 5.6B). In a phylogram, branch lengths are drawn in a scale proportional to evolutionary distances. Birds and crocodiles share a common ancestor: they are sister taxa. A clade can be distinguished that joins all the mammals (i.e., platypus, opossum, mouse, rat, human, cow, whale, and seal), which indicates – as expected – their common

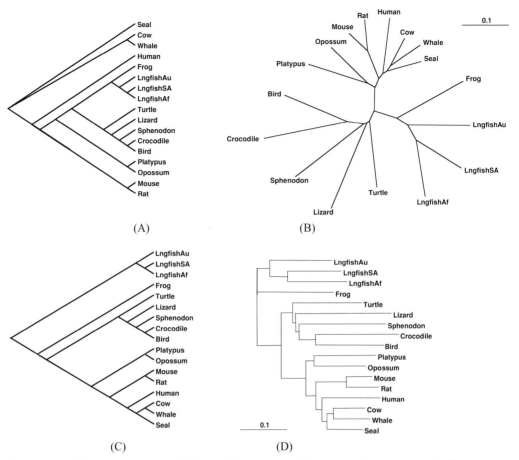

Figure 5.6   Different TreeView displays of the NJ tree of the mtDNA data set. Genetic distances are calculated using the F84 model with an empirical transition/transversion ratio of 2. (A) Slanted cladogram. (B) Unrooted phylogram. The top scale measures genetic distances in substitutions per nucleotide. (C) Slanted cladogram rooted with the three "lungfish" sequences. (D) Phylogram rooted with the three "lungfish" sequences. The bottom scale measures genetic distances in substitutions per nucleotide.

ancestry. From systematic studies based on morphological characters, it is known that lungfish belong to a clearly distinct phylogenetic lineage with respect to the amniote vertebrates (e.g., reptiles, birds, and mammals). Thus, the three lungfish sequences (i.e., LngfishAf, LngfishSA, and LngfishAu) can be chosen as outgroups and the tree rerooted so that an evolutionary direction will be given.

1. Choose Define outgroup...from the Tree menu and select the three lungfish sequences (LngFishAu, LngFishSA, LngFishAf).
2. Choose Root with outgroup...from the Tree menu.

The rooted tree can be viewed as a cladogram (Figure 5.6C) by selecting `Slanted cladogram` from the `Tree` menu, or as a phylogram with branch lengths proportional to evolutionary distances (Figure 5.6D) by selecting `Phylo-gram` in the `Tree` menu. The tree can be printed by selecting `Print...` from the `File` menu.

Choosing an **outgroup** to infer the root of a tree can be a tricky task. The chosen outgroup must belong to a clearly distinct lineage with respect to the **ingroup** sequences (i.e., the sequences in the data set under investigation), but it should not be so divergent that it cannot be aligned unambiguously against them. Therefore, before aligning the outgroup with the ingroup sequences, and using such alignment for phylogenetic analysis, it is good practice to obtain a few **dot plots** (see Sections 3.6.1 and 3.10) that compare the potential outgroup with some of the ingroup sequences. If the dot plot does not show a clear diagonal, as in Figure 3.8B, it is better to choose a different outgroup or to align outgroup and ingroup sequences only in the genome region in which a clear diagonal is visible in the dot plot (see also Section 9.5 for other alternatives implemented in TREECON).

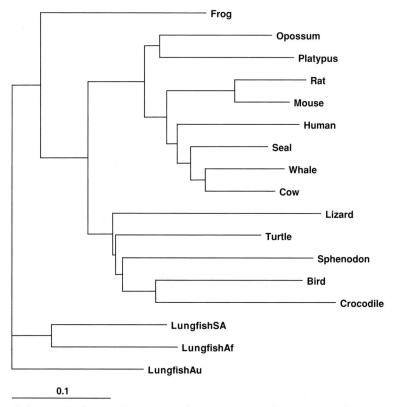

Figure 5.7 Fitch-Margoliash tree of the mtDNA data set. Genetic distances are calculated using the F84 model with an empirical transition/transversion ratio of 2.

5.8 Inferring a Fitch-Margoliash tree for the mtDNA data set

The Fitch-Margoliash tree is calculated with the program `Fitch.exe` employing the same ***mtDNA distance matrix*** used for estimating the NJ tree. The only option to change is Option G (select `Yes`), which slows down the computation but increases the probability of finding a tree minimizing the squared difference between tree distance and patristic distance (see Section 5.2.3). Again, the tree written into a `treefile` can be rerooted and printed with the `TreeView` program (Figure 5.7).

5.9 Inferring an NJ tree for the HIV-1 data set

Similar trees can be obtained for the HIV-1 data set (`hivALN.phy` file). Figure 5.8 shows the NJ tree estimated with F84 distances (ti/tv = 2.0). So far, three main lineages of HIV-1 are known: Group M, Group N, and Group O. HIV-1 Group M, the dominant group in the human pandemic, is further classified in several subtypes (A, B, C, etc.....). The example data set consists of 11 HIV-1 Group M strains, 2 SIVcpz sequences isolated from African chimpanzees, and 1 HIV-1

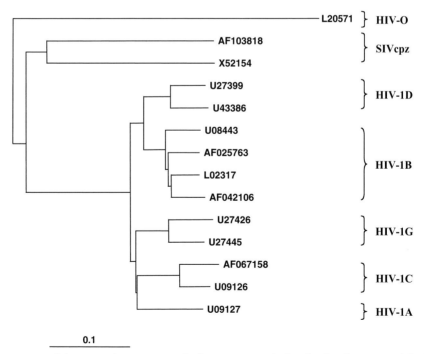

Figure 5.8 NJ tree of the HIV-1 data set. Genetic distances are calculated using the F84 model with an empirical transition/transversion ratio of 2. The bottom scale measures genetic distances in substitutions per nucleotide. HIV-O is used as outgroup.

Group O strain, which was used as the outgroup to root the tree. The different HIV-1 subtypes are *monophyletic*, and they share a common ancestor with the SIVcpz sequences. Additionally, the most external branch of the tree is a separate HIV-1 Group O lineage still infecting humans. Therefore, the tree suggests an interspecies transmission event from African chimpanzees to humans somewhere in the past at the origin of the HIV-1 Group M subtypes.

5.10 Bootstrap analysis with `PHYLIP`

The mtDNA data set discussed herein was originally obtained to support the common origin of birds and crocodiles versus an alternative hypothesis that suggested mammals as the birds' closest lineage (Hedges, 1994). Both the NJ and the Fitch-Margoliash tree show birds and crocodiles clustering together. However, the two trees differ about the clustering of turtles and lizards (compare Figures 5.6 and 5.7). It is not unusual to obtain slightly different tree topologies from different tree-building algorithms. In this example, the reliability of the crocodile-bird relationship in the phylogenetic trees estimated previously must be evaluated. As discussed in Section 5.3, one of the most widely used methods to evaluate the reliability of specific branches in a tree is bootstrap analysis, which can be carried out with `PHYLIP` as follows:

1. Run the program `Seqboot.exe` using as the `infile` the aligned mtDNA sequences. First, the program asks for a random-number seed. The number, which should be odd, is used to feed a random-number generator needed by the program to generate replicates of the data set with random resampling of the alignment's columns (see Section 5.3). Type a 5, for example, and click `enter`; the following menu will appear:

```
Bootstrapped sequences algorithm, version 3.5c

Settings for this run:
  D      Sequence, Morph, Rest., Gene Freqs?  Molecular
                                                 sequences
  J        Bootstrap, Jackknife, or Permute?  Bootstrap
  R                      How many replicates?  100
  I              Input sequences interleaved?  Yes
  0      Terminal type (IBM PC, VT52, ANSI)?   ANSI
  1      Print out the data at start of run?   No
  2  Print indications of progress of run?     Yes

Are these settings correct? (type Y or the letter for
    one to change)
```

Option J selects the kind of analysis (bootstrap by default). Option R determines the number of replicates. Type R and enter 1000. Enter Y: the program will compute 1000 replicates of the original alignment and save them in the outfile.

2. Rename the `outfile` as `infile` and run `DNAdist.exe`. By selecting Option M, the program asks how many different data sets have to be analyzed. Enter 1000; the following option should appear in the menu:

   ```
   M          Analyze multiple data sets?  Yes, 1000 data sets
   ```

   Enter Y. `DNAdist.exe` will compute 1000 distance matrixes from the 1000 replicates of the original mtDNA alignment and write them in the `outfile`.

3. Rename the `outfile` as `infile`. It is now possible to calculate, using the new `infile`, 1000 NJ or Fitch-Margoliash trees with `Neighbor.exe` or `Fitch.exe` by selecting Option M from their menus. As expected, the trees will be written in the `treefile`. Because the computation of 1000 Fitch-Margoliash trees can be very slow, especially if Option G is selected (see Section 5.5), enter 200 in Option M so that only the first 200 replicates in the `infile` will be analyzed by `Fitch.exe`.

4. The collection of trees in the `treefile` from the previous step is the input data to be used with the program `Consense.exe`. Rename `treefile` as `infile`, run `Consense.exe`, and enter Y. Note that there is no indication of the calculation progress provided by `Consense.exe`, which – depending on the size of the data set – can take from a few seconds to a few minutes. The consensus tree (see Section 5.3), written in the `treefile` in the typical Newick format (see Figure 5.5), can be viewed with `TreeView`. The outfile also contains detailed information about the bootstrap analysis.

The bootstrap values can be viewed with `TreeView` by selecting `Show internal edges labels` from the `Tree` menu. Figure 5.9 shows the NJ bootstrap consensus tree as it would be displayed by `TreeView`. In 994 of our 1000 replicates, birds and crocodiles cluster together (i.e., bootstrap value 99.4%). A slightly different value could be obtained if a different set of bootstrap replicates were analyzed, for example, by feeding a different number to the random-number generator; but as far as a sufficient number of replicates (≥ 500) is analyzed, such differences will not be statistically significant. Ninety-nine point four percent is an excellent support that strengthens confidence in the monophyletic origin of these two groups. Notice that there is no bootstrap support for the earliest internal node in the tree joining the outgroup with all the other OTUs. Clearly, a bootstrap value there would not have any meaning because that node is, of course, always present in any tree from any replicate. The `outfile` produced by `Consense.exe` also contains a representation of the tree using ASCII characters (Figure 5.10). Although the tree in Figure 5.9 (and 5.10) is drawn as rooted, it is in fact an unrooted tree. The 1000 bootstrap value on the node leading to the "opossum" branch in Figure 5.10 has no meaning. By default, the program draws the tree as rooted,

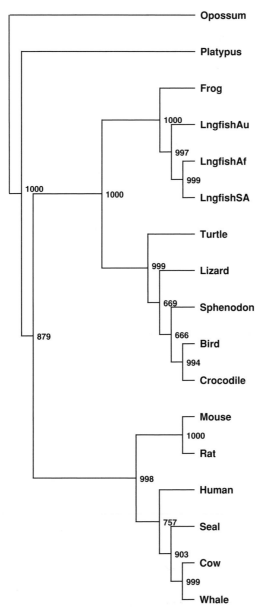

Figure 5.9 NJ consensus tree of the mtDNA data set as displayed in `TreeView`. The bootstrap values are located closest to the node of the clade that they support.

arbitrarily using one of the taxa as outgroup, and writes 1000 as bootstrap support for the most external branch leading to that taxon.

It is important to distinguish between the consensus tree and the "real" tree constructed using the NJ method because the two trees do not necessarily have the same topology (Box 5.4). When they are different, the topology of the consensus

CONSENSUS TREE:
The numbers at the forks indicate the number
of times the group consisting of the species
that are to the right of that fork occurred
among the trees, out of 1000.00 trees.

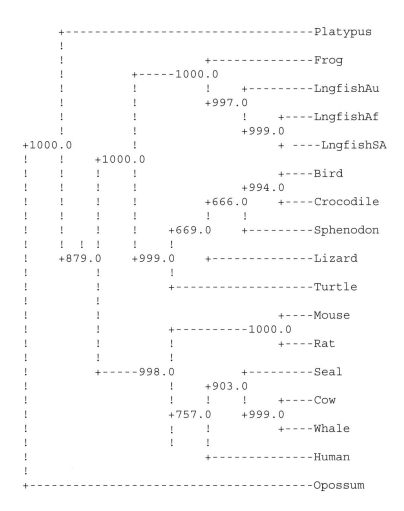

Remember: This is an unrooted tree!

Figure 5.10   NJ consensus tree as written in the outfile by the Consense program. Bootstrap values are located on the node of the clade that they support. Note that the 1,000 bootstrap value located on the earliest node leading to the "opossum" branch does not have any meaning.

Box 5.4 Bootstrap analysis with PHYLIP

Consense.exe outfile

Part of the output file produced by the program Consense.exe for the 1000 bootstrap replicates of the mtDNA data set discussed in Section 5.6 is shown here.

```
Majority-rule and strict consensus tree program, version
   3.573c
```

Species in order:

Opossum
Platypus
Rat
Mouse
Whale
Cow
Seal
Human
Lizard
Bird
Crocodile
Turtle
Sphenodon
Frog
LngfishSA
LngfishAf
LngfishAu

Sets included in the consensus tree

Set (species in order)	How many times out of 1000.00
....****	1000.00
..**......	1000.00
........** *******	1000.00
....**....	999.00
....**.	999.00
........** ***....	999.00
..******..	998.00
....***	997.00
..........* *......	994.00
....***...	903.00
..******** *******	879.00
....****..	757.00
........** *.*....	669.00
.........* *.*....	666.00

```
Sets NOT included in consensus tree:

Set (species in order)              How many times out of 1000.00

.........* ***....                  245.00
..*****... .......                  170.00
.........** *......                 159.00
........*. ..*....                  112.00
.*******.. .......                  88.00
.........* **.....                  77.00
....**.*.. .......                  75.00
..**...*.. .......                  66.00
.........** **.....                 46.00
.*.......** *******                 33.00
........*. .*......                 24.00
......**.. .......                  12.00
..**..*... .......                  9.00
.......... *.*....                  5.00
.......... ...*..*                  3.00
..****.... .......                  3.00
..****.*.. .......                  3.00
.......... .**....                  2.00
..**..**.. .......                  2.00
.........*. *......                 1.00
....*****_*_ *******                1.00
.......*** *******                  1.00
.....**... .......                  1.00
.........* *******                  1.00
.......... ....**                   1.00
```

The outfile first lists the taxa included in the tree (Species in order). The second section of the file, Sets included in the consensus tree, lists the clades that are present in more than 50% of the bootstrap replicates and, therefore, are included in the consensus tree. The clades are indicated as follows: each "." represents a taxon in the same order as it appears in the list and taxa belonging to the same clade are represented by "*". For example, "..**...... .... 1000.00" means that the clade joining rat and mouse (i.e., the third and fourth species in the list) is present in all 1000 trees estimated from the bootstrap replicates. It is possible that a particular clade present in the original tree is not included in the consensus tree because a different topology with other clades was better supported by the bootstrap test. In this case, the bootstrap value of that particular clade can be found in the third section, Sets NOT included in consensus tree.

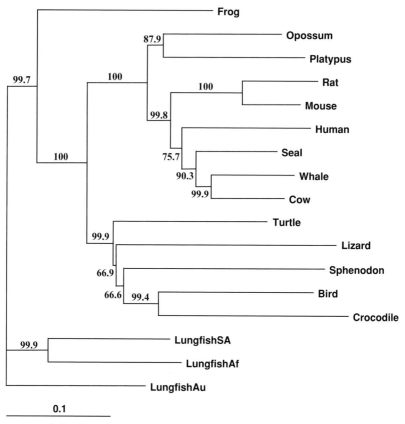

Figure 5.11   NJ tree of the mtDNA data set with bootstrap values. Genetic distances are calculated using the F84 model with an empirical transition/transversion ratio of 2. The bottom scale measures genetic distances in substitutions per nucleotide.

tree must be disregarded because it is not based on the real data set but rather on bootstrap resamplings of the original data. For publication, the bootstrap values must be displayed on a tree based on the original alignment, which usually includes branch lengths proportional to genetic distances. For example, Figure 5.11 shows the NJ tree of the mtDNA data set (the same as in Figure 5.6D) in which the bootstrap values from the consensus tree in Figure 5.9 have been added. The tree from TreeView can be exported as a graphic file by selecting Save as graphic... from the File menu. This file can then be imported in most drawing programs and bootstrap values can be added on the branches as shown in Figure 5.11.

An example of a common use of bootstrap analysis is the classification of new HIV-1 Group M strains into one of the known subtypes. A viral isolate can be assigned to one of the known subtypes when it clusters with the reference sequences of that subtype available in the Los Alamos database (see Chapter 2) with high bootstrap support (i.e., more than 80%). However, although the classification is based on bootsrapping, the actual guidelines for HIV-1 subtyping are slightly more complicated because recombination between different HIV-1 subtypes is quite common (Robertson et al., 2000) (see Chapter 14).

5.11   Other programs

Several other programs are available to compute phylogenetic trees with distance-based methods; some, such as PAUP* and TREECON, are discussed in the following chapters. A rather complete list of the phylogeny software available is maintained by Joe Felsenstein at **http://evolution.genetics.washington.edu/PHYLIP/ software.html**, and most packages can be downloaded following the links on his Web site. The packages contain complete installation and operation documentation; however, programmers typically assume that users have a background in phylogeny and molecular evolution. For those programs not discussed in this book, it may be necessary to read the original documentation to understand the details or the meaning of the implemented computation.

REFERENCES

Bruno, W. J., N. D. Socci, and A. L. Halpern (2000). Weighted neighbor-joining: A likelihood-based approach to distance-based phylogeny reconstruction. *Molecular Biology and Evolution*, 17, 189–197.

Buneman, P. (1971). The recovery of trees from measures of dissimilarity. In: *Mathematics in the Archeological and Historical Sciences*, eds. F. Hodson et al. Edinburgh University.

Cavalli-Sforza, L. L., and A. W. F. Edwards (1967). Phylogenetic analysis: Models and estimation procedures. *Evolution*, 32, 550–570.

Charleston, M. A., M. D. Hendy, and D. Penny (1994). The effect of sequence length, tree topology, and number of taxa on the performance of phylogenetic methods. *Journal of Computational Biology*, 1, 133–151.

Efron, B. and G. Gong (1983). A leisurely look at the bootstrap, the jackknife, and cross-validation. *American Statistician*, 37, 36–48.

Farris, J. S. (1977). On the phenetic approach to vertebrate classification. In: *Major Patterns in Vertebrate Evolution*, eds. M. K. Hecht, P. C. Goody, and B. M. Hecht, pp. 823–850. Plenum Press, New York.

Felsenstein, J. (1978). The number of evolutionary trees. *Systematic Zoology*, 27, 27–33.

Felsenstein, J. (1982). Numerical methods for inferring evolutionary trees. *Quarterly Review of Biology*, 57, 379–404.

Felsenstein, J. (1985). Confidence limits on phylogenies: An approach using the bootstrap. *Evolution*, 39, 783–791.

Felsenstein, J. (1988). Phylogenies from molecular sequences: Inference and reliability. *Annual Review of Genetics*, 22, 521–565.

Felsenstein, J. (1993). *PHYLIP (Phylogeny Inference Package) Version 3.5c*. Distributed by the author. Department of Genetics, University of Washington, Seattle.

Felsenstein, J. and H. Kishino (1993). Is there something wrong with the bootstrap on phylogenies? A reply to Hillis and Bull. *Systematic Biology*, 42, 193–200.

Fitch, W. M. and E. Margoliash (1967). Construction of phylogenetic trees: A method based on mutation distances as estimated from cytochrome c sequences is of general applicability. *Science*, 155, 279–284.

Gascuel, O. (1997). An improved version of the NJ algorithm based on a simple model of sequence data. *Molecular Biology and Evolution*, 14, 685–695.

Hedges, S. B. (1992). The number of replications needed for accurate estimation of the bootstrap P value in phylogenetic studies. *Molecular Biology and Evolution*, 9, 366–369.

Hedges, S. B. (1994). Molecular evidence for the origin of birds. *Proceedings of the National Academy of Sciences of the USA*, 91, 2621–2624.

Hillis, D. M. and J. J. Bull (1993). An empirical test of bootstrapping as a method for assessing confidence in phylogenetic analysis. *Systematic Biology*, 42, 182–192.

Huelsenbeck, J. P. and D. M. Hillis (1993). Success of phylogenetic methods in the four-taxon case. *Systematic Biology*, 42, 247–264.

Jukes, T. H. and C. R. Cantor (1969). Evolution of protein molecules. In: *Mammalian Protein Metabolism*, eds. H. H. Munro, pp. 21–132. Academic Press, New York.

Kidd, K. K. and L. A. Sgaramella-Zonta (1971). Phylogenetic analysis: Concepts and methods. *American Journal of Human Genetics*, 23, 235–252.

Kimura, M. (1980). A simple method for estimating evolutionary rates of base substitutions through comparative studies of nucleotide sequences. *Journal of Molecular Evolution*, 16, 111–120.

Klotz, L. C., R. L. Blanken, N. Komar, and R. M. Mitchell (1979). Calculation of evolutionary trees from sequence data. *Proceedings of the National Academy of Sciences of the USA*, 76, 4516–4520.

Kuhner, M. K. and J. Felsenstein (1994). A simulation comparison of phylogeny algorithms under equal and unequal evolutionary rates. *Molecular Biology and Evolution*, 11, 459–468.

Kumar, S. (1996). A stepwise algorithm for finding minimum-evolution trees. *Molecular Biology and Evolution*, 13, 584–593.

Kumar, S., K. Tamura, and M. Nei (1993). *MEGA: Molecular Evolutionary Genetics Analysis, Version 1.01*. Pennsylvania State University, University Park.

Li, W-H. (1981). Simple method for constructing phylogenetic trees from distance matrixes. *Proceedings of the National Academy of Sciences of the USA*, 78, 1085–1089.

Li, W-H. (1997). *Molecular Evolution*. Sunderland, MA: Sinauer Associates.

Lockhart, P. J., M. A. Steel, M. D. Hendy, and D. Penny (1994). Recovering evolutionary trees under a more realistic model of sequence evolution. *Molecular Biology and Evolution*, 11, 605–612.

Nei, M., S. Kumar, and K. Takahashi (1998). The optimization principle in phylogenetic analysis tends to give incorrect topologies when the number of nucleotides or amino acids used is small. *Proceedings of the National Academy of Sciences of the USA* 95, 12390–12397.

Olsen, G. J. (1987). Earliest phylogenetic branchings: Comparing rRNA-based evolutionary trees inferred with various techniques. *Cold Spring Harbor Symposia on Quantitative Biology*, LII, 825–837.

Page, R. D. M. and E. C. Holmes (1998). *Molecular Evolution: A Phylogenetic Approach*. Oxford: Blackwell Science.

Pearson, W. R., G. Robins, and T. Zhang (1999). Generalized neighbor-joining: More reliable phylogenetic tree reconstruction. *Molecular Biology and Evolution*, 16, 806–816.

Robertson, D. L., J. P. Anderson, J. A. Bradac et al. (2000). HIV-1 nomenclature proposal [letter]. *Science*, 288, 5–6.

Russo, C. A. M., N. Takezaki, and M. Nei (1996). Efficiencies of different genes and different tree-building methods in recovering a known vertebrate phylogeny. *Molecular Biology and Evolution*, 13, 525–536.

Rzhetsky, A. and M. Nei (1992). A simple method for estimating and testing minimum-evolution trees. *Molecular Biology and Evolution*, 9, 945–967.

Rzhetsky, A. and M. Nei (1993). Theoretical foundation of the minimum-evolution method of phylogenetic inference. *Molecular Biology and Evolution*, 10, 1073–1095.

Saitou, N. and M. Nei (1987). The neighbor-joining method: A new method for reconstructing phylogenetic trees. *Molecular Biology and Evolution*, 4, 406–425.

Saitou, N. and T. Imanishi (1989). Relative efficiencies of the Fitch-Margoliash, maximum-parsimony, maximum-likelihood, minimum-evolution, and neighbor-joining methods of phylogenetic tree construction in obtaining the correct tree. *Molecular Biology and Evolution*, 6, 514–525.

Sneath, P. H. A. and R. R. Sokal (1973). *Numerical Taxonomy*. San Francisco: W. H. Freeman.

Sokal, R. R. and C. D. Michener (1958). A statistical method for evaluating systematic relationships. *University of Kansas Scientific Bulletin*, 28, 1409–1438.

Sourdis, J. and C. Krimbas (1987) Accuracy of phylogenetic trees estimated from DNA sequence data. *Molecular Biology and Evolution, 4*, 159–166.

Studier, J. A. and K. J. Keppler (1988). A note on the neighbor-joining algorithm of Saitou and Nei. *Molecular Biology and Evolution*, 5, 729–731.

Swofford, D. L., G. J. Olsen, P. J. Waddell, and D. M. Hillis (1996). Phylogenetic inference. In: *Molecular Systematics*, eds. D. M. Hillis, C. Moritz, and B. K. Mable, pp. 407–514, Sunderland, MA: Sinauer Associates.

Swofford, D. L. (2002). *PAUP** (*Phylogenetic Analysis Using Parsimony*) (**and other methods*), *Version 4*. Sunderland, MA (USA): Sinauer Associates.

Van de Peer, Y. and R. De Wachter (1994). TREECON for Windows: A software package for the construction and drawing of evolutionary trees for the Microsoft Windows environment. *Computer Applications in the Biosciences*, 10, 569–570.

Van de Peer, Y., S. Rensing, U.-G. Maier, and R. De Wachter (1996). Substitution rate calibration of small ribosomal subunit RNA identifies chlorarachniophyte endosymbionts as remnants of green algae. *Proceedings of the National Academy of Sciences of the USA*, 93, 7732–7736.

Van de Peer, Y., P. De Rijk, J. Wuyts, T. Winkelmans, and R. De Wachter (2000a). The European small subunit ribosomal RNA database. *Nucleic Acid Research*, 28, 175–176.

Van de Peer, Y., S. Baldauf, W. F. Doolittle, and A. A. Meyer (2000b). An updated and comprehensive phylogeny of the rRNA crown eukaryotes based on rate-calibrated evolutionary distances. *Journal of Molecular Evolution* (in press).

Xia, X. (2000). *Data Analysis in Molecular Biology and Evolution (DAMBE)*. Boston: Kluwer Academic Publishers.

Zharkikh, A. and W-H. Li (1992a). Statistical properties of bootstrap estimation of phylogenetic variability from nucleotide sequences. I. Four taxa with a molecular clock. *Molecular Biology and Evolution*, 9, 1119–1147.

Zharkikh, A. and W-H. Li (1992b). Statistical properties of bootstrap estimation of phylogenetic variability from nucleotide sequences. II. Four taxa without a molecular clock. *Journal of Molecular Evolution*, 35, 356–366.

Phylogeny inference based on maximum-likelihood methods with TREE–PUZZLE

THEORY

Arndt von Haeseler and Korbinian Strimmer

6.1 Introduction

The concept of likelihood refers to situations that typically arise in natural sciences in which given some data **D**, a decision must be made about an adequate explanation of the data. Thus, a specific model and a hypothesis are formulated in which the model as such is generally not in question. In the phylogenetic framework, one part of the model is that sequences actually evolve according to a tree. The possible hypotheses include the different tree structures, the branch lengths, the parameters of the model of sequence evolution, and so on. By assigning values to these elements, it is possible to compute the probability of the data and to make statements about their plausibility. If the hypothesis varies, the result is that some hypotheses produce the data with higher probability than others. Coin-tossing is a standard example. After flipping a coin $n = 100$ times, $h = 21$ heads and $t = 79$ tails were observed. Thus, $\mathbf{D} = (21, 79)$ constitutes the data. The model then states that with some probability, $\theta \in [0, 1]$ heads appear when the coin is flipped. Moreover, it is assumed that the outcome of each experiment is independent of the others, that θ does not change during the experiment, and that the experiment has only two outcomes (head or tail). The model is now fully specified. Because both heads and tails were obtained, θ must be larger than zero and smaller than 1. Moreover, any probability textbook explains that the probability to observe exactly $H = h$ heads in n experiments can be calculated according to the following binomial distribution:

$$\Pr[H = h] = \binom{n}{h} \theta^h (1 - \theta)^{n-h} \qquad (6.1)$$

Equation 6.1, the so-called binomial formula, can be read in two ways. First, it is assumed that θ is known; then the probability of $h = 0, \ldots, n$ heads in n tosses can

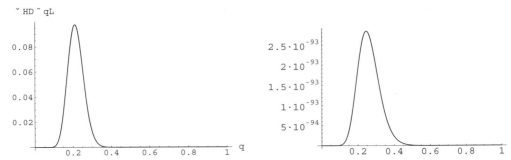

Figure 6.1   Left: Likelihood function of a coin-tossing experiment showing 21 heads and 79 tails. Right: Likelihood function of the Jukes-Cantor model of sequence evolution for a sequence with length 100 and 21 observed differences.

be computed. Second, Equation 6.1 can be seen as a function of θ, where n and k are given; this defines the so-called **likelihood function**

$$L(\theta) = \Pr[H = h] = \binom{n}{h} \theta^h (1 - \theta)^{n-h}$$ (6.2)

From Figure 6.1, which illustrates the likelihood function for the coin-tossing example, it can be seen that some hypotheses (i.e., choices of θ) generate the observed data with a higher probability than others. In particular, Equation 6.2 becomes maximal if $\theta = \frac{21}{100}$. This value also can be computed analytically. For ease of computation, first compute the logarithm of the likelihood function, which results in sums rather than products:

$$\log[L(\theta)] = \log\binom{n}{h} + h\log\theta + (n - h)\log(1 - \theta)$$ (6.3)

The problem is now to find the value of θ maximizing the function. From elementary calculus, it is known that the maximum of a function $y = f(x)$ – when it exists – is given by the value x for which the first derivative of the function equals zero. Differentiation of Equation 6.3 with respect to θ yields the following:

$$\frac{\partial \log[L(\theta)]}{\partial \theta} = \frac{h}{\theta} + \frac{n - h}{1 - \theta}$$ (6.4)

This derivative is equal to zero if $\theta = \frac{h}{n}$, positive for smaller values of θ, and negative for larger values, so that $\log[L(\theta)]$ attains its maximum when $\hat{\theta} = h/n$ (the hat notation on θ indicates an estimate rather than the real unknown value). Thus, $\hat{\theta} = h/n$ is the **maximum-likelihood estimate** (**MLE**) of the probability of observing a heads in a single coin toss. In general, when the value of θ that maximizes Equation 6.3 is selected, the observed data are produced with the highest likelihood, which is precisely the maximum-likelihood principle. However, the resulting like-lihoods are usually small (e.g., $L(21/100) \approx 0.0975$); conversely, the likelihoods of competing hypotheses can be compared by computing the odds ratio. Note that the hypothesis that the coin is a fair one ($\theta = 1/2$) results in a likelihood of $L(1/2) \approx 1.61 \cdot 10^{-9}$; thus, the MLE of $\theta = 0.21$ is $6 \cdot 10^7$ times more likely to produce the data than $\theta = 0.50$! This comparison of odds ratios leads to the statistical test procedure discussed in Chapters 7 and 10.

In evolution, point mutations are considered chance events, just like tossing a coin. Therefore, at least in principle, the probability of finding a mutation along one branch in a *phylogenetic tree* can be calculated by using the same maximum-likelihood framework discussed previously. The main idea behind phylogeny inference with maximum likelihood is to determine the tree topology, branch lengths, and parameters of the evolutionary model (e.g., transition/transversion ratio, base frequencies, rate variation among sites) (see Chapter 4) that maximize the probability of observing the sequences at hand. In other words, the likelihood function is the conditional probability of the data (i.e., sequences) given a hypothesis (i.e., a model of substitution with a set of parameters θ and the tree τ, including branch lengths):

$$L(\tau, \theta) = \text{Prob}(\text{Data} \mid \tau, \theta) = \text{Prob}(\text{Aligned sequences} \mid \text{tree,}$$
$$\text{model of evolution}) \quad (6.5)$$

The MLEs of τ and θ are those making the likelihood function as large as possible:

$$\hat{\tau}, \hat{\theta} = \arg\max_{\tau, \theta} L(\tau, \theta) \quad (6.6)$$

Before proceeding to the next section, some cautionary notes are necessary. First, the likelihood function must not be confused with a probability. It is defined in terms of a probability, but it is the probability of the observed event, not of the unknown parameters. The parameters have no probability because they do not depend on chance. Second, the probability of getting the observed data has nothing to do with the probability that the underlying model is correct. For example, if the model states that the sequences evolve according to a tree, although they have recombined, then the final result will still be a tree that gives rise to the maximum-likelihood value (see also Chapter 14). The probability of the data given the MLE of the parameters does not provide any hints that the model assumptions are true. One can only compare the maximum-likelihood values with other likelihoods for model parameters that are elements of the model. To determine whether the hypothesis of tree-like evolution is reasonable, the type of relationship allowed among sequences must be enlarged; this is discussed in Chapter 12.

6.2   The formal framework

Before entering the general discussion about maximum-likelihood tree reconstruction, the simplest example (i.e., reconstructing a maximum-likelihood tree for two sequences) is considered. A tree with two **taxa** has only one branch connecting the two sequences; the sole purpose of the exercise is reconstructing the branch length that produces the data with maximal probability.

6.2.1   The simple case: Maximum-likelihood tree for two sequences

In what follows, it is assumed that the sequences are evolving according to the Jukes and Cantor model (see Chapter 4). Each position evolves independently from the remaining sites and with the same **evolutionary rate**. The alignment has length l for the two sequences $S_i = (s_i(1), \ldots, s_i(j))$, $(i = 1, 2)$, where $s_i(j)$ is the nucleotide, the amino acid, or any other letter from a finite alphabet at sequence position j in sequence i. The likelihood function is, then, according to Equation 4.31:

$$L(d) = \prod_{j=1}^{l} \pi_{s_i(j)} P_{s_1(j)s_2(j)}\left(-\frac{d}{3/4}\right) \tag{6.7}$$

where d, the number of substitutions per site, is the parameter of interest and $P_{xy}(t)$ is the probability of observing nucleotide y if nucleotide x was originally present. Following Equations 4.12a and 4.12b, the following is obtained:

$$P_{xy}\left(-\frac{4}{3}d\right) = \begin{cases} \frac{1}{4}\left(1 + 3\exp\left[-\frac{4}{3}d\right]\right) \equiv \tilde{P}_{xx}(d), & \text{if } x = y \\[2ex] \frac{1}{4}\left(1 - \exp\left[-\frac{4}{3}d\right]\right) \equiv \tilde{P}_{xy}(d), & \text{if } x \neq y \end{cases} \tag{6.8}$$

To infer d, the relevant statistic is the number of identical pairs of nucleotides (l_0) and the number of different pairs (l_1), where $l_0 + l_1 = l$. Therefore, the alignment is summarized as $\mathbf{D} = (l_0, l_1)$ and the score is computed as:

$$\log[L(d)] = C + l_0 \log\left[\tilde{P}_{xx}(d)\right] + l_1 \log\left[\tilde{P}_{xy}(d)\right] \tag{6.9}$$

which is maximized if

$$d = -\frac{3}{4}\log\left[1 - \frac{4}{3} \cdot \frac{l_1}{l_1 + l_0}\right] \tag{6.10}$$

In practice, the MLE of the number of substitutions per site equals the method-of-moments estimate (see Equation 4.15a). Therefore, the maximum-likelihood tree relating the sequences S_1 and S_2 is a straight line of length d, with the sequences as endpoints.

```
L20571     ...AAAGTAATGAAGAAGAACAACAGGAAGTCATGGAGCTTATACATA...

AF10138    ...ATGGAGAAGAAGAAG---------AGACTCTGGCTAAGTTATTGT...

X52154     ...ATGGAGAAGAAGAAG---------AGAGACTGGAACAGCTTATCC...

U09127     ...ATGGGGATAGAGAGGAATTATCCTTGCTGGTGGACATGGGGGATT...

U27426     ...AGGGGGATACAGATGAATTGGCAACACTTGTGGAAATGGGGAACT...

U27445     ...AAGGGGATACGGACGAATTGGCAACACTTCTGGAGATGGGGAACT...

U067158    ...AGGGGGACACTGAGGAATTATCAACAATGGTGGATATGGGGCGTC...

U09126     ...GAGGGGATACAGAGGAATTGGAAACAATGGTGGATATGGGGCATC...

U27399     ...AGGGAGATGAGGAGGAATTGTCAGCATTTGTGGGGATGGGGCACC...

U43386     ...AGGGAGATGCAGAGGAATTATCAGCATTTATGGAAATGGGGCATC...

L02317     ...AAGGAGATCAGGAAGAATTATCAGCACTTGTGGAGATGGGGCACC...

AF025763   ...AAGGGGATCAGGAAGAATTGTCAGCACTTGTGGAGATGGGGCATG...

U08443     ...AAGGAGATGAGGAAGCATTGTCAGCACTTATGGAGAGGGGGCACC...

AF042106   ...AAGGGGATCAGGAAGAATTATCGGCACTTGTGGACATGGGGCACC...
```

Figure 6.2   Part of the mtDNA sequence alignment used as a relevant example throughout the book.

This example was completely computable because it is the simplest model of sequence evolution and, more importantly, because only two sequences – which can only be related by one tree – were considered. The following sections set up the formal framework to study more sequences.

6.2.2 The complex case

When the data set consists of n ($n > 2$) aligned sequences, rather than computing the probability $P(t)$ of observing two nucleotides at a given site in two sequences, the probability of finding a certain column or pattern of nucleotides in the data set is computed. Let D_j denote the nucleotide pattern of site $j = 1, \ldots, l$ in the alignment (Figure 6.2). The unknown probability obviously depends on the model of sequence evolution, M, and the tree, τ, relating the n sequences with the number of substitutions along each branch of the tree (i.e., the branch lengths). In theory, each site could be assigned its own model of sequence evolution according to the general time reversible model (see Chapter 4) and its own set of branch lengths. Then, however, the goal to reconstruct a tree from an alignment becomes almost computationally intractable and several simplifications are needed. First, it is assumed that each site s in the alignment evolves according to the same model M; for example, the Tamura Nei (TN) model (see Equations 4.32a, b, c) (i.e., γ, κ, and π are the same for each site in the alignment). The assumption also implies that all sites evolve at the same rate μ (see Equation 4.24). To overcome this simplification, the rate at a site is modified by a rate-specific factor, $\rho_j > 0$. Thus, the ingredients for the probability of a certain site pattern are available, and

$$\Pr[D_s, \tau, M, \rho_j], \ j=1, \ldots, l \tag{6.11}$$

specifies the probability to observe pattern D_j. If it is also assumed that each sequence site evolves independently (i.e., according to τ, M, with a site specific rate ρ_j), then the probability of observing the alignment (data) $\mathbf{D} = (D_1, \ldots, D_l)$ equals the product of the probabilities at each site, as follows:

$$L(\tau, M, \rho) \equiv \Pr[\mathbf{D}, \tau, M, \rho] = \prod_{j=1}^{l} \Pr[D_j, \tau, M, \rho_j] \tag{6.12}$$

When the data are fixed, Equation 6.12 is again a likelihood function (see Equations 6.2 and 6.5), which allows for the two ways of looking at it (see the previous section). First, for a fixed choice of τ, M, and the site rate vector ρ, the probability to observe the alignment $\mathbf{D}$ can be computed with Equation 6.11. Second, for a given alignment $\mathbf{D}$, Equation 6.12 can be used to find the MLEs.

In what follows, the two issues are treated separately. However, to simplify the matter, it is assumed that the site-specific rate factor ρ_j is drawn from a Γ-*distribution* with expectation 1 and variance $1/\alpha$ (Uzzel and Corbin, 1971; Wakeley, 1993, where α defines the shape of the distribution) (see also Section 4.6.1).

6.3 Computing the probability of an alignment for a fixed tree

Consider the tree τ with its branch lengths (i.e., number of substitutions), the model of sequence evolution M with its parameters (e.g., transition/transversion ratio, stationary base composition), and the site-specific rate factor $\rho_j = 1$ for each site j. The goal is to compute the probability of observing one of the 4^n possible patterns in an alignment of n sequences. The tree displayed in Figure 6.3 illustrates

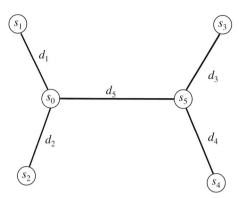

Figure 6.3   Four-sequence tree, with branch lengths d_1, d_2, d_3, and d_4 leading to sequences S_1, S_2, S_3, and S_4 and branch length d_5 connecting the "ancestral" sequences S_0 and S_5.

the principle for four sequences ($n = 4$). Because the model M belongs to the GTR class – that is, a time-reversible model (see Chapter 4) – it is assumed that evolution started from sequence S_0 and then proceeded along the branches of tree τ with branch lengths d_1, d_2, d_3, d_4, and d_5. To compute $\Pr[\,D_j,\ T,\ M,\ 1]$ for a specific site j, where $D_j = (s_1, s_2, s_3, s_4)$ are the nucleotides observed, it is necessary to know the ancestral states s_0 and s_5. The conditional probability of the data given the ancestral states then will be as follows:

$$\Pr[\,D_j,\ \tau,\ M,\ 1|s_0,\ s_5] = P_{s_0 s_1}(d_1) \cdot P_{s_0 s_2}(d_2) \cdot P_{s_0 s_5}(d_5) \cdot P_{s_5 s_3}(d_3) \cdot P_{s_5 s_4}(d_4)$$

(6.13)

The computation follows immediately from the considerations in Chapter 4. However, in almost any realistic situation, the ancestral sequences are not available. Therefore, one sums over all possible combinations of ancestral states of nucleotides. As discussed in Section 4.4, nucleotide substitution models assume *stationarity*; that is, the relative frequencies of A, C, G, and T (π_A, π_C, π_G, π_T) are at equilibrium. Thus, the probability for nucleotide s_0 will equal its stationary frequency $\pi(s_0)$, from which it follows that

$$\Pr[\,D_j,\ \tau,\ M,\ 1] = \sum_{s_0}\sum_{s_5} \pi(s_0) P_{s_0 s_1}(d_1) \cdot P_{s_0 s_2}(d_2) \cdot P_{s_0 s_5}(d_5)$$
$$\cdot P_{s_5 s_3}(d_3) \cdot P_{s_5 s_4}(d_4)$$

(6.14)

Although this equation looks like exponentially, many summands need to be computed exponentially, the sum can be efficiently assessed by evaluating the likelihoods moving from the end nodes of the tree to the root (Felsenstein, 1981). In each step, two nodes from the tree are removed and replaced by a single node. This process bears some similarity to the computation of the minimal number of substitutions on a given tree in the ***maximum parsimony*** framework (Fitch, 1971) (see Chapter 7). However, contrary to maximum parsimony, the distance (i.e., number of substitutions) between the two nodes is considered. Under the maximum parsimony framework, if two sequences share the same nucleotide, then the most recent common ancestor also carries this nucleotide (see Chapter 7). In the maximum-likelihood framework, this nucleotide is shared by the ancestor only with a certain probability, which gets smaller if the sequences are only very remotely related.

6.3.1 Felsenstein's pruning algorithm

Equation 6.14 shows how to compute the likelihood of a tree for a given position in a sequence alignment. To generalize this equation for more than four sequences, it is necessary to sum all the possible assignments of nucleotides at the $n - 2$ inner nodes of the tree. Unfortunately, this straightforward computation is not feasible, but the amount of computation can be reduced considerably by noticing the following recursive relationship in a tree. Let $D_j = (s_1, s_2, s_3, \ldots, s_n)$ be a pattern at a site j, with tree τ and a model M fixed. Nucleotides at inner nodes of the tree are abbreviated as x_i, $i = n + 1, \ldots, 2n - 2$. For an inner node i with offspring o_1 and o_2, the vector $\mathbf{L}_j^i = (L_j^i(A), L_j^i(C), L_j^i(G), L_j^i(T))$ is defined recursively as

$$
L_j^i(s) = \left[\sum_{x \in \{A,C,G,T\}} P_{s,x}(d_{o_1}) L_j^{o_1}(x) \right] \cdot \left[\sum_{x \in \{A,C,G,T\}} P_{s,x}(d_{o_2}) L_j^{o_2}(x) \right],
$$
$$
s \in \{A, C, G, T\} \quad (6.15)
$$

with

$$
L_j^i(s) = \begin{cases} 1, & \text{if } s = s_i \\ 0 & \text{otherwise} \end{cases} \quad (6.16)
$$

where d_{o_1} and d_{o_2} are the number of substitutions connecting node i and its descendants in the tree. Without loss of generality, it is assumed that the node $2n - 2$ has three offspring: o_1, o_2, and o_3, respectively. For this node, Equation 6.15 is modified accordingly. These two equations allow a much more efficient computation of the likelihoood for each position of an alignment by realizing that

$$
\Pr[D_j, \tau, M, 1] = \sum_{s \in \{A,C,G,T\}} \pi_s L_j^{2n-1}(s) \quad (6.17)
$$

Equation 6.17 then can be used to compute the likelihood of the full alignment with the aid of Equation 6.12. In practice, the calculation of products is avoided, moving instead to log likelihoods; that is, Equation 6.12 becomes

$$\text{Log}[L(\tau, M, 1)] = \text{Log}\left[\prod_{j=1}^{l} \Pr[D_j, \tau, M, 1]\right] = \sum_{j=1}^{l} \text{Log}\left[\Pr[D_j, \tau, M, 1]\right]$$

(6.18)

6.4 Finding a maximum-likelihood tree

Equations 6.15 through 6.18 show how to compute the likelihood of an alignment, if everything were known. In practice, however, branch lengths of the tree are unknown. Branch lengths are computed numerically by maximizing Equation 6.18; that is, by finding those branch lengths for tree τ maximizing the log-likelihood function, which is accomplished by applying Newton's method or other numerical routines. Such a computation is usually time-consuming and sometimes the result depends on the numerical method. Nevertheless, maximizing the likelihood for a single tree is not the major challenge in phylogenetic reconstruction; the daunting task is to actually find the tree among all possible tree structures that maximizes the global likelihood. Unfortunately, for any method that has an explicit optimality criterion (e.g., maximum parsimony, distance methods, and maximum likelihood), no efficient algorithms are known that guarantee the localization of the best tree(s) in the huge space of all possible tree topologies. The naive approach to simply compute the maximum-likelihood value for each tree topology is prohibited by the huge number of tree structures, even for moderately sized data sets, that can be computed for n sequences according to

$$t_n = \frac{(2n-5)!}{2^{n-3}(n-3)!} = \prod_{i=3}^{n}(2i-5)$$

(6.19)

When computing the maximum-likelihood tree, the model parameters and branch lengths have to be computed for each tree, and then the tree that yields the highest likelihood has to be selected. Because of the numerous tree topologies, testing all possible trees is impossible, and it is also not computationally feasible to estimate the model parameters for each tree. Thus, various heuristics are used to suggest reasonable trees, including stepwise addition (e.g., used in Felsenstein's PHYLIP package: program DNAML.exe) (Felsenstein, 1993), and ***star decomposition*** MOLPHY (Adachi and Hasegawa, 1995), as well as the ***neighbor-joining*** (***NJ***) algorithm (Saitou and Nei, 1987). These methods are discussed in Chapter 7 because they are implemented in the PAUP* program and can be used when searching for the best tree according to maximum parsimony (Swofford et al., 1995). What follows focuses on the description of another heuristic to find a plausible

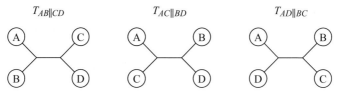

Figure 6.4   The three different informative tree topologies for the quartet $q = (A, B, C, D)$.

candidate for the maximum-likelihood tree: the **quartet-puzzling method** implemented in the TREE–PUZZLE program (Strimmer and von Haeseler, 1996a; Strimmer, Goedman, and von Haeseler, 1987).

6.4.1   The quartet-puzzling algorithm

Given a set of n aligned nucleotide (or amino-acid) sequences, any group of four of them is called a quartet. The quartet-puzzling algorithm analyzes all possible quartets in a data set, taking advantage of the fact that for a quartet, just three unrooted tree topologies are possible (Figure 6.4). In essence, the algorithm is a three-step procedure. The first step, the so-called **maximum-likelihood step**, computes for each of the $\binom{n}{4} = \frac{n!}{4!(n-4)!}$ possible quartets the maximum-likelihood values L_1, L_2, and L_3 for the three possible four-sequence trees T_1, T_2, T_3, assuming

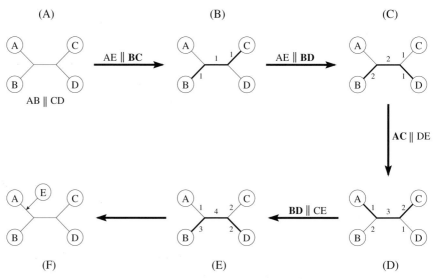

Figure 6.5   Schematic view of the quartet-puzzle step. Sequence E is to be inserted into the already reconstructed subtree. The location of the insertion is guided by the result of the maximum-likelihood reconstruction of all quartets comprising three sequences from the subtree and sequence E. A penalty is given to the branches supporting a different clustering with respect to the quartets where E was included. The branch showing the smallest penalty value is selected as the place of insertion.

a user-defined model M of sequence evolution (see Equation 6.12). The resulting list of $3 \cdot \binom{n}{4}$ likelihoods is then used in the **quartet-puzzle step** to compute an intermediate tree by inserting sequences sequentially in an already reconstructed **subtree**. Figure 6.5 illustrates the insertion procedure, in which a tree with sequences A, B, C, and D is already reconstructed, and sequence E is inserted according to the four-sequence trees that contain sequence E and three sequences from the partial tree. Eventually, sequence E is inserted at the branch with minimal penalty. Figure 6.6 shows all intermediate trees for a small example of five sequences. For large data it is not feasible to compute all intermediate trees. Thus the quartet-puzzle step is repeated at least a thousand times for various input orders of sequences

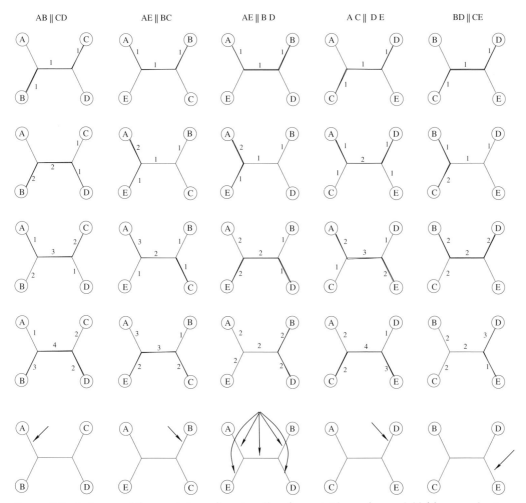

Figure 6.6 Outcome of the quartet-puzzling step, if each quartet is used as an initial four-species tree. The arrows indicate where to place the fifth sequence.

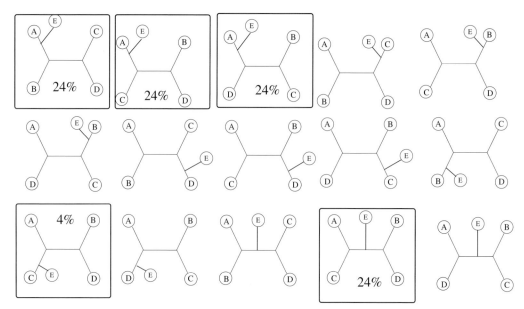

Figure 6.7   Frequency (in %) of the three topologies generated by the quartet-puzzling algorithm for the five-sequences example in Figure 6.6.

to avoid reconstruction artifacts due to the ordering of the sequences and to get a representative collection of trees. Figure 6.7 shows how often each of the 15 five-sequence topologies is reconstructed during the repeated application of the quartet-puzzling algorithm.

Finally, in step three, the **majority-rule consensus** (Margush, 1981) is computed from the resulting intermediate trees. The resulting tree is called the **quartet-puzzling tree**. The **consensus step** provides information about the number of times a particular grouping occurred in the intermediate trees. This so-called **reliability value**, or **support value**, measures (in %) how frequently a group of sequences occurs among all intermediate trees. All groups that occur in more than 50% of the collection of intermediate trees are represented in the majority-rule **consensus tree**, which is not necessarily the maximum-likelihood tree. In the example in Table 6.1, the TREE–PUZZLE program would output tree clustering sequences A and E (support value 72%) and B and D (support value 52%). In this case, the tree is fully resolved. If the data do not provide a good resolution of phylogenetic relationships for a subgroup of sequences, the consensus tree will display a small reliability value for the corresponding internal branch of the tree. The reliability value should not be confused with the usual **bootstrap values**; whereas reliability values are an intrinsic result of the quartet-puzzling algorithm, bootstrapping is an external procedure that can be applied to any tree-building method (see Chapter 5).

Table 6.1 Outcome of the computation of the majority-rule consensus tree for the five-sequence example (in %)

Cluster	T-1	T-2	T-3	T-4	T-5	Total frequency
(A, B)	0	0	0	0	0	0
(A, C)	0	0	0	0	24	24
(A, D)	0	0	0	0	0	0
(A, E)	24	24	24	0	0	72
(B, C)	0	0	24	0	0	24
(B, D)	0	24	0	4	24	52
(B, E)	0	0	0	0	0	0
(C, D)	24	0	0	0	0	24
(C, E)	0	0	0	4	0	4
(D, E)	0	0	0	0	0	0

6.5 Estimating the model parameters with maximum likelihood

When inferring the parameters of a model that allows for substitution-rate heterogeneity across sites, which include the α parameter of the discrete Γ-distribution (see Sections 4.6.1 and 4.10.1), it is necessary to consider an overall tree because the relationship of the sequences is important for the estimation (Sullivan et al., 1996). In theory, it is possible to optimize both the parameters of the model and the tree topology with maximum likelihood by simultaneously finding the phylogenetic tree and the set of parameters maximizing the likelihood function. However, for large data sets (i.e., more than ten to fifteen sequences), this approach becomes computationally non-feasible (discussed previously). Therefore, the accepted strategy is to infer a "reasonable" tree topology with faster – although less reliable – reconstruction methods and use that tree to estimate the parameters. Eventually, a maximum-likelihood tree can be reestimated with the new set of parameters. This approach assumes that parameter estimates are not greatly disturbed when using a slightly incorrect topology. Among the fast distance-based tree reconstruction methods, NJ (Saitou and Nei, 1987) exhibits the best performance in terms of accuracy and speed. Although its probability to find the true underlying tree decreases rather quickly with the number of taxa, it has been shown that the NJ tree is always similar to the true tree, even for numerous sequences (Strimmer and von Haeseler, 1996b).

The following iterative procedure is implemented in the TREE–PUZZLE program:

1. Based on reasonable pairwise "genetic distance, estimates" an NJ tree is computed.
2. Then, maximum-likelihood branch lengths are computed for this tree topology and parameters of the sequence evolution are estimated.
3. Based on these estimates, a new NJ tree is computed and Step 2 is repeated.

Steps (2) and (3) are repeated until the estimates of the model parameters are stable. Eventually, given the unrooted tree topology and a set of optimized parameters, maximum-likelihood branch lengths can be computed using the likelihood function (Equation 6.12), which involves simultaneous optimization of $2n - 3$ branch lengths for a tree with n sequences. The computing time can be accelerated by estimating an approximate likelihood. Adachi and Hasegawa (1996) applied a least-squares fit of pairwise maximum-likelihood distances to a given tree topology and showed that the resulting suboptimal set of branch-length estimates is close to the maximum likelihood. TREE–PUZZLE employs this idea to obtain approximate estimates of model parameters, saving computation time and still serving as an efficient tool to estimate model parameters.

6.6   Likelihood-mapping analysis

The chapter so far has discussed the problem of reconstructing a phylogenetic tree. A maximum-likelihood approach also may be used to study the amount of *evolutionary information* contained in a data set. The analysis is based on the maximum-likelihood values for the three possible trees with four sequences. If L_1, L_2, and L_3 are the likelihoods of trees T_1, T_2, and T_3, then it is possible to compute the posterior probabilities of each tree T_i as $p_i = L_i/(L_1 + L_2 + L_3)$, where the p_i terms sum to 1 and $0 < p_i < 1$ for each i. The probabilities p_1, p_2, and p_3 can be reported simultaneously as a point P lying inside an equilateral triangle, each corner of the triangle representing one of the three possible tree topologies (Figure 6.8A). If P is close to one corner – for example, the corner T_1 – the tree T_1 receives the highest support. In a maximum-likelihood analysis, the tree T_i, which satisfies $p_i = \max\{p_1, p_2, p_3\}$, is selected as the MLE. However, this decision is questionable if P is close to the center of the triangle. In that case, the three likelihoods are of similar magnitude; in such situations, a more realistic representation of the data is a star-like tree rather than an artificially **strictly bifurcating tree** (see Section 1.6).

Therefore, the **likelihood-mapping method** (Strimmer and von Haeseler, 1997) partitions the area of the equilateral triangle into seven regions (Figure 6.8B). The three trapezoids at the corners represent the areas supporting strictly bifurcating trees (i.e., Areas 1, 2, and 3 in Figure 6.8B). The three rectangles on the sides represent regions where the decision between two trees is not obvious (i.e., Areas 4, 5, and 6 in Figure 6.8B for trees 1 and 2, 2 and 3, and 3 and 1). The center of the triangle represents sets of P vectors where all three trees are poorly supported (i.e., Area 7 in Figure 6.8B). Given a data set of n aligned sequences, the likelihood-mapping analysis works as follows. The three likelihoods for the three tree topologies of each possible quartet (or of a random sample of the quartets when their number is

(A)

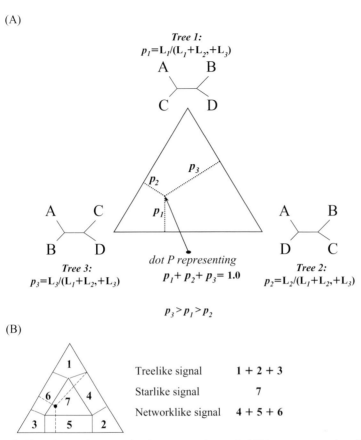

Figure 6.8 Likelihood mapping. (A) The three posterior probabilities p_1, p_2, and p_3 for the three possible unrooted trees of four taxa are reported as a dot (P) inside an equilateral triangle, where each corner represents a specific tree topology with likelihood L_1, L_2, and L_3, respectively. (B) Seven main areas in the triangle supporting different evolutionary information.

too great; e.g., 10,000) are reported as a dot in an equilateral triangle like the one in Figure 6.8A. The distribution of dots in the seven major areas of the triangle (see Figure 6.8B) gives an overall impression of the tree-likeness of the data. Note that because the method evaluates random quartets drawn from N sequences, which one of the three topologies is supported by any corner of the triangle is not relevant. What is informative about the mode of evolution of the sequences under investigation is the percentage of dots belonging to the three main different areas in the equilateral triangle. The three corners (Areas $1 + 2 + 3$; see Figure 6.8B) represent fully resolved tree topologies; that is, the presence of **tree-like phylogenetic signal** in the data. The center is the area of **star-like phylogeny** (Area 7; see Figure 6.8B); the three areas on the sides (Areas $4 + 5 + 6$; see Figure 6.8B) represent **network-like phylogeny**, in which the data support conflicting tree topologies (see also Chapter 12).

From the biological standpoint, a likelihood mapping showing more than 20–30% of dots in the star-like or network-like area suggests that either the data are noisy – and, therefore, not reliable for phylogenetic inference – or *recombination* is at work. In the latter case, it would also be useful to perform an analysis with methods that explore and display conflicting trees, such as *split decomposition* or *bootscanning* (see Chapters 12 and 14). A more detailed study is given in Nieselt-Struwe and von Haeseler (2001).

PRACTICE

Arndt von Haeseler and Korbinian Strimmer

6.7 Software packages

There are several software packages that compute maximum-likelihood trees from DNA or amino-acid sequences, including PHYLIP (see Chapters 4 and 5). A detailed list is found at Joe Felsenstein's Web site, **http://evolution.genetics. washington.edu/PHYLIP/software.html**. Because program packages are emerging at a rapid pace, the reader is advised to visit this Web site for regular updates.

6.8 An illustrative example of quartet-puzzling tree reconstruction

In what follows, the `hivALN.phy` file will be analyzed with the latest version of TREE-PUZZLE (5.0). Place the `hivALN.phy` file in the TREE-PUZZLE folder and run the executable; the following text appears:

```
WELCOME TO TREE-PUZZLE 5.0.

Please enter a file name for the sequence data:
Type the filename hivALN.phy and press Enter.
Input data set contains 14 sequences of length 2781
(consists very likely of nucleotides)

GENERAL OPTIONS
 b                     Type of analysis? Tree reconstruction
 k             Tree search procedure? User-defined trees
 z   Compute clocklike branch lengths? No
 e               Parameter estimates? Approximate (faster)
 x        Parameter estimation uses? 1st input tree
SUBSTITUTION PROCESS
 d       Type of sequence input data? Auto: Nucleotides
 h           Codon positions selected? Use all positions
 m              Model of substitution? HKY (Hasegawa et al.,
                   1985)
 t Transition/transversion parameter? Estimate from data set
 f              Nucleotide frequencies? Estimate from data set
RATE HETEROGENEITY
 w        Model of rate heterogeneity? Uniform rate

Quit [q], confirm [y], or change [menu] settings:
```

Each option can be selected and/or edited by typing the corresponding letter. For example, if the user types b repeatedly, then the option will change from `tree`

reconstruction to Likelihood mapping to tree reconstruction again. The letter k cycles from quartet puzzling to user defined
trees to pairwise distances only (no tree). For example, a typical
run to infer a tree based on DNA sequences would start with the following setting:

```
b                    Type of analysis? Tree reconstruction
k               Tree search procedure? Quartet puzzling
v      Approximate quartet likelihood? Yes
u             List unresolved quartets? No
n             Number of puzzling steps? 10000
j             List puzzling step trees? No
o                   Display as outgroup? L20571
z     Compute clocklike branch lengths? No
e                   Parameter estimates? Approximate (faster)
x            Parameter estimation uses? Neighbor-joining tree
SUBSTITUTION PROCESS
d          Type of sequence input data? Auto: Nucleotides
h             Codon positions selected? Use all positions
m                Model of substitution? HKY (Hasegawa et al.,
                                       1985)
t     Transition/transversion parameter? Estimate from data set
f             Nucleotide frequencies? Estimate from data set
RATE HETEROGENEITY
w           Model of rate heterogeneity? Gamma-distributed
                                        rates
a    Gamma distribution parameter alpha? Estimate from data set
c        Number of Gamma rate categories? 8

Quit [q], confirm [y], or change [menu] settings:
```

By entering y, the program computes a quartet-puzzling tree based on 10,000 intermediate trees, where the likelihoods for the quartets are only approximate. The
parameter estimates are also approximate and are based on an NJ tree that TREE-
PUZZLE computes automatically at the beginning of the optimization routine. The
model of sequence evolution assumed is HKY (see Sections 4.6 and 4.9), and it is possible to estimate distances with Γ-distributed rate heterogeneity over sites, where the
parameter is estimated with the aid of eight discrete categories (see Section 4.6.1).
If these settings are confirmed, the following output will appear on the screen:

```
Optimizing missing substitution process parameters
Optimizing missing rate heterogeneity parameters
Optimizing missing substitution process parameters
Optimizing missing rate heterogeneity parameters
Optimizing missing substitution process parameters
Optimizing missing rate heterogeneity parameters
Optimizing missing substitution process parameters
Optimizing missing rate heterogeneity parameters
```

```
Writing parameters to file outfile
Writing pairwise distances to file outdist
Computing quartet maximum likelihood trees
Computing quartet puzzling tree
Computing maximum likelihood branch lengths (without clock)

All results written to disk:
          Puzzle report file: outfile
        Likelihood distances: outdist
           PHYLIP tree file: outtree

The computation took 299 seconds (= 5.0 minutes = 0.1 hours)
       including input 1003 seconds (= 16.7 minutes = 0.3 hours)
```

The outfile, the most important file, summarizes the complete phylogenetic analyses. Outdist contains the matrix of pairwise distances inferred from the model parameters. outtree describes the resulting consensus tree in the Newick notation and can be displayed with the TreeView program (see Chapter 5 and Figure 5.5). Because the content of the outfile is self-explanatory, it is not discussed here.

A new feature recently implemented in TREE–PUZZLE Version 5.0 enables output of all different tree topologies found during puzzling steps. In the HIV example, 685 different intermediate trees were found, in which the most frequent tree occurred about 6.6%. Therefore, the quartet-puzzling algorithm can be used as a tool to generate a collection of plausible candidate trees, and this collection can subsequently be employed to search for the most likeli tree. The consensus tree does not necessarily coincide with the maximum-likelihood tree, especially when the consensus is not fully resolved. If the user is interested in the maximum-likelihood tree, option j should be changed to unique topologies, which will be output in the outptorder file. A typical line of this file looks like the following:

```
[              1.              657              6.57              14
685 10000](L20571,((AF10138,X52154),(U09127,(((U27426,U27445),
(U067158,U09126)),
((U27399,U43386),(((L02317,AF042106),AF025763),U08443))))));
```

The first column is a simple numbering scheme, in which each tree is numbered according to its frequency (i.e., second column and third column). Column four (14) gives the first time among 10,000 (column 6) puzzling steps, when the tree was found. Column five shows how many different trees were found (685). To compute the maximum-likelihood tree among all intermediate trees, rename outptorder to intree and run TREE–PUZZLE again with the following settings:

```
GENERAL OPTIONS
  b                    Type of analysis? Tree reconstruction
  k            Tree search procedure? User-defined trees
```

```
z   Compute clocklike branch lengths? No
e                 Parameter estimates? Approximate (faster)
x           Parameter estimation uses? Neighbor-joining tree
SUBSTITUTION PROCESS
d           Type of sequence input data? Auto: Nucleotides
h             Codon positions selected? Use all positions
m               Model of substitution? HKY (Hasegawa et al.,
                1985)
t   Transition/transversion parameter? Estimate from data set
f               Nucleotide frequencies? Estimate from data set
RATE HETEROGENEITY
w           Model of rate heterogeneity? Gamma distributed rates
a Gamma distribution parameter alpha? Estimate from data set
c     Number of Gamma rate categories? 8

Quit [q], confirm [y], or change [menu] settings: y
```

Now the program will compute the maximum-likelihood values for all intermediate trees using the parameter estimates of the sequence-evolution model from
the iterative procedure based on the NJ tree. The computation takes time, but it
provides more insight about the data and reliability of the tree. The resulting outfile shows the likelihood computation for each tree topology and summarizes
results at the end of the file.

6.9   Likelihood-mapping analysis of the HIV data set

Option b in the GENERAL OPTIONS menu of TREE-PUZZLE allows switching
from Tree reconstruction to Likelihood mapping. When the number
of possible quartets for the data set is less than 10,000, the program computes the
P vectors (see Section 6.6) for all of them; otherwise, only the P vectors of 10,000
random quartets are computed. In the latter case, the user can decide how many
quartets to evaluate by selecting from the GENERAL OPTIONS menu option n
number of quartets and typing the number. Typing 0 forces TREE-PUZZLE
to analyze all the possible quartets; however, this is not efficient for large data sets
because the computation can be very slow – a random sample of quartets works
just as well.

Box 6.1 shows part of the outfile from the likelihood-mapping analysis of
the 1,001 possible quartets for the HIV data set. Most of the points gather in one of
the corners of the triangle, but some are also found in the center and at the sides.
Specifically, 4.4% of all quartets lie in the rectangles and 2.9 % in the central triangle:
they represent the unresolved part in the data. Because most of the quartets favor
one tree and only a fraction of about 7% does not support a unique phylogeny, an
overall phylogenetic tree with a good resolution is expected to fit the data well. The
HIV quartet-puzzling consensus tree calculated in the previous section is mostly

Box 6.1 Likelihood-mapping analysis with TREE-PUZZLE

TREE-PUZZLE writes results of the likelihood-mapping analysis at the end of the outfile. For example, for the HIV data set (file: hivALN.phy; substitution model: HKY with Γ-distances, parameters estimated via maximum likelihood):

```
LIKELIHOOD MAPPING STATISTICS
Occupancies of the three areas 1, 2, 3:
```

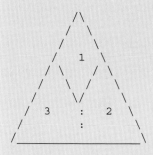

```
Number of quartets in region 1: 359 (= 35.9%)
Number of quartets in region 2: 328 (= 32.8%)
Number of quartets in region 3: 314 (= 31.4%)


Occupancies of the seven areas 1, 2, 3, 4, 5, 6, 7:
```

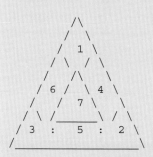

```
Number of quartets in region 1: 333 (= 33.3%)  left:    172  right: 161
Number of quartets in region 2: 299 (= 29.9%)  bottom: 153  top:    146
Number of quartets in region 3: 296 (= 29.6%)  bottom: 146  top:    150
Number of quartets in region 4: 15 (= 1.5%)  bottom: 5   top:    10
Number of quartets in region 5: 18 (= 1.8%)  left:    5   right: 13
Number of quartets in region 6: 11 (= 1.1%)  bottom: 8   top:    3
Number of quartets in region 7: 29 (= 2.9%)
```

TREE-PUZZLE also outputs a drawing of the likelihood-mapping triangle in encapsulated postscript format (file: outlm.eps) that can be printed or imported in a graphic application to be edited.

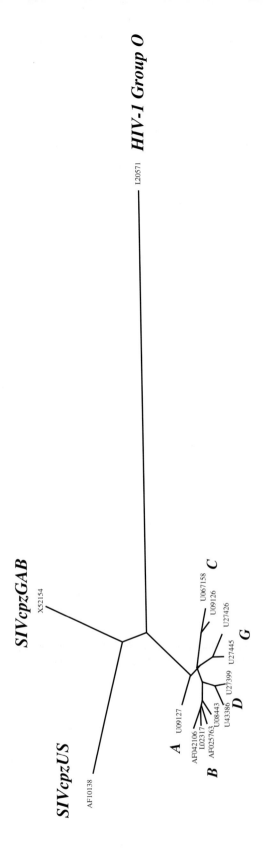

Figure 6.9 Quartet-puzzling consensus tree for the HIV/SIV data set. The major groups of HIV-1 and the Group M subtypes are indicated in bold.

resolved, but contains a ***politomy*** (see Section 1.6) joining Subtypes G, C, and the clade B/D together (Figure 6.9). This is also not surprising in light of the noise in the data revealed by likelihood mapping. Such a result may suggest a ***star-like radiation*** at the origin of the HIV-1 Group M subtypes, but it could also imply the presence of ***inter-subtype recombination*** in HIV-1. These issues are analyzed further in Chapters 12 and 14, which discuss general methods for detecting and investigating recombination and conflicting phylogenetic signals in molecular data.

REFERENCES

Adachi, J. and M. Hasegawa (1995). *MOLPHY: Programs for Molecular Phylogenetics, Version 2.3.* Tokyo: Institute of Statistical Mathematics.

Felsenstein, J. (1981). Evolutionary trees from DNA sequences: A maximum-likelihood approach. *Journal of Molecular Evolution*, 17, 368–376.

Felsenstein, J. (1993). *PHYLIP: Phylogenetic Inference Package, Version 3.5c.* Seattle: Department of Genetics, University of Washington.

Fitch, W. M. (1971). Toward defining the course of evolution: Minimum change for a specific tree topology. *Systematic Zoology*, 20, 406–416.

Margush, T. (1981). Consensus n-trees. *Bulletin of Mathematical Biology*, 43, 239–244.

Nieselt-Struwe, K. and A. von Haeseler (2001). Quartet-mapping, a generalization of the likelihood-mapping procedure. *Molecular Biology and Evolution*, 18, 1204–1219.

Saitou, N. and M. Nei (1987). The neighbor-joining method: A new method for reconstructing phylogenetic trees. *Molecular Biology and Evolution*, 4, 406–425.

Strimmer, K. and A. von Haeseler (1996a). Quartet-puzzling: A quartet maximum-likelihood method for reconstructing tree topologies. *Molecular Biology and Evolution*, 13, 964–969.

Strimmer, K. and A. von Haeseler (1996b). Accuracy of neighbor-joining for n-taxon trees. *Systematic Biology*, 45, 514–521.

Strimmer, K. and A. von Haeseler (1997). Likelihood mapping: a simple method to visualize phylogenetic content in a sequence alignment. *Proc. Natl. Ac. Sc.* 94, 6815–6819.

Strimmer, K., N. Goldman, and A. von Haeseler (1997). Bayesian probabilities and quartet-puzzling. *Molecular Biology and Evolution*, 14, 210–211.

Sullivan, J., K. E. Holsinger, and C. Simon (1996). The effect of topology on estimates of among sites rate variation. *Journal of Molecular Evolution*, 42, 308–312.

Swofford, D. L., G. J. Olsen, P. J. Waddell, and D. M. Hillis (1995). Phylogenetic inference. In: *Molecular Systematics*, eds. D. M. Hillis, C. Moritz, and B. K. Mable, pp. 407–514. Sunderland, MA, Sinauer Associates.

Uzzel, T. and K. W. Corbin (1971). Fitting discrete probability distributions to evolutionary events. *Science*, 172, 1089–1096.

Wakeley, J. (1993). Substitution rate variation among sites in hypervariable region 1 of human mitochondrial DNA. *Journal of Molecular Evolution*, 37, 613–623.

Phylogeny inference based on parsimony and other methods using PAUP*

THEORY

David L. Swofford and Jack Sullivan

7.1 Introduction

Methods for inferring evolutionary trees can be divided into two broad categories: those that operate on a matrix of discrete characters that assigns one or more attributes or character states to each **taxon** (i.e., sequence or gene-family member); and those that operate on a matrix of pairwise distances between **taxa**, with each distance representing an estimate of the amount of divergence between two taxa since they last shared a common ancestor (see Chapter 1). The most commonly employed discrete-character methods used in molecular phylogenetics are **parsimony** and **maximum-likelihood** methods. For molecular data, the character-state matrix is typically an aligned set of DNA or protein sequences, in which the states are the nucleotides A, C, G, and T (i.e., DNA sequences) or symbols representing the 20 common amino acids (i.e., protein sequences); however, other forms of discrete data such as restriction-site presence/absence and gene-order information also may be used.

Parsimony, maximum-likelihood, and some **distance methods** are examples of a broader class of phylogenetic methods that rely on the use of **optimality criteria**. Methods in this class all operate by explicitly defining an **objective function** that returns a score for any input tree topology. This tree score thus allows any two or more trees to be ranked according to the chosen optimality criterion. Ordinarily, phylogenetic inference under **criterion-based methods** couples the selection of a suitable optimality criterion with a search for an optimal tree topology under that criterion. Because the number of tree topologies grows exponentially with the number of taxa (see Table 1.3), criterion-based methods are necessarily slower than algorithmic approaches, such as **UPGMA** or **neighbor-joining** (**NJ**), which simply cluster taxa according to a prescribed set of rules and operations (see Chapter 5). However, we believe that criterion-based methods have a strong advantage in that the basis for preferring one tree over another is made mathematically precise, unlike the case for algorithmic methods. For example, NJ was originally described (Saitou and Nei, 1987) as a method for approximating a tree that minimizes the sum of least-squares branch lengths – the *minimum-evolution* criterion (see Chapter 5). However, it rarely achieves this goal for data sets of nontrivial size, and rearrangements of the NJ tree that yield a lower minimum-evolution score can usually be found. This result makes it difficult to defend the presentation of an NJ tree as the most reasonable estimate of a phylogeny. With criterion-based methods, it can at least be said that a given tree topology was the best that could be found according to the criterion. If others want to dispute that tree, they are free to criticize the criterion or search for better trees according to the criterion, but it is clear why the tree was chosen. It is true that *bootstrapping* or *jackknifing methods* (see Chapter 5) can be used to quantify uncertainty regarding the groupings implied by an NJ tree, but fast approximations are also available for criterion-based methods.

Although it contains additional capabilities, the PAUP* program (Swofford, 2002) is primarily a program for estimating phylogenies using criterion-based methods. It includes support for parsimony, maximum-likelihood (nucleotide data), and distance methods. This chapter discusses the use of PAUP* as a tool for phylogenetic inference. First, some theoretical background is provided for parsimony analysis, which – unlike distance and maximum-likelihood methods (see Chapters 5 and 6) – is not treated elsewhere in this book. Next, strategies are described for searching for optimal trees that are appropriate for any of the existing optimality criteria. Finally, in the Practice section, many of the capabilities of PAUP* are illustrated using the real-data examples common to other chapters in this book.

7.2 Parsimony analysis – background

Although the first widely used methods for inferring phylogenies were pairwise distance methods, parsimony analysis has been the predominant approach used to construct phylogenetic trees from the early 1970s until relatively recently; despite some limitations, it remains an important and useful technique. The basic idea underlying parsimony analysis is simple: one seeks the tree, or collection of trees, that minimizes the amount of evolutionary change (i.e., transformations of one character state into another) required to explain the data (Kluge and Farris, 1969; Farris, 1970; Fitch, 1971).

The goal of minimizing evolutionary change is often defended on philosophical grounds. One line of argument is the notion that when two hypotheses provide equally valid explanations for a phenomenon, the simpler one should always be preferred. This position is often referred to as "Ockham's Razor": shave away all that is unnecessary. To use simplicity as a justification for parsimony methods in phylogenetics, one must demonstrate a direct relationship between the number of character-state changes required by a tree topology and the complexity of the corresponding hypotheses. The connection is usually made by asserting that each instance of **homoplasy** (i.e., sharing of identical character states that cannot be explained by inheritance from the common ancestor of a group of taxa) constitutes an ad hoc hypothesis, and that the number of such ad hoc hypotheses should be minimized. Related arguments have focused on the concepts of falsifiability and corroboration most strongly associated with the writings of Karl Popper, suggesting that parsimony is the only method consistent with a hypothetico-deductive framework for hypothesis testing. However, as argued recently by de Queiroz and Poe (2001), careful interpretation of Popper's work does not lead unambiguously to parsimony as the method of choice. Furthermore, the linkage between parsimony and simplicity is tenuous, as highlighted by the recent work of Tuffley and Steel (1997). It demonstrates that parsimony and likelihood become equivalent under an extremely parameter-rich likelihood model that assigns a separate parameter for each character (site) on every branch of the tree, which is hardly a "simple" model. So, despite more than 20 years of ongoing debate between those who advocate the exclusive use of parsimony methods and those who favor maximum-likelihood and related model-based statistical approaches, the issue remains unsettled and the camps highly polarized. Although we place ourselves firmly on the statistical side, we believe that parsimony methods will remain part of a complete phylogenetic-analysis toolkit for some time because they are fast and have been demonstrated to be quite effective in many situations (e.g., Hillis, 1996). In this sense, parsimony represents a useful "fallback" method when model-based methods cannot be used due to computational limitations.

Although parsimony methods are most effective when rates of evolution are slow (i.e., the expected amount of change is low), it is often asserted incorrectly that this is an "assumption" of parsimony methods. In fact, parsimony can perform extremely well under high rates of change as long as the lengths of the branches on the true underlying tree do not exhibit certain kinds of pathological inequalities (Hillis et al., 1994) (see also **long-branch attraction**, Section 5.3). Nonetheless, it is difficult to state what the assumptions of parsimony analysis actually are. Conditions can be specified in which parsimony does very well in recovering the true evolutionary tree; however, alternative conditions can be found where it fails horribly. For example, in the so-called Felsenstein zone, standard parsimony methods converge to the wrong tree with increasing certainty as more data are accumulated. This is because a greater proportion of identical character states will be shared by chance between unrelated taxa than are shared by related taxa due to common ancestry

(Felsenstein, 1978; Swofford et al., 1996). Because parsimony requires no explicit assumptions other than the standard one of independence among characters (and some might even argue with that), about all one can say is that it assumes that the conditions that would cause it to fail do not apply to the current analysis. Fortunately, these conditions are now relatively well understood. The combined use of model-based methods – less susceptible to long-branch attraction (see Section 5.3) and related artifacts but limited by computational burden – with faster parsimony methods that permit greater exploration of alternative tree topologies provides a mechanism for making inferences maintaining some degree of protection against the circumstances that could cause parsimony estimates alone to be misleading.

7.3 Parsimony analysis – methodology

The problem of finding optimal trees under the parsimony criterion can be separated into two subproblems: (1) determining the amount of character change, or tree length, required by any given tree; and (2) searching over all possible tree topologies for the trees that minimize this length. The first problem is easy and fast, whereas the second is slow due to the extremely large number of possible tree topologies for anything more than a small number of taxa.

7.3.1 Calculating the length of a given tree under the parsimony criterion

For n taxa, an unrooted binary tree (i.e., a fully bifurcating tree) contains n terminal nodes representing those sequences, $n - 2$ internal nodes, and $2n - 3$ branches (edges) that join pairs of nodes. Let τ represent some particular tree topology, which could be, for example, an arbitrarily chosen tree from the space of all possible trees. The length of this tree is given by

$$L(\tau) = \sum_{j=1}^{N} l_j \tag{7.1}$$

where N is the number of sites (characters) in the alignment and l_j is the length for a single site j. This length l_j is the amount of character change implied by a most parsimonious reconstruction that assigns a character-state x_{ij} to each node i for each site j. For terminal nodes, the character-state assignment is fixed by the input data. Thus, for binary trees:

$$l_j = \sum_{k=1}^{2N-3} c_{a(k),b(k)} \tag{7.2}$$

where $a(k)$ and $b(k)$ are the states assigned to the nodes at either end of branch k, and c_{xy} is the cost associated with the change from state x to state y. In the simplest form of parsimony (Fitch, 1971), this cost is simply 1 if x and y are different or 0 if they are identical. However, other cost schemes may be chosen. For example, one common scheme for nucleotide data is to assign a greater cost to transversions than to transitions (see Chapter 1), reflecting the observation that the former occur more frequently in many genes and, therefore, are accorded less weight. The cost scheme can be represented as a *cost matrix*, or *step matrix*, that assigns a cost for the change between each pair of character states. In general, the cost matrix is symmetric (e.g., $c_{AG} = c_{GA}$), with the consequence that the length of the tree is the same regardless of the position of the root. If the cost matrix contains one or more elements for which $c_{xy} \neq c_{yx}$, then different rootings of the tree may imply different lengths, and the search among trees must be done over rooted trees rather than unrooted trees.

Direct algorithms for the determination of the l_j are available and are described briefly in this section. First, however, it will be instructive to examine the calculation of tree length using the brute-force approach of evaluating all possible character-state reconstructions. In general, with an alphabet size of r (e.g., $r = 4$ states for nucleotides, $r = 20$ states for amino acids) and T taxa, the number of these reconstructions for each site is equal to r^{T-2}. Consider the following example:

$$j$$

W	... ACAGGAT ...
X	... ACACGCT ...
Y	... GTAAGGT ...
Z	... GCACGAC ...

Suppose that the tree $((W, Y),(X, Z))$ is being evaluated (see Figure 5.5 about the Newick representation of *phylogenetic trees*), that the lengths for the first $j-1$ sites have been calculated, and that the length of site j is to be determined next. Because there are four sequences, the number of reconstructions to be evaluated is $4^{(4-2)} = 16$. The lengths implied by each of these reconstructions under two different cost schemes are shown in Figure 7.1. With equal costs, the minimum length is two steps, and this length is achievable in three ways (i.e., internal nodes assignment "A–C", "C–C", and "G–C"). If a similar analysis for the other two trees is conducted, both of the trees $((W, X),(Y, Z))$ and $((W, Z),(Y, X))$ are also found to have lengths of two steps. Thus, this character does not discriminate among the three tree topologies and is said to be parsimony-uninformative under this cost scheme. With 4:1 transversion:transition weighting, the minimum length is five steps, achieved by two reconstructions (i.e., internal node assignments "A–C" and "G–C"). However, similar evaluation of the other two trees (not shown) finds a minimum of eight steps on both trees (i.e., two transversions are required rather

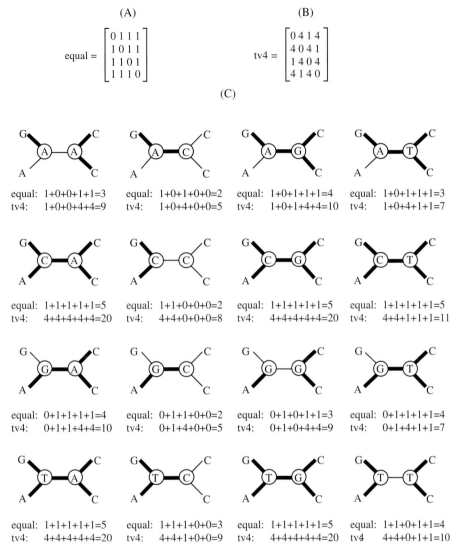

$$\text{equal} = \begin{bmatrix} 0\ 1\ 1\ 1 \\ 1\ 0\ 1\ 1 \\ 1\ 1\ 0\ 1 \\ 1\ 1\ 1\ 0 \end{bmatrix}$$

(A)

$$\text{tv4} = \begin{bmatrix} 0\ 4\ 1\ 4 \\ 4\ 0\ 4\ 1 \\ 1\ 4\ 0\ 4 \\ 4\ 1\ 4\ 0 \end{bmatrix}$$

(B)

(C)

equal: 1+0+0+1+1=3
tv4: 1+0+0+4+4=9

equal: 1+0+1+0+0=2
tv4: 1+0+4+0+0=5

equal: 1+0+1+1+1=4
tv4: 1+0+1+4+4=10

equal: 1+0+1+1+1=3
tv4: 1+0+4+1+1=7

equal: 1+1+1+1+1=5
tv4: 4+4+4+4+4=20

equal: 1+1+0+0+0=2
tv4: 4+4+0+0+0=8

equal: 1+1+1+1+1=5
tv4: 4+4+4+4+4=20

equal: 1+1+1+1+1=5
tv4: 4+4+1+1+1=11

equal: 0+1+1+1+1=4
tv4: 0+1+1+4+4=10

equal: 0+1+1+0+0=2
tv4: 0+1+4+0+0=5

equal: 0+1+0+1+1=3
tv4: 0+1+0+4+4=9

equal: 0+1+1+1+1=4
tv4: 0+1+4+1+1=7

equal: 1+1+1+1+1=5
tv4: 4+4+4+4+4=20

equal: 1+1+1+0+0=3
tv4: 4+4+1+0+0=9

equal: 1+1+1+1+1=5
tv4: 4+4+4+4+4=20

equal: 1+1+0+1+1=4
tv4   4+4+0+1+1=10

Figure 7.1   Determination of the length of a tree by brute-force consideration of all possible state assignments to the internal nodes. Calculations are for one site of one of the possible trees for the four taxa W, X, Y, and Z (character states: A, C, G, T). (A) Cost matrix that assigns equal cost to all changes from one nucleotide to another. (B) Cost matrix that assigns four times as much weight to transversions as to transitions (rows and columns are ordered A, C, G, T). (C) All sixteen possible combinations of state assignments to the two internal nodes and the lengths under each cost scheme. Minimum lengths are two steps for equal costs and five steps for 4:1 tv:ti weighting.

than one transition plus one transversion). Under these unequal costs, the character becomes informative in the sense that some trees have lower lengths than others, which demonstrates that the use of unequal costs may provide more information for phylogeny reconstruction than equal-cost schemes.

The method used in this example, in principle, could be applied to every site in the data set, summing these lengths to obtain the total tree length for each possible tree, and then choosing the tree that minimizes the total length. Obviously, for real applications, a better way is needed for determining the minimum lengths that does not require evaluation of all r^{n-2} reconstructions. A straightforward dynamic programming algorithm (Sankoff and Rousseau, 1975) provides such a method for general cost schemes; the methods of Farris (1970), Fitch (1971), and others handle special cases of this general system with simpler calculations. Sankoff's algorithm is illustrated using the example in Box 7.1; the original papers may be consulted for a full description of the algorithm. Dynamic programming operates by solving a set of subproblems and assembling those solutions in a way that guarantees optimality for the full problem. In this instance, the best length that can be achieved for each subtree – given each of the possible state assignments to each node – is determined, moving from the tips toward the root of the tree. Upon arriving at the root, an optimal solution for the full tree is guaranteed. A nice feature of this algorithm is that at each stage, only one pair of vectors of conditional subtree lengths above each node needs to be referenced; knowing how those subtree lengths were obtained is unnecessary.

For simple cost schemes, the full dynamic programming algorithm described in Box 7.1 is not necessary. Farris (1970) and Fitch (1971) described algorithms for characters with ordered and unordered states, respectively, which can be proven to yield optimal solutions (Hartigan, 1973; Swofford and Maddison, 1987); these two algorithms are equivalent for binary (i.e., two-state) characters. Fitch's algorithm, illustrated in Box 7.2, is relevant for sequence data when the cost of a change from any state to any other state is assumed to be equal.

7.4   Searching for optimal trees

Having specified a means for calculating the score of a tree under our chosen criterion, the more difficult task of searching for an optimal tree can be confronted. The methods described in the following sections can be used for parsimony, *least-squares distance criteria*, and maximum likelihood. Regrettably, this search is complicated by the huge number of possible trees for anything more than a small number of taxa.

Box 7.1 Calculation of the minimum tree length under general cost schemes using Sankoff's algorithm

For symmetric cost matrixes, we can root an unrooted tree arbitrarily to determine the minimum tree length. Then, for each node i (labeled in boldface italics), we compute a conditional-length vector S_{ij} containing the minimum possible length above i, given each of the possible state assignments to this node for character j (for simplicity, we drop the second subscript because only one character is considered in turn). Thus, s_{ik} is the minimum possible length of the subtree descending from node i if it is assigned state k. For the tip sequences, this length is initialized to 0 for the state(s) actually observed in the data or to infinity otherwise. The algorithm proceeds by working from the tips toward the root, filling in the vector at each node based on the values assigned to the node's children (i.e., immediate descendants).

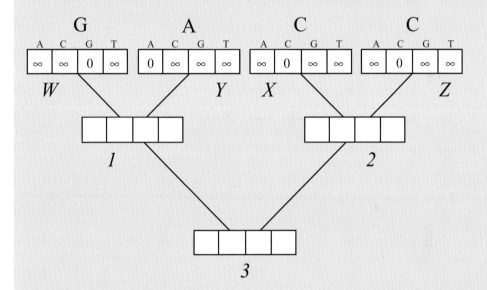

We now visit node 1. For each element k of this vector, we consider the costs associated with each of the four possible assignments to each of the child nodes W and Y, and the cost needed to reach these states from state k, which is obtained from the cost matrix (C in this example, we use the cost matrix from Figure 7.1B, which represents a 4:1 tv:ti weighting). This calculation is trivial for nodes ancestral to two terminal nodes because only one state needs to be considered for each child. Thus, if we assign state A to node 1, the minimum length of the subtree above node 1 given this assignment is the cost of a change from A to G in the left branch, plus the cost of a (non) change from A to A in the right branch: $s_{1A} = c_{AG} + c_{AA} = 1 + 0 = 1$. Similarly, s_{1C} is the sum of c_{CG} (left branch) and c_{CA} (right branch), or 8. Continuing in this manner, we obtain

Box 7.1 (continued)

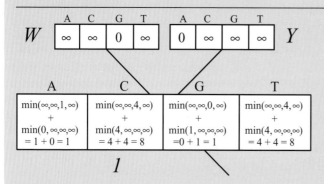

for the subtree of node 1.
The calculation for node 2 proceeds analogously:

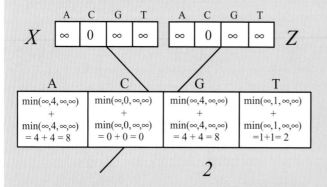

The calculation for the root node (node 3) is somewhat more complicated: for each state k at this node, we must explicitly consider each of the four state assignments to each of the child nodes 1 and 2. For example, when calculating the length conditional on the assignment of state A to node 3, for the left branch we consider in turn all four of the assignments to node 1. If node 1 is assigned state A as well, the length would be the sum of 1 (for the length above node 1) plus 0 (for the nonchange from state A to state A). If we instead choose state C for node 1, the length contributed by the left branch would be 8 (for the length above node 1) plus 4 (for the change from A to C). The same procedure is used to determine the conditional lengths for the right branch. By summing these two values for each state k, we obtain the entire conditional-length vector for node 3:

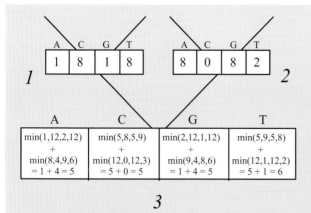

A	C	G	T
1	8	1	8

1

A	C	G	T
8	0	8	2

2

A	C	G	T
min(1,12,2,12)	min(5,8,5,9)	min(2,12,1,12)	min(5,9,5,8)
+	+	+	+
min(8,4,9,6)	min(12,0,12,3)	min(9,4,8,6)	min(12,1,12,2)
= 1 + 4 = 5	= 5 + 0 = 5	= 1 + 4 = 5	= 5 + 1 = 6

3

Since we are now at the root of the tree, the conditional-length vector s_3 provides the minimum possible lengths for the full tree given each of the four possible state assignments to the root, and the minimum of these values is the tree length we seek. Observe that this length, 5, is the same value we obtained using the brute-force enumeration shown in Figure 7.1. As an exercise, the reader may wish to verify that other rootings yield the same length.

This algorithm provides a means of calculating the length required by any character on any tree under any cost scheme. We can obtain the length of a given tree by repeating the procedure outlined here for each character and summing over characters. In principle, we can then find the most parsimonious tree by generating and evaluating all possible trees, although this exhaustive-search strategy would only be feasible for a relatively small number of sequences (i.e., eleven in the current version of PAUP*).

Box 7.2 Calculation of the minimum tree length using Fitch's algorithm for equal costs

As for the general case, we can root an unrooted tree arbitrarily to determine the minimum tree length. We will assign a **state set** X_i to each node i; this set represents the set of states that can be assigned to each node so that the minimum possible length of the subtree above that node can be achieved. Also, for each node, we will store an accumulated length s_i, which represents the minimum possible (unconditional) length in the subtree descending from node i.

The state sets for the terminal nodes are initialized to the states observed in the data, and the accumulated lengths are initialized to zero. For the example of Figure 7.1, this yields:

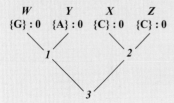

Box 7.2 continued

For binary trees, the state-set calculation for each internal node i follows two simple rules, based on the state sets and accumulated lengths of the two descendant child nodes, denoted $L(i)$ and $R(i)$:

1. Form the intersection of the two child state sets: $X_{L(i)} \cap X_{R(i)}$. If this intersection is nonempty, let X_i equal this intersection. Set the accumulated length for this node to be the sum of the accumulated lengths for the two child nodes: $s_i = s_{L(i)} + s_{R(i)}$.
2. If the intersection was empty (i.e., $X_{L(i)}$ and $X_{R(i)}$ are disjoint), let X_i equal the union of the two child state sets: $X_{L(i)} \cup X_{R(i)}$. Set the accumulated length for this node to be the sum of the accumulated lengths for the two child nodes *plus one*: $s_i = s_{L(i)} + s_{R(i)} + 1$.

Proceeding to node 1 in our example, we find that the intersection of $\{G\}$ and $\{A\}$ is the empty set ($\emptyset$); therefore, by Rule 2, we let $X_1 = \{A\} \cup \{G\} = \{A, G\}$ and $s_1 = 0 + 0 + 1 = 1$:

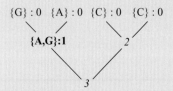

For node 2, the intersection of $\{C\}$ and $\{C\}$ is (obviously) $\{C\}$; therefore, by Rule 1, we obtain:

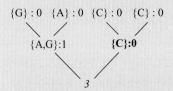

Finally, for node 3, the intersection of $\{A, G\}$ and $\{C\} = \emptyset$, so the state set at the root is equal to the union of these sets, and the corresponding accumulated length equals $1 + 0 + 1 = 2$:

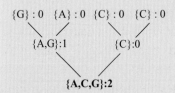

Thus, the length required by the tree is equal to two steps, as we obtained previously using the brute-force approach. As an exercise, the reader may wish to verify that other rootings of the tree yield the same length.

7.4.1 Exact methods

For eleven or fewer taxa, a brute-force **exhaustive search** is feasible; an algorithm is needed that can guarantee generation of all possible trees for evaluation using the previously described methods. The procedure outlined in Box 7.3 is used in PAUP*. This algorithm recursively adds the tth taxon in a stepwise fashion to all possible trees containing the first $t - 1$ taxa until all n taxa have been joined. The algorithm is easily modified for rooted trees by including one additional artificial taxon that locates the root of each tree. In this case, the first three trees generated represent each of the three possible rootings of an unrooted three-taxon tree, and the algorithm proceeds as in the unrooted case. Thus, the number of rooted trees for n taxa is equal to the number of unrooted trees for $n + 1$ taxa.

It is clear from the description of the algorithm for generating all possible trees in Box 7.3 that the number of possible trees grows by a factor that increases by two with each additional taxon, as expressed in the following relationship:

$$B(t) = \prod_{i=3}^{t} (2i - 5) \tag{7.3}$$

where $B(t)$ is the number of unrooted trees for t taxa. For example, the number of unrooted trees for 7 taxa is $1 \times 3 \times 5 \times 7 \times 9 = 945$ and the number of unrooted trees for 20 taxa is over 2×10^{20}. Clearly, the exhaustive-search method can be used for only a relatively small number of taxa. An alternative exact procedure, the **branch-and-bound method** (Hendy and Penny, 1982), is useful for data sets containing from 12 to 25 or so taxa, depending on the "messiness" of the data. This method operates by implicitly evaluating all possible trees, but cutting off paths of the search tree when it is determined that they cannot possibly lead to optimal trees. The branch-and-bound method is illustrated for a hypothetical six-taxon data set in Figure 7.2. We present the example as if parsimony is the optimality criterion, but this choice is not important to the method. The algorithm effectively traces the same route through the search tree as would be used in an exhaustive search (see Box 7.3), but the length of each tree encountered at a node of the search tree is evaluated even if it does not contain the full set of taxa. Throughout this traversal, an upper bound on the length of the optimal tree(s) is maintained; initially, this upper bound can simply be set to infinity. The traversal starts by moving down the left branch of the search tree successively connecting taxa D and E to the initial tree, with lengths of 221 and 234 steps, respectively. Then, connecting taxon F provides the first set of full-tree lengths. After this connection, it is known that a tree of 241 steps exists, although it is not yet known whether this tree is optimal. This number, therefore, is taken as a new upper bound on the length of the optimal tree (i.e., the optimal tree length cannot be longer than 241 steps because a tree at this

Box 7.3 Generation of all possible trees

The approach for generation of all possible unrooted trees is straightforward. Suppose that we have a data set containing six sequences (i.e., taxa). We begin with the only tree for the first three taxa in the data set, and connect the fourth taxon to each of the three branches on this tree.

This generates all three of the possible unrooted trees for the first four taxa. Now, we connect the fifth taxon, E, to each branch on each of these three trees. For example, we can join taxon E to the tree on the right above:

By connecting taxon E to the other two trees in a similar manner, we generate all 15 of the possible trees for the first five taxa. Finally, we connect the sixth taxon, F, to all locations on each of these 15 trees (7 branches per tree), yielding a total of 105 trees. Thus, the full-search tree can be represented as follows (only the paths toward the full 6-taxon trees are shown):

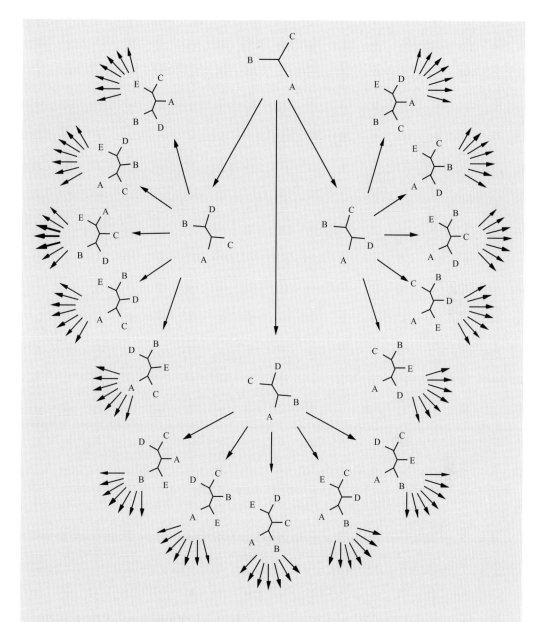

The lengths (or likelihoods or distance scores) of each of these 105 trees can now be evaluated, and the set of optimal trees identified.

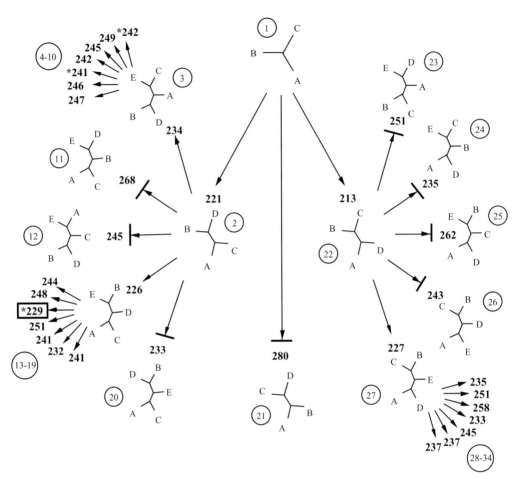

Figure 7.2 The branch-and-bound algorithm for the exact solution of the problem of finding an optimal parsimony tree. The search tree is the same as shown in Box 7.3, with tree lengths for a hypothetical data set shown in boldface type. If a tree lying at a node of this search tree (thus joining fewer taxa) has a length that exceeds the current lower bound on the optimal tree length, this path of the search tree is terminated (indicated by a cross-bar), and the algorithm backtracks and takes the next available path. When a tip of the search tree is reached (i.e., when we arrive at a tree containing the full set of taxa), the tree is either optimal (and therefore retained) or suboptimal (and rejected). When all paths leading from the initial three-taxon tree have been explored, the algorithm terminates, and all most parsimonious trees will have been identified. Asterisks indicate points at which the current lower bound is reduced. See text for additional explanation; circled numbers represent the order in which phylogenetic trees are visited in the search tree.

length has already been identified). Now, the algorithm backtracks on the search tree and takes the second path out of the tree 221-step, 4-taxon tree. The 5-taxon tree containing taxon E obtained by following this path requires 268 steps. Thus, there is no point in evaluating the seven trees produced by connecting taxon F to this tree because they cannot possibly require fewer than 268 steps, and a tree of 241 steps has already been found. By cutting off paths in this way, large portions of the search tree may be avoided and a considerable amount of computation time saved. The algorithm proceeds to traverse the remainder of the search tree, cutting off paths where possible and storing optimal trees when they are found. In this example, a new optimal tree is found at a length of 229 steps, allowing the upper bound on the tree length to be further reduced. Then, when the 233-step tree containing the first five taxa is encountered, the seven trees that would be derived from it can be immediately rejected because they would also require at least 233 steps. The algorithm terminates when the root of the search tree has been visited for the last time, at which time all optimal trees will have been identified.

Several refinements to the branch-and-bound method improve its performance considerably. Briefly, these include (1) using a heuristic method such as stepwise addition (discussed later in this section) or NJ (see Chapter 5) to find a tree whose length provides a smaller initial upper bound, which allows earlier termination of search paths in the early stages of the algorithm; (2) ordering the sequential addition of taxa in a way that promotes earlier cutoff of paths (rather than just adding them in order of their appearance in the data matrix); and (3) using techniques such as pairwise character incompatibility to improve the *lower* bound on the minimum length of trees that can be obtained by continuing outward traversal of the search tree (allowing earlier cutoffs). All of these refinements are implemented in PAUP* (see the program documentation for further information).

The branch-and-bound strategy may be used for any optimality criterion whose objective function is guaranteed to be nondecreasing as additional taxa are connected to the tree. Obviously, this is true for parsimony; increasing the variability of the data by adding additional taxa cannot possibly lead to a decrease in tree length. It is also true for maximum-likelihood and some distance criteria, including least-squares methods that score trees by minimizing the discrepancy between observed and path-length distances (see Chapter 5). However, it does not work for the minimum-evolution distance criterion. In minimum evolution, one objective function is optimized for the computation of branch lengths (i.e., least-squares fit), but a different one is used to score the trees (i.e., sum of branch lengths). Unfortunately, the use of these two objective functions makes it possible for the minimum-evolution score to decrease when a new taxon is joined to the tree, invalidating the use of the branch-and-bound method in this case.

7.4.2 Approximate methods

When data sets become too large to use the exact searching methods described in the previous section, it becomes necessary to resort to the use of heuristics: approximate methods that attempt to find optimal solutions but provide no guarantee of success. In PAUP* and several other programs, a two-phase system is used to conduct approximate searches. In the simplest case, an initial starting tree is generated using a "greedy" algorithm that builds a tree sequentially according to some set of rules. In general, this tree will not be optimal, because decisions are made in the early stages without regard for their long-term implications. After this starting tree is obtained, it is submitted to a round of perturbations in which neighboring trees in the perturbation scheme are evaluated. If some perturbation yields a better tree according to the optimality criterion, it becomes the new "best" tree and it is in turn submitted to a new round of perturbations. This process continues until no better tree can be found in a full round of the perturbation process.

One of the earliest and still most widely used greedy algorithms for obtaining a starting tree is **stepwise addition** (e.g., Farris, 1970), which follows the same kind of search tree as the branch-and-bound method described previously. However, unlike the exact exhaustive enumeration and branch-and-bound methods, stepwise addition commits to a path out of each node on the search tree that looks most promising at the moment, which is not necessarily the path leading to a global optimum. In the example in Figure 7.2, Tree 22 is shorter than Trees 2 or 21; thus, only trees derivable from Tree 22 remain as candidates. Following this path ultimately leads to selection of a tree of 233 steps (Figure 7.3), which is only a local rather than a global optimum. The path leading to the optimal 229-step tree was rejected because it appeared less promising at the 4-taxon stage. This tendency to become "stuck" in local optima is a common property of greedy heuristics, and they are often called **local-search methods** for that reason.

Because stepwise addition rarely identifies a globally optimal tree topology for real data and any nontrivial number of taxa, other methods must be used to improve the solution. One such class of methods involves tree-rearrangement perturbations known as **branch-swapping**. These methods all involve cutting off one or more pieces of a tree (subtrees) and reassembling them in a way that is locally different from the original tree. Three kinds of branch-swapping moves are used in PAUP*, as well as in other programs. The simplest type of rearrangement is a **nearest-neighbor interchange** (**NNI**), illustrated in Figure 7.4. For any binary tree containing T terminal taxa, there are $T - 3$ internal branches. Each branch is visited, and the two topologically distinct rearrangements that can be obtained by swapping a subtree connected to one end of the branch with a subtree connected to the other end of the branch are evaluated. This procedure generates a relatively small number of perturbations whose lengths or scores can be compared to the original tree. A more extensive rearrangement scheme is **subtree pruning and regrafting** (**SPR**), illustrated in Figure 7.5, which involves clipping off all possible subtrees from the main tree and reinserting them at all possible locations, but avoiding pruning and grafting operations that would generate the same tree

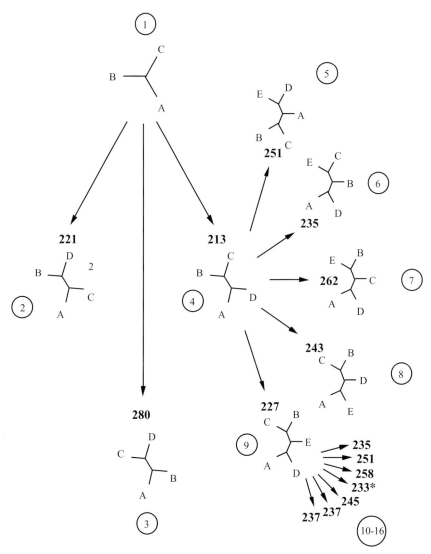

Figure 7.3   A greedy stepwise-addition search applied to the example in Figure 7.2. The best four-taxon tree is determined by evaluating the lengths of the three trees obtained by joining Taxon D to Tree 1 containing only the first three taxa. Taxa E and F are then connected to the five and seven possible locations, respectively, on Trees 4 and 9, with only the shortest trees found during each step being used for the next step. In this example, the 233-step tree obtained is not a global optimum (see Figure 7.2). Circled numbers indicate the order in which phylogenetic trees are evaluated in the stepwise-addition search.

redundantly. The most extensive rearrangement strategy available in PAUP* is *tree bisection and reconnection* (*TBR*), illustrated in Figure 7.6. TBR rearrangements involve cutting a tree into two subtrees by cutting one branch, and then reconnecting the two subtrees by creating a new branch that joins a branch on one subtree

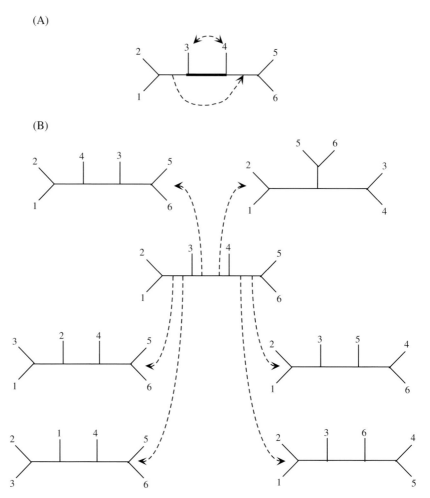

Figure 7.4 Nearest-neighbor interchange (NNI) rearrangements. (A) An NNI around the central branch. (B) All six of the possible NNIs for the tree in (A).

to a branch on the other. All possible pairs of branches are tried, again avoiding redundancies.

Note that the set of possible NNIs for a tree is a subset of the possible SPR rearrangements, and that the set of possible SPR rearrangements is in turn a subset of the possible TBR rearrangements. For TBR rearrangements, a "reconnection distance" can be defined by numbering the branches from zero starting at the cut branch (Figures 7.6C and 7.6E). The reconnection distance is then equal to the sum of numbers of the two branches that are reconnected. The reconnection distances then have the following properties: (1) NNIs are the subset of TBRs that have a reconnection distance of 1; (2) SPRs are the subset of TBRs so that exactly one of the two reconnected branches is numbered zero; and (3) TBRs that are neither NNIs

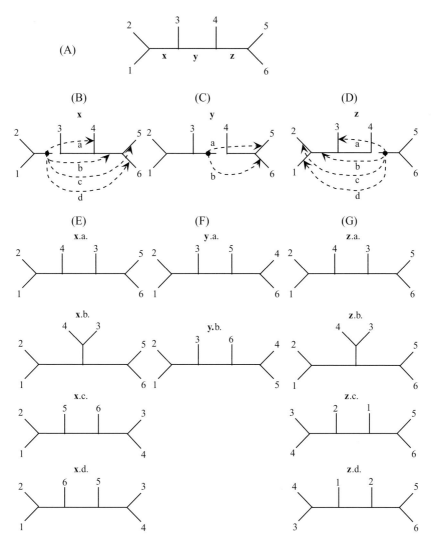

Figure 7.5   Subtree pruning-regrafting (SPR) rearrangements. (A) A tree to be rearranged. (B, C, D) SPRs resulting from pruning of branches **x**, **y**, and **z**, respectively. In addition to these rearrangements, all terminal taxa (i.e., leaves) would be pruned and reinserted elsewhere on the tree. (E, F, G) Trees resulting from regrafting of branches **x**, **y**, and **z**, respectively, to other parts of the tree.

nor SPRs are those for which both reconnected branches have nonzero numbers. The reconnection distance can be used to limit the scope of TBR rearrangements tried during the branch-swapping procedure.

The default strategy used for each of these rearrangement methods is to visit branches of the "current" tree in some arbitrary and predefined order. At each branch, all of the nonredundant branch swaps are tried and the score of each

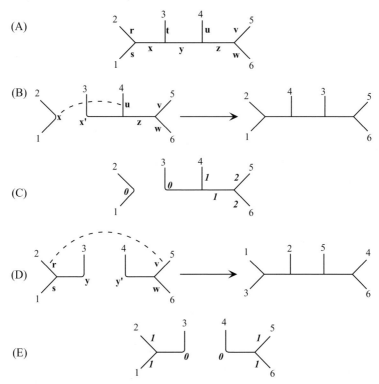

Figure 7.6 Tree bisection-reconnection (TBR) rearrangements. (A) A tree to be rearranged. (B) Bisection of branch **x** and reconnection to branch **u**; other TBRs would connect **x** to **z**, **v**, and **w**, respectively. (C) Branch-numbering for reconnection distances involving branch **x** (see text). (D) Bisection of branch **y** and reconnection of branch **r** to **v**; other TBRs would connect **r** to **w**, **r** to **y'**, **s** to **v**, **s** to **w**, **s** to **y'**, **y** to **v**, and **y** to **w**, respectively. (E) Branch-numbering for reconnection distances involving branch **y** (see text). All other branches, both internal and external, also would be cut in a full round of TBR swapping.

resulting tree is obtained (e.g., using the methods described in Figure 7.3 for parsimony). If a rearrangement is successful in finding a shorter tree, the previous tree is discarded and the rearrangement process is restarted on this new tree. If all possible rearrangements have been tried without success in finding a better tree, the swapping process terminates. Optionally, when trees are found that are equal in score to the current tree (e.g., equally parsimonious trees or trees that have identical likelihoods within round-off error), they are appended to a list of optimal trees. In this case, when the arrangement of one tree finishes, the next tree in the list is obtained and input to the branch-swapping algorithm. If a rearrangement of this next tree yields a better tree than any found so far, all trees in the current list are discarded and the entire process is restarted using the newly discovered tree. The algorithm then terminates when every possible rearrangement has been tried on each of the stored trees. In addition to identifying multiple and equally good trees, this strategy often identifies better trees than would be found if only a single tree were stored at any one time. This can happen when all of the trees within one re-arrangement of the current tree are no better than the current tree; however, some of the adjacent trees can in turn be rearranged to yield trees that are better.

Although they are often quite effective, **hill-climbing algorithms**, such as the branch-swapping methods implemented in PAUP*, are susceptible to the problem of entrapment in local optima. By only accepting proposed rearrangements that are equal to or better than the current best tree, these algorithms eventually reach the peak of the slope on which they start; however, the peak may not represent a global optimum. One generally successful method for improving the chances of obtaining a globally optimal solution is to begin the search from a variety of starting points in the hope that at least one of them will climb the right hill. An option available in PAUP* is to start the search from randomly chosen tree topologies. However, in practice, these randomly chosen trees usually fit the data so poorly that they end up merely climbing a foothill on a rugged landscape, and usually fail to find trees that are as good as those resulting from using a starting tree obtained by stepwise addition. An alternative method takes advantage of the fact that for data sets of nontrivial size and complexity, varying the sequence in which taxa are added during stepwise addition may produce different tree topologies that each fit the data reasonably well. Starting branch-swapping searches from a variety of random-addition-sequence replicates thereby provides a mechanism for performing multiple searches that each begin at a relatively high point on some hill, increasing the probability that the overall search will find an optimal tree.

Random-addition-sequence searches are also useful in identifying multiple **islands** of trees (Maddison, 1991) that may exist. Each island represents all of the trees that can be obtained by a sequence of rearrangements, starting from any tree in the island, keeping and rearranging all optimal trees that are discovered. If two optimal trees exist so that it is impossible to reach one tree by a sequence of rearrangements starting from the other without passing through trees that are suboptimal, these trees are on different islands. Because trees from different islands tend to be topologically dissimilar, it is important to detect multiple islands when they exist.

The methods described previously are generally effective for data sets containing up to 100 or so taxa. However, for larger data sets, they are not as efficient as some newer methods that use a variety of stochastic-search and related algorithms that are better able to avoid entrapment in local optima (Lewis, 1998; Goloboff, 1999; Nixon, 1999; Moilanen, 2001). Nixon's (1999) "parsimony ratchet" can be implemented in PAUP* using the PAUPRat program (Sikes and Lewis, 2001).

PRACTICE

David L. Swofford and Jack Sullivan

The basic capabilities of PAUP* (Swofford, 2002; **http://paup.csit.fsu.edu/**) are illustrated by analyzing the same three data sets used elsewhere in this book (see Section 2.9). Because PAUP* does not currently support model-based analyses of amino-acid sequence data, only parsimony analysis and basic tree searching are described using the glycerol-3-phosphate dehydrogenase data set. Distance methods and a variety of additional capabilities are demonstrated using the amniote *mtDNA* data set. Finally, the flexibility of PAUP* for evaluating and comparing maximum-likelihood models is described through the use of the HIV data set. In doing this, we illustrate the convenience and power of being able to switch easily between different optimality criteria in a single program. Versions of PAUP* are available with a full graphical user interface (Macintosh), a partial graphic user interface (Microsoft Windows), and a command-line-only interface (UNIX/Linux and Microsoft Windows console). Because the command-line interface is available on all platforms, it is used exclusively for the analyses that follow. For each example, it is assumed that PAUP* has been successfully installed and is invoked by typing paup at the operating-system prompt, or double-clicking the executable application in the PAUP* folder (Mac and Windows versions). Instructions on how to purchase and install PAUP* for different platforms can be found at **http://paup.csit.fsu.edu/**.

PAUP* uses the NEXUS format for input data files and command scripts. The details of this format, a command-reference manual, and quick-start tutorial describing data-file formats are available at **http://paup.csit.fsu.edu/downl.html**. Box 7.4 covers these topics and outlines how to convert the example data sets from PHYLIP to NEXUS format.

7.5 Analyzing data with PAUP* through the command-line interface

A typical phylogenetic analysis with PAUP* might follow these steps:

1. Choose a phylogenetic optimality criterion: parsimony (default when the program starts), maximum likelihood (Set criterion=likelihood) or distance (Set criterion=distance).
2. Choose the appropriate settings for the selected parsimony, likelihood, or distance analysis with the PSet, LSet, or DSet command, respectively.
3. Compute the tree topology with the HSearch (all optimality criteria) or NJ (distance) command (see Section 7.7).

Box 7.4 The PAUP* program

The PAUP* (Phylogenetic Analysis Using Parsimony* and other methods) program is distributed by Sinauer Associates and purchasing information can be found at **http://paup.csit.fsu.edu/**. The PAUP* installer creates a PAUP folder on the local computer containing the executable application plus a number of other files, including sample data files in NEXUS format and an extensive documentation in pdf format. Versions of the program are available for Macintosh, MS Windows, and UNIX/Linux. The Mac version is controlled using either a full graphical interface or a command-line interface; only the command-line interface is available for the other platforms. Because it is intuitively easy to switch to the Mac graphic interface for the user familiar with the PAUP* commands, this book focuses on how to use the program through the command-line interface. What follows is a brief introduction to the NEXUS format and the main PAUP* commands. More details can be found in the quick-start tutorial available at **http://paup.csit.fsu.edu/downl.html** and in the pdf documentation in the PAUP folder.

The NEXUS format

PAUP* reads input files in NEXUS format. A NEXUS file must begin with the string

#NEXUS

on the first line. The actual sequences plus some additional information about the data set are then written in the next lines within a so-called data block. A data block begins with the line

Begin data;

and it must end with the line

End;

Here is an example of a NEXUS file containing four sequences ten nucleotides long:

(Dimensions ntax=4 nchar=10):

```
#NEXUS
Begin data;
    Dimensions ntax=4 nchar=10;
    Format datatype=nucleotide gap=- missing=? interleave;
    Matrix
L20571          ATGACAG-AA
AF10138         ATGAAAG-AT
X52154          AT?AAAGTAT
U09127          ATGA??GTAT
;
End;
```

The sequences are DNA with gaps indicated by a – symbol, and missing characters indicated by a ? symbol (Format datatype=nucleotide gap=- missing=? matchchar=. Interleave). The aligned sequences are reported on the line after the keyword Matrix. Each

Box 7.4 continued

sequence starts on a new line with its name followed by a few blank spaces and then the nucleotide (or amino-acid) sequence itself. A semicolon (;) must be placed in the line after the last sequence (to indicate the end of the Matrix block).

Running PAUP* (MS Windows version)

By double-clicking the PAUP executable inside the PAUP folder (the version available at the time of this writing is called win-paup4b10.exe), an Open window appears asking to select a file to be edited or executed. By choosing edit, the file is opened in a text-editor window, so that it is possible, for example, to convert its format to NEXUS by manual editing. If the file is already in NEXUS format, it can be executed by selecting Execute in the Open window. The PAUP* Display window appears showing information about the data in the NEXUS file (e.g., number of sequences, number of characters) which are now ready to be analyzed by entering PAUP* commands in the command line at the bottom of the PAUP* Display window. By clicking on cancel, an empty PAUP* Display window appears. To quit the program, just click the x in the upper right corner of the PAUP* Display window, or choose exit from the File menu.

During a phylogenetic analysis with PAUP*, it is good practice to keep a record of all the different steps, displayed in the program Display window, on a log file. To do that at the beginning of a new analysis, choose Log output to disk from the File menu. The log file can be viewed later with any text editor, including PAUP* working in edit mode (described previously). The program can convert aligned sequences from PHYLIP and other formats to NEXUS. In the following example, the file hivALN.phy (see Section 2.9) is converted from PHYLIP to NEXUS format:

1. Place the file hivALN.phy in the PAUP folder.
2. Run PAUP* by double-clicking the executable icon and click on cancel.
3. Type in the command line the following:

    toNexus fromFile=hivALN.phy toFile=hivALN.nex;

4. Click execute below the command line or press Enter.

A file called hivALN.nex will appear in the PAUP folder.

The command toNexus converts files to the NEXUS format. The option fromFile= indicates the input file with the sequences in format other than NEXUS. toFile= is used to name the new NEXUS file being created. Note that PAUP* is case-insensitive, which means that, for example, the commands toNexus and TONEXUS are exactly the same. In what follows, uppercase and lowercase in commands or options are only used for clarity reasons. The file gdp.phy, containing the protein example data set (see Section 2.9) in PHYLIP interleaved format (see Box 2.2), is translated into NEXUS with the following command block:

toNexus fromFile=gdp.phy toFile=gdp.nex Datatype=protein
    Interleaved=yes;

Note that it is necessary to specify protein for the option Datatype and to set Interleaved to yes (the starting default option is no).

Commands and options

As discussed in this chapter, PAUP* can perform distance-, parsimony-, and maximum-likelihood-based phylogenetic analyses of molecular sequence data. Such analyses are carried out by entering commands on the command line. For most commands, one or more options may be supplied. Each option has a default value, but the user can assign an alternative value among those allowed for that particular option by typing *option=value*. Any command and its options with assigned values can be typed in the command line at the bottom of the PAUP* Display window and executed by clicking execute below the command line or pressing Enter. A semicolon (;) is used to indicate the end of a particular command. If desired, multiple commands can be entered, each separated from the next by a semicolon, in which case the commands are executed in the order they appear in the command line. The complete list of available options for a given command can be obtained by entering

```
command-name ?;
```

in the PAUP* command line, where *command-name* represents the name of a valid PAUP* command. In this way, it is also possible for the user to check the current value of each option. Try, for example, to type and execute Set ? (when only one command is executed, the semicolon can be omitted). The following list will be displayed in the PAUP* window:

```
Usage: Set [options...] ;
Available options:
Keyword ---- Option type ----------------------- Current default -
Criterion    Parsimony|Likelihood|Distance        Parsimony
MaxTrees     <integer-value>                       100
```

This list is an abridged version of the possible options and their current values for the Set command. Set is used to choose whether to perform a parsimony-, maximum-likelihood-, or distance-based analysis by assigning the appropriate value – selected among the Option type in the list – to the Criterion option. By default, PAUP* performs parsimony-based analysis. By typing set Criterion=Likelihood or set Criterion=Distance in the command line, the analysis criterion switches to maximum likelihood or distance, respectively.

MaxTrees is another option that can be used in conjunction with the Set command. MaxTrees controls the maximum number of trees that will be saved by PAUP* when a heuristic tree search with parsimony or maximum likelihood is performed (see Sections 7.4.2 and 7.6). The default value is 100, but the user can enter a different integer number (e.g., 1000) by typing set MaxTrees=1000.

4. Evaluate the reliability of the inferred tree(s) with bootstrapping (Bootstrap command). Also, for maximum likelihood, the zero-branch-length test is available (see Section 7.8).
5. Examine the tree in the PAUP* Display window with the DescribeTrees command; or save it in a tree file with the SaveTrees command by typing savetrees file=*tree-file-name*).

A note of caution: By default, the SaveTrees command saves trees in NEXUS format without including branch lengths. To include branch lengths, type:

```
savetrees file=tree-file-name brlens=yes;
```

Trees in NEXUS format can be opened with TreeView (see Chapter 5) or reimported in PAUP* (after executing the original sequence data file) with the following command:

```
gettrees file=tree-file-name;
```

Complete details can be found in the .pdf manual in the PAUP folder.

7.6   Basic parsimony analysis and tree-searching

Assuming that the file gpd.nex is located in the current directory, start PAUP* and execute the file. The number of taxa in this data set (i.e., 12) is effectively the maximum for which an exhaustive search is feasible. This can be accomplished using the command:

```
alltrees fd=histogram;
```

The fd=histogram option requests output of the frequency distribution of tree lengths in the form of a histogram with the default number (i.e., 20) of class intervals. The resulting output follows:

```
Exhaustive search completed:
    Number of trees evaluated = 654729075
    Score of best tree found = 1304
    Score of worst tree found = 1865
    Number of trees retained = 2
    Time used = 02:04:49.2


Frequency distribution of tree scores:

                mean=1726.617435 sd=68.439984 g1=-0.840669
                    g2=0.570847
1304.00000 /---------------------------------------------------
1332.05000 + (490)
1360.10000 + (4451)
1388.15000 + (25977)
1416.20000 + (98229)
1444.25000 + (291779)
1472.30000 + (716148)
1500.35000 +# (1679478)
1528.40000 +## (3379015)
1556.45000 +### (6530925)
```

```
1584.50000 +###### (11994903)
1612.55000 +######### (20450808)
1640.60000 +############## (31955251)
1668.65000 +#################### (48539228)
1696.70000 +########################### (67297301)
1724.75000 +################################### (85253204)
1752.80000 +#############################################
           (101495260)
1780.85000 +###############################################
           ###(116910918)
1808.90000 +##############################################
           (106579209)
1836.95000 +#################### (46954684)
1865.00000 +## (4571817)
           \---------------------------------------------------
```

Thus, it is known with certainty that the shortest trees require 1,304 steps and that there are two trees of this length. Furthermore, although there are many trees slightly longer than the shortest tree, these "good" trees represent only a small fraction of the more than 6.5×10^8 possible trees for 10 taxa. This search requires significant computation time (i.e., approximately 2 hours on a moderately fast Pentium III machine). If the distribution of tree lengths is not of interest, the branch-and-bound algorithm may be used to find an exact solution more quickly by typing:

```
bandb;
```

This search takes less than one second and finds the same two trees. If the data set had been much larger, finding exact solutions using the exhaustive-search or branch-and-bound methods would not have been feasible, in which case the HSearch command could be used to perform a heuristic search. For example,

```
hsearch/addseq=random nreps=100;
```

would request a search from 100 starting trees generated by stepwise addition with random-addition sequences, followed by TBR (the default) branch-swapping on each starting tree. In this small example, every one of the random-addition-sequence replicates finds the same two trees obtained using the exact methods, but this often will not be the case with heuristic searches on larger, more complex data sets.

To examine the trees, the ShowTrees or DescribeTrees commands are available. ShowTrees simply outputs a diagram representing each tree topology, but provides no other information:

```
showtrees all;
```

The two resulting trees are shown in the output as follows:

```
Tree number 1 (rooted using default outgroup)
```

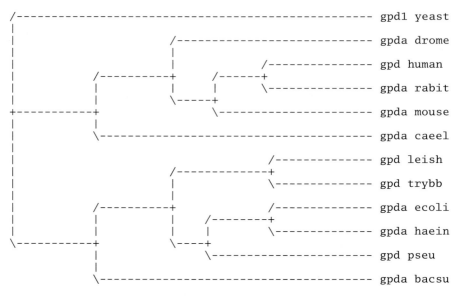

```
Tree number 2 (rooted using default outgroup)
```

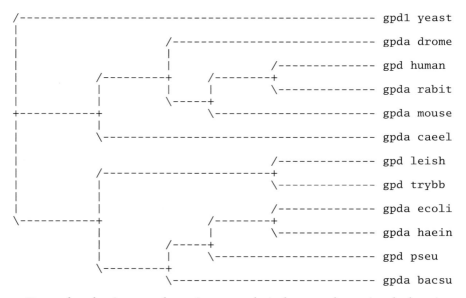

Remember that in general, parsimony analysis does not determine the location of the root. By default, PAUP* assumes that the first taxon in the data matrix is the only **outgroup taxon** and roots the tree at the adjacent internal node leading to a basal **polytomy** as indicated previously. In this example, a more appropriate outgroup would be the four bacterial sequences, which can be specified as follows:

```
outgroup gpdaecoli gpdahaein gpdabacsu gpdpseu;
```

or, alternatively, by

```
outgroup 9-12;
```

Because the trees are represented internally as **unrooted** trees, they may now be output according to the new outgroup specification by reissuing the ShowTrees command. Alternatively, the DescribeTrees command can be used to request output of the tree in *phylogram* format, where the branches of the tree are drawn proportionally to the number of changes (see Chapter 5) assigned under the parsimony criterion. Other information is also available through additional DescribeTrees options. For example, the command

```
describetrees 1/plot=phylogram diagnose;
```

outputs a drawing of the first tree in phylogram format (now reflecting the new rooting), plus a table showing the number of changes required by each character on the first tree, the minimum and maximum possible number of changes on any tree, and several goodness-of-fit measures derived from the following values:

```
Tree number 1 (rooted using user-specified outgroup)
Warning: Tree can not be rooted such that specified in-
group is monophyletic.
Tree length = 1304
Consistency index (CI) = 0.8627
Homoplasy index (HI) = 0.1373
CI excluding uninformative characters = 0.8442
HI excluding uninformative characters = 0.1558
Retention index (RI) = 0.7635
Rescaled consistency index (RC) = 0.6587
```

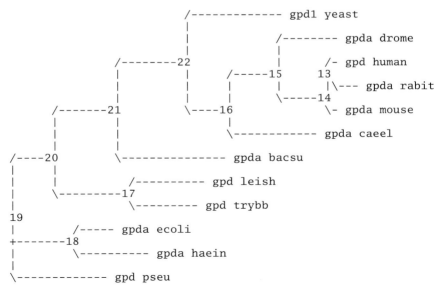

```
Character diagnostics:
```

Character	Range	Min steps	Tree steps	Max steps		CI	RI	RC	HI	G-fit
21	1	1	1	1	1.000	0/0	0/0	0.000	1.000
23	1	1	1	1	1.000	0/0	0/0	0.000	1.000
.										
.										
.										
33	5	5	5	6	1.000	1.000	1.000	0.000	1.000
34	5	5	6	7	0.833	0.500	0.417	0.167	0.750
35	5	5	5	6	1.000	1.000	1.000	0.000	1.000

To assess whether this result is strongly supported, bootstrap or jackknife analysis (see Chapter 5) can be performed using the Bootstrap or Jackknife commands. For example, the command

```
bootstrap nreps=1000;
```

requests a bootstrap analysis with 1,000 pseudoreplicate samples using the current optimality criterion and associated settings.

Results of the bootstrap analysis can be summarized using a majority-rule *consensus tree* that indicates the frequency in which each bipartition, or division of the taxa into two groups, was present in the trees obtained for each bootstrap replicate (see Box 5.3). Interpretation of *bootstrap values* is complicated (e.g., Hillis and Bull, 1993) (see also Chapter 5), but clearly clades with close to 100% support can be considered very robust.

```
Bootstrap 50% majority-rule consensus tree
```

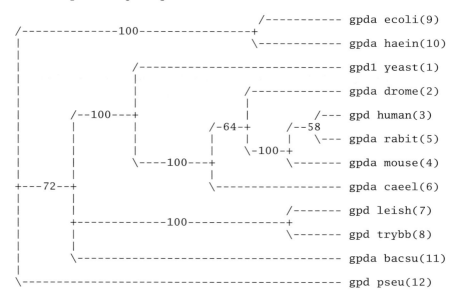

In addition to the consensus tree, the output includes a table showing the frequency of all groups that were found at a frequency of 5% or higher, so that support for groups not included in the consensus also can be evaluated as follows:

```
Bipartitions found in one or more trees and frequency of
    occurrence (bootstrap support values):
          1 1
123456789012            Freq        %
------------------------------
........**..      1000.00 100.0%
......**....      1000.00 100.0%
......******      1000.00 100.0%
..***.......      1000.00 100.0%
.*****......       997.67  99.8%
........**.*       717.93  71.8%
.****.......       639.45  63.9%
..*.*.......       578.98  57.9%
........****       409.44  40.9%
......****.*       398.03  39.8%
...**.......       290.05  29.0%
.*...*......       197.91  19.8%
..****......       162.64  16.3%
........***.       137.13  13.7%
..**........       130.96  13.1%
......*****.        88.57   8.9%
......**...*        64.57   6.5%
......****..        58.73   5.9%
..........**        51.85   5.2%
```

(Note: the format of this output follows the same convention of the output file from the program Consense.exe of the PHYLIP package; see Box 5.4). Another measure of nodal support is the **decay index**, or **Bremer support** (Bremer, 1988; Donoghue et al., 1992). If the shortest tree inconsistent with the monophyly of some group is three steps longer than the most parsimonious tree (on which the group is monophyletic), the decay index is equal to 3. These values can be calculated in two ways. For groups that decay quickly, all trees within one step, two steps, and so on, of the shortest tree can be stored and a strict consensus tree calculated for each length. The following commands can be used to find groups that occur in all trees within 1, 2, and 3 steps, respectively, of the shortest tree:

```
hsearch keep=1305; contree;

hsearch keep=1306; contree;

hsearch keep=1307; contree;
```

For example, a group consisting of *Drosophila* plus three mammals (e.g., human, mouse, rabbit) is found in all trees of length 1,306 and shorter, but this group disappears from the consensus of all trees of length 1,307 and shorter, indicating that there are some trees at 1,307 steps that do not contain this 4-taxon clade. The decay index is, therefore, $1,307 - 1,304 = 3$. Evaluating the meaning of decay indexes is probably even more difficult than bootstrap values, but any group for which the decay index is less than 4 should probably be viewed with suspicion (DeBry, 2001).

An alternative method for calculating decay indexes is to use the ***converse constraints*** feature of PAUP*. In general, a ***constraint tree*** is a user-defined tree that limits the search space to those trees that are either compatible or incompatible with the constraint tree. For monophyly constraints, a tree T is compatible with the constraint tree if it is equal to the constraint tree or if the constraint tree can be obtained by collapsing one or more branches of T. For calculating decay indexes, the monophyly of the group of interest is indicated and a search is performed using converse constraints. For the example discussed previously, the constraint tree can be defined using the command

```
constraints mamdros = ((2,3,4,5));
```

The tree resulting from this definition can then be viewed:

```
showconstr;
Constraint-tree "mamdros":

/----------------------------------------------- gpd1 yeast
|
|                                /-------------------- gpda drome
|                                |
|                                +-------------------- gpd human
+----------------------+
|                                +-------------------- gpda mouse
|                                |
|                                \-------------------- gpda rabit
|
+----------------------------------------------- gpda caeel
|
+----------------------------------------------- gpd leish
|
+----------------------------------------------- gpd trybb
|
+----------------------------------------------- gpda ecoli
|
+----------------------------------------------- gpda haein
|
+----------------------------------------------- gpda bacsu
|
\----------------------------------------------- gpd pseu
```

Next, a constrained search is conducted as follows:

```
HSearch enforce constraints=mamdros converse
    addseq=random nreps=100;
```

It is especially desirable to use multiple random-addition-sequence starting points when searching under converse constraints to improve the chances of identifying the shortest trees that are incompatible with the constraint tree. In this case, 4 trees of length 1,307 are found in the constrained search, yielding a decay index of 3, as was determined previously. This second approach is much more useful for determining larger **decay values**, where saving all trees within n steps of the shortest tree becomes more unwieldy as n increases. As it is often but not always the case, the low decay index in the example mirrors a small bootstrap value for the same group.

7.7 Analysis using distance methods

The same search strategies discussed previously may be used for searching trees under distance optimality criteria. Because distance methods were covered extensively in Chapter 5, what follows only illustrates how to perform some commonly used analyses with PAUP*, using the vertebrate mtDNA data set as an example.

The DSet command is used to set the type of distance to be computed:

```
dset distance=jc;
```

This will set the distances to Jukes-Cantor distances (see Chapter 5); other distance options available in PAUP* (as well as the current settings) can be viewed by typing Dset ?. To calculate a standard NJ tree (Saitou and Nei, 1987) according to the current distance transformation, use the command

```
nj;
```

Before proceeding with further analyses, note that the vertebrate mtDNA data set exhibits heterogeneous base frequencies among taxa, as can be seen from the output of the command

```
basefreq;
```

In the presence of such heterogeneity, an appropriate distance transformation is the LogDet distance (see Section 4.12), which can be selected using the command

```
dset distance=logdet;
```

Criterion-based searches are performed using the same commands as were previously described for parsimony analysis. However, because the default optimality criterion is parsimony, the following command must be issued to force use of distance-based optimality criteria.

```
set criterion=distance;
```

By default, distance analyses use an NJ method to obtain a starting tree (addseq= nj). This can be overridden to use random addition sequences, as follows:

```
hsearch/addseq=random nreps=100;
```

The resulting tree can then be output as a phylogram:

```
describe/plot=phylogram;
```

```
Minimum evolution score = 2.47415
```

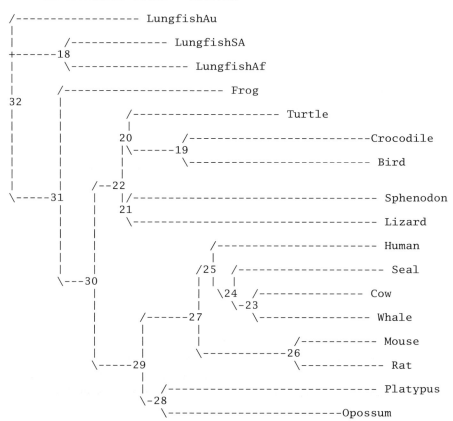

```
/----------------- LungfishAu
|
|         /------------- LungfishSA
+------18
|         \----------------- LungfishAf
|
|         /---------------------- Frog
32        |
|         |         /------------------- Turtle
|         |         |
|         |        20         /-----------------------------Crocodile
|         |        |\------19
|         |         |         \------------------------ Bird
|         |         |
|         |     /--22
\-----31   |     |/------------------------------------ Sphenodon
|         |     21
|         |     \----------------------------- Lizard
|         |
|         |                   /--------------------- Human
|         |                   |
|         |          /25      /------------------- Seal
|         |          | | |
\---30    |          | \24    /-------------- Cow
|         |          |  \-23
|         |    /------27      \-------------- Whale
|         |    |           
|         |    |                     /----------- Mouse
|         |    |         \-----------26
\-----29  |                       \----------- Rat
|
|   /------------------------- Platypus
\-28
\-----------------------Opossum
```

This tree places turtles with archosaurs. Constraints may be used to find the best tree consistent with the conventional placement of turtle (i.e., basal reptile):

```
constraints turtlebasal = ((Crocodile,Bird,
    Sphenodon, Lizard),Turtle);
```

```
hsearch enforce constraints=turtlebasal;
```

The score of the resulting tree is 2.47570, which is barely worse than the best un-constrained tree. The low support for the placement of the turtle can be confirmed using ***bootstrap analysis***:

```
bootstrap nreps=1000;
```

The bootstrap support results, in fact, only about 51%.

Minimum evolution (see Chapter 5) is the default method used for distance analyses in PAUP*. This method uses least-squares fit to determine the branch lengths, but then uses the sum of these branch lengths as the objective function for computing the score of each tree. The unweighted least-squares and Fitch-Margoliash (weighted least-squares) variants are also available by replacing n with 0 or 2, respectively, in the following command

```
dset objective=ls power=n;
```

prior to beginning a distance-criterion search.

One particularly problematic aspect of distance analysis involves the handling of negative branch lengths, which often occur in unconstrained optimization of the least-squares fit between path-length and observed distances. Because nega-tive branch lengths have no obvious biological meaning, they probably should be avoided. A constrained optimization is available in PAUP* to perform the least-squares optimization under the restriction of branch-length nonnegativity, which can be requested as follows:

```
dset negbrlen=prohibit;
```

We believe that this option provides the most appropriate way to deal with the negative-branch length problem, but analyses run significantly more slowly when this option is specified. The default setting in PAUP* is to allow negative branch lengths when performing the least-squares fit calculations, but to set negative values to zero before computing the score of the tree (`negbrlen=setzero`). Thus, if the least-squares fit does not require negative lengths, the tree scores obtained in this way are identical, whereas trees for which the fit is improved by allowing negative branch lengths are still penalized accordingly. Negative branch lengths also may be set to their absolute value (i.e., `negbrlen=setabsval`) (Kidd and Sgaramella-Zonta, 1971) or simply used "as is" (i.e., `negbrlen=allow`) (Rzhetsky and Nei, 1992).

It is also possible to input distances directly into PAUP* using a NEXUS format `Distances` block. This feature permits mixing the tree-searching capabilities of PAUP* with distance transformations that are not available within the program (e.g., model-based protein distances). Examples for using user-supplied distance matrixes are given in the command-reference document in the PAUP folder.

7.8   Analysis using maximum-likelihood methods

Maximum-likelihood methods are discussed in Chapter 6, which describes the *quartet-puzzling algorithm* implemented in TREE–PUZZLE as a heuristic strategy for a maximum-likelihood tree search. PAUP* is also able to carry out a number of maximum-likelihood-based analyses on nucleotide-sequence data. In particular, the program can perform exhaustive, branch-and-bound, and various types of *heuristic searches*, described previously, on aligned nucleotide sequences. The following example assumes that the HIV data set in NEXUS format (hivALN.nex) (see Box 7.4) has already been executed in PAUP*.

The first task in any maximum-likelihood analysis is to choose an appropriate model. The default model in PAUP* is the HKY model (see Chapter 4) with a transition:transversion ratio of 2; however, this model should never be used uncritically. Although automated model-selection techniques are available and useful (Posada and Crandall, 1998) (see Chapter 10), we find that manual model selection can be highly informative and may identify good models that might otherwise be missed. Unfortunately, we cannot recommend a precise solution that is best for all situations (if such a solution were available, then it could simply be automated). Instead, we outline the following general strategy, which basically follows the "top-down dynamical" approach of Chapter 10, but without enforcing any particular a priori specification of models to be compared.

1. Start with some reasonable tree for the data. This tree need not be optimal in any sense, but it should at least be a "good" tree under some criterion.
2. Set the likelihood model to the most complex one available in PAUP* (six-substitution-type general time-reversible model with some fraction of invariable sites and rates at variable sites following a gamma distribution (=GTR+I+Γ)).
3. Estimate all model parameters (i.e., relative substitution rates, base frequencies, proportion of invariable sites, gamma shape parameter) and calculate the likelihood score of the tree using the LScores command.
4. Examine the parameter estimates carefully in an attempt to identify possible model simplifications that will not greatly reduce the fit of the model. Evaluate the simplified model and decide whether it is acceptable. If the attempted simplification is rejected, return to the previous model.
5. Repeat Step 4 until no further acceptable simplifications can be found.
6. Proceed to search for an optimal tree using the model and parameter estimates selected previously.

Some of these steps deserve further elaboration. The reason that the tree used for Step 1 is not critical is that parameter estimates do not vary much from tree to tree, as long as the trees are reasonably good for the data (i.e., those that are much better than randomly chosen trees and include clades that are well supported under any optimality criterion) (Yang, 1994; Sullivan et al., 1996; unpublished observations). The determination of whether to reject a simplified model in Step 4 can be accomplished using likelihood-ratio tests (LRTs), the AIC or BIC criteria (see Chapter 10), or more subjective methods. Model-selection criteria are somewhat controversial. LRTs offer the appeal of statistical rigor, but the choice of an appropriate alpha level for the test (e.g., 0.05, 0.01, 0.001) is arbitrary. Furthermore, even if a simple model is strongly rejected in favor of a more complex model, the simpler model may still be better for tree inference because it requires the estimation of fewer parameters. As models become more complex, more parameters must be estimated from the same amount of data, so that the variance of the estimates increases. Conversely, use of an oversimplified model generally leads to biased estimates. We follow Burnham and Anderson (1998) and others in preferring models that optimize the trade-off between bias and variance as a function of the number of parameters in the model. Unfortunately, it seems impossible to avoid some degree of subjectivity when selecting models, but this should not be taken as a reason to avoid model-based methods entirely (Sullivan and Swofford, 2001).

We illustrate our model-selection strategy using the HIV data set. First, we obtain a starting tree, which can be, for example, a parsimony, an NJ, or a maximum-likelihood tree obtained under a simple model such as Jukes-Cantor. In practice, the choice of this tree rarely makes a difference in the model selected (unpublished observations). In the absence of further information, the LogDet transformation represents a good all-purpose distance, and an NJ tree computed from the LogDet distance matrix can serve as the starting tree:

```
dset distance=logdet;
nj;
```

Steps 2 and 3 are then accomplished using the LSet and/or LScores commands:

```
lset nst=6 rmatrix=estimate basefreq=estimate
    rates=gamma shape=estimate pinvar=estimate;

lscores;
```

Or, equivalently:

```
lscores/nst=6 rmatrix=estimate basefreq=estimate
    rates=gamma shape=estimate pinvar=estimate;
```

NST represents the number of substitution types. Instead of two (i.e., transitions versus transversions), it is assumed that all six of the pairwise substitution rates are potentially different. Γ-distributed rates for variable sites (see Chapter 4) are specified via rates=gamma. Rather than taking default values, all parameters are estimated from the data. The resulting output follows. Note that PAUP* reports the tree score as the negative log likelihood rather than the log likelihood itself;

therefore, smaller likelihood scores are better. This is done so that all optimizations performed by the program can be treated as minimization problems.

```
Likelihood scores of tree(s) in memory:
  Likelihood settings:
    Number of substitution types = 6
    Substitution rate-matrix parameters estimated via ML
    Nucleotide frequencies estimated via ML
    Among-site rate variation:
      Assumed proportion of invariable sites = estimated
      Distribution of rates at variable sites = gamma
        (discrete approximation)
        Shape parameter (alpha) = estimated
        Number of rate categories = 4
        Representation of average rate for each category = mean
    These settings correspond to the GTR+G+I model
    Number of distinct data patterns under this model = 1356
    Molecular clock not enforced
    Starting branch lengths obtained using Rogers-Swofford
      approximation method
    Branch-length optimization = one-dimensional Newton-Raphson
      with pass limit=20, delta=1e-06
    -ln L (unconstrained) = unavailable due to missing-data
      and/or ambiguities


Tree                    1
-------------------
-ln L    20642.70321
Base frequencies:
  A           0.366605
  C           0.181585
  G           0.214291
  T           0.237519
Rate matrix R:
  AC          2.00873
  AG          4.12877
  AT          0.72819
  CG          1.15669
  CT          4.09635
  GT          1.00000
P_inv         0.132576
Shape         0.802011
```

Searching for simplifications, it is apparent that the substitution rates for the two transitions, r_{AG} and r_{CT}, are nearly equal (i.e., approximately 4.1). Thus, an obvious first try at simplifying the model involves equating these two rates and reducing the number of parameters by one. This is accomplished by specifying a particular submodel of the GTR model using the RClass option, which allows entries in the rate matrix to be pooled into larger classes. The classification of substitution types follows the order r_{AC}, r_{AG}, r_{AT}, r_{CG}, r_{CT}, r_{GT}. All rates assigned the same alphabetic character are pooled into a single category. Thus, the following command pools the two transitions (r_{AG} and r_{CT}) into one class, but leaves each transversion distinct, and reestimates all other parameters (all options from the previous command are still in effect):

```
lscores/rclass=(a b c d b e);
```

The resulting output shows that the tree score is only trivially worse than the full GTR+I+Γ model (20642.708 versus 20642.703):

```
Tree                    1
-------------------
-ln L    20642.70755
Base frequencies:
   A          0.366803
   C          0.181386
   G          0.214502
   T          0.237309
Rate matrix R:
   AC           2.01173
   AG           4.11724
   AT           0.72854
   CG           1.15761
   CT           4.11724
   GT           1.00000
P_inv         0.132566
Shape         0.802072
```

Thus, it is appropriate to accept the simpler model with the two transitions pooled into one category.

Often, models with either Γ-distributed rates or invariable sites alone fit the data nearly as well as those with both types of rate variation. To examine this possibility for the HIV data set, the reduced model is evaluated, but with all sites variable (pinv=0) and following a gamma distribution (still the current setting):

```
lscores/pinv=0;
```

The likelihood score obtained for this model is 20647.857, or 5.149 units worse. An LRT can be performed by multiplying this score difference by 2 and comparing to a chi-squared distribution with 1 degree of freedom (see Chapter 10). The P-value for this test is 0.0013, suggesting that the invariable-sites plus gamma model (I+Γ)

fits significantly better than the Γ model alone. The AIC (see Chapter 10) also favors the I+Γ rate-variation model.

Next, we evaluate the model with some sites invariable, and all variable sites evolving at the same rate:

```
lsc/rates=equal pinv=est;
```

The likelihood score for this model, 20854.240, is even worse, so it also is rejected in favor of the I+Γ model.

After failing to simplify the model by eliminating among-site rate-variation parameters, attention can be returned to the substitution-rate matrix. Inspection of this matrix reveals that the relative rate for the CG substitution ($r_{CG} = 1.158$) is not much different than $r_{GT} = 1$. Thus, these two substitutions also can be pooled and the resulting likelihood score calculated:

```
lsc/rclass=(a b c d b d) rates=gamma shape=estimate
    pinv=estimate;
```

The resulting likelihood score and parameter estimates follow:

```
Tree                    1
-------------------
-ln L     20643.18670
Base frequencies:
    A         0.366830
    C         0.182505
    G         0.214685
    T         0.235981
Rate matrix R:
    AC        1.89422
    AG        3.88229
    AT        0.68654
    CG        1.00000
    CT        3.88229
    GT        1.00000
P_inv         0.134026
Shape         0.804921
```

The likelihood score (20643.187) is not much worse than the previous acceptable model (20642.708). The difference in scores is equal to 0.479, and an LRT using twice this value with 1 degree of freedom yields a P-value of 0.33. Thus, the simpler model is preferred. Although we do not show the results, further attempts to simplify the model by additional substitution-pooling (e.g., rclass=(a b c c b c)) or assuming equal base frequencies (i.e., basefreq=equal) yield likelihood scores that are significantly worse than this model. It is interesting that the model chosen here, to our knowledge, has not been named and would not have been found using current automated model-selection procedures.

As an aside, the RClass option may be used to select other specific models. For example, lset rclass=(a b a a c a); can be used to specify the Tamura-Nei model (see Chapter 4), which assumes one rate for transversions, a second rate for purine transitions, and a third rate for pyrimidine transitions. RClass is meaningful only when the substitution-rate matrix is being estimated (rmatrix=estimate) and the number of substitution types is set to six (nst=6). Specific (relative) rates also may be specified using the RMatrix option. For example, the command

```
lset rmatrix=(1 5 1 1 2.5);
```

also specifies a Tamura-Nei model in which purine and pyrimidine transitions occur at a rate of 5 and 2.5 times (respectively) that of transversions. Note that these rates are assigned relative to the G–T rate, which is assigned a value of 1 and is not included in the list of RMatrix values.

Having chosen a model, we can now proceed to search for an optimal tree. To set the criterion to maximum likelihood, execute the following command:

```
set criterion=likelihood;
```

Because the criterion is now likelihood, the HSearch command will perform a maximum-likelihood tree search using the default or current model settings. To reset the model chosen, the following command can be issued:

```
lset nst=6 rclass=(a b c d b d) rmatrix=estimate
    basefreq=estimate rates=gamma shape=estimate
    pinv=estimate;
```

With these settings, PAUP* will estimate all model parameters on each tree evaluated in the search. Whereas in general this is the ideal method, it is too slow for data sets containing more than a small number of taxa. A more computationally tractable approach uses a successive-approximations strategy that alternates between parameter estimation and tree search. First, the parameter values are estimated on a starting tree (e.g., the same NJ tree used for the model-selection procedure). These values are then fixed at their estimated values for the duration of a heuristic search. If the search finds a new tree that is better than the tree for which the parameter values were estimated, then the parameter values are reestimated on this new tree. The iteration continues until the same tree is found in two consecutive searches. To implement this iterative strategy for the HIV example, the parameter values must first be fixed to the values estimated during the model-selection procedure. An easy way to do this is as follows (assuming that the NJ tree is still in memory from before):

```
lscores;
```

```
lset rmatrix=previous basefreq=previous shape=previous
    pinv=previous;
```

Each parameter set to "previous" in this command is assigned the value that was most recently estimated for that parameter (i.e., in the preceding LScores command when the parameters were set to "estimate"). A standard heuristic search using simple stepwise addition followed by TBR rearrangements (see Section 7.4.2) is performed using the command:

```
hsearch;
```

A tree of likelihood score 20631.986 is found, which is an improvement over the NJ starting tree. Consequently, the model parameters are reestimated on this new tree:

```
lset rmatrix=estimate basefreq=estimate shape=estimate
    pinv=estimate;
```

With the newly optimized parameter values, the likelihood score of the tree found by the previous search improves slightly to 20631.940:

```
Tree                    1
--------------------
-ln L     20631.93993
Base frequencies:
   A           0.367310
   C           0.182714
   G           0.213862
   T           0.236114
Rate matrix R:
   AC           1.87478
   AG           3.85110
   AT           0.66716
   CG           1.00000
   CT           3.85110
   GT           1.00000
P_inv          0.129789
Shape          0.794980
```

The parameter estimates are close to those obtained using the original NJ tree (discussed previously), which is why the likelihood score did not improve appreciably. Nonetheless, to make sure that the change in parameter values will not lead to a new optimal tree, the search should be repeated, fixing the parameter values to these new estimates:

```
lset rmatrix=previous basefreq=previous
    shape=previous pinv=previous;
```

```
hsearch;
```

This search finds the same tree as the previous one did, so the iterations can stop. In our experience, this iterative, successive-approximations strategy nearly always converges to the same final solution regardless of the starting tree. A more formal evaluation of its effectiveness is under way at the time of this writing.

Although it is not important for the relatively small HIV data set, it may be desirable for data sets containing larger numbers of taxa to reduce the scope of the rearrangements using the ReconLimit option. If ReconLimit is set to *x*, then any rearrangement for which the reconnection distance (see Section 7.4.2) exceeds *x* is not evaluated. For data sets of 30 or more taxa, use of values such as ReconLimit=12 can often substantially reduce the time required for a TBR search without greatly compromising its effectiveness.

As a starting tree for the heuristic search, it is possible to use an NJ tree (instead of a tree obtained by stepwise addition) with the command HSearch Start=NJ;. Although the program finds the same maximum-likelihood tree for the HIV example data set, this will not necessarily be true for other data sets. In general, the use of different starting trees permits a more thorough exploration of the tree space, and it is always a good idea to compare results of alternative search strategies. Use of random-addition sequences (i.e., HSearch AddSeq=Random;) is especially desirable. The default number of random-addition-sequence replicates is rather small (10); the option NReps=*n* can be used to specify a different number of replicates *n*.

As for parsimony and distance methods, PAUP* can perform bootstrapping and jackknifing analyses under the likelihood criterion, although these analyses may be very time-consuming. In addition, a zero-branch-length test can be performed on the maximum-likelihood tree by setting the ZeroLenTest option of the LSet command to full and then executing the DescribeTrees command:

```
lset zerolentest=full; describetrees;
```

If more than one tree is in memory, the result of the test is printed on the screen (and/or in the log file: see Box 7.4) for each tree. In the zero-branch-length test, the statistical support for each branch of a previously estimated maximum-likelihood tree is obtained as follows. After collapsing the branch under consideration (by constraining its length to zero), the likelihood of the tree is recalculated and compared with the likelihood of the original tree. If the likelihood of the former is significantly worse than that of the latter according to an LRT, the branch is considered statistically supported. As a result of executing these commands, PAUP* prints out a list of all pairs of neighboring nodes in the tree (including terminal nodes), each pair with a corresponding *p-value* giving the support for the branch connecting the two nodes:

```
Branch lengths and linkages for tree  #1 (unrooted)

              Connected   Branch     Standard   --- L.R.    test ---
    Node       to node    length      error     lnL diff      P*
    ----------------------------------------------------------------
    L20571 (1)     26      1.24987    0.08869    413.769     <0.001
    15             26      0.13232    0.02919      8.696     <0.001
    AF10138 (2)    15      0.39919    0.03438    133.805     <0.001
    X52154 (3)     15      0.41887    0.03499    130.733     <0.001
    25             26      0.26245    0.03244     30.645     <0.001
    U09127 (4)     25      0.09300    0.00992     41.974     <0.001
    24             25      0.03462    0.00862      6.020     <0.001
    16             24      0.08143    0.00893    114.759     <0.001
    U27426 (5)     16      0.08363    0.00793    142.627     <0.001
    U27445 (6)     16      0.05983    0.00696     87.930     <0.001
    23             24      0.02223    0.00570     14.460     <0.001
    17             23      0.09663    0.00941    175.513     <0.001
    U067158 (7)    17      0.07276    0.00732    135.090     <0.001
    U09126 (8)     17      0.05423    0.00649     77.533     <0.001
    22             23      0.04219    0.00707     40.833     <0.001
    18             22      0.07051    0.00803    110.980     <0.001
    U27399 (9)     18      0.05609    0.00653     95.309     <0.001
    U43386 (10)    18      0.07063    0.00724    132.291     <0.001
    21             22      0.05521    0.00717     76.209     <0.001
    20             21      0.01345    0.00399     10.580     <0.001
    19             20      0.00971    0.00326      9.305     <0.001
    L02317 (11)    19      0.04976    0.00555    147.044     <0.001
    AF042106 (14)  19      0.06781    0.00646    224.290     <0.001
    AF025763 (12)  20      0.06151    0.00619    206.095     <0.001
    U08443 (13)    21      0.06074    0.00636    149.236     <0.001
    ----------------------------------------------------------------
    Sum                    3.61865
```

   * Probability of obtaining a likelihood ratio as large or larger than the observed ratio under the null hypothesis that a branch has zero length (full reoptimization after forcing a branch length to zero)

-Ln likelihood = 20631.93993

In this case, all the branches in the maximum-likelihood tree seem to be robustly supported by this test. However, the test only assesses whether a given resolution of a trichotomy is a significant improvement relative to leaving the trichotomy un-resolved. For example, it is possible for incompatible groups on different trees to both receive significant support using this test. The fact that not all of the clades received high bootstrap values in the parsimony and NJ trees, as well as in maximum-likelihood bootstrap analyses (not shown), highlights the need for caution in interpreting the results of this test.

REFERENCES

Bremer, K. (1988). The limits of amino-acid sequence data in angiosperm phylogenetic recon-struction. *Evolution* 42: 795–803.

Burnham, K. P., and D. R. Anderson (1998). *Model Selection and Inference: A Practical Information-Theoretic Approach.* New York: Springer-Verlag.

de Queiroz, K., and S. Poe (2001). Philosophy and phylogenetic inference: A comparison of likelihood and parsimony methods in the context of Karl Popper's writings on corroboration. *Systematic Biology* 50: 305–321.

DeBry, R. W. (2001). Improving interpretation of the decay index for DNA sequence data. *Systematic Biology* 50: 742–752.

Donoghue, M. J., R. G. Olmstead, J. F. Smith, and J. D. Palmer (1992). Phylogenetic relationships of Dipsacales based on rbcL sequences. *Annals of the Missouri Botanical Garden* 79: 333–345.

Farris, J. S. (1970). Estimating phylogenetic trees from distance matrixes. *American Nature* 106: 645–668.

Felsenstein, J. (1978). Cases in which parsimony and compatibility methods will be positively misleading. *Systematic Zoology* 27, 401–410.

Fitch, W. M. (1971). Toward defining the course of evolution: Minimum change for a specific tree topology. *Systematic Zoology* 20: 406–416.

Goloboff, P. (1999). Analyzing large data sets in reasonable times: Solutions for composite optima. *Cladistics* 15: 415–428.

Hartigan, J. A. (1973). Minimum mutation fits to a given tree. *Biometrics* 29: 53–65.

Hendy, M. D., and D. Penny (1982). Branch-and-bound algorithms to determine minimal evo-lutionary trees. *Mathematical Biosciences* 59: 277–29.

Hillis, D. M. (1996). Inferring complex phylogenies. *Nature* 383: 130.

Hillis, D. M., and J. J. Bull (1993). An empirical test of bootstrapping as a method for assessing confidence in phylogenetic analysis. *Systematic Biology* 42: 182–192.

Hillis, D. M., J. P. Huelsenbeck, and D. L. Swofford (1994). Hobgoblin of phylogenetics. *Nature* 369: 363–364.

Kidd, K. K., and L. A. Sgaramella-Zonta (1971). Phylogenetic analysis: Concepts and methods. *American Journal of Human Genetics* 23: 235–252.

Kluge, A. G., and J. S. Farris (1969). Quantitative phyletics and the evolution of anurans. *Systematic Zoology* 18: 1–32.

Lewis, P. O. (1998). A genetic algorithm for maximum-likelihood phylogeny inference using nucleotide-sequence data. *Molecular Biology and Evolution* 15: 277–283.

Maddison, D. R. (1991). The discovery and importance of multiple islands of most parsimonious trees. *Systematic Zoology* 40: 315–328.

Moilanen, A. (2001). Simulated evolutionary optimization and local search: Introduction and application to tree search. *Cladistics* 17: S12–S25.

Nixon, K. C. (1999). The parsimony ratchet, a new method for rapid parsimony analysis. *Cladistics* 15: 407–414.

Posada, D., and K. A. Crandall (1998). MODELTEST: Testing the model of DNA substitution. *Bioinformatics* 14: 817–818.

Rzhetsky, A., and M. Nei (1992). A simple method for estimating and testing minimum-evolution trees. *Molecular Biology and Evolution* 9: 945–967.

Saitou, N., and M. Nei (1987). The neighbor-joining method: A new method for reconstructing phylogenetic trees. *Molecular Biology and Evolution* 4: 406–425.

Sankoff, D., and P. Rousseau (1975). Locating the vertixes of a Steiner tree in an arbitrary metric space. *Math. Progr.* 9: 240–276.

Sikes, D. S., and P. O. Lewis (2001). *Beta Software, Version 1. PAUPRat: PAUP* Implementation of the Parsimony Ratchet.* Distributed by the authors. Storrs: University of Connecticut, Department of Ecology and Evolutionary Biology. (**http://viceroy.eeb.uconn.edu/paupratweb/pauprat.htm**).

Sullivan, J., K. E. Holsinger, and C. Simon (1996). The effect of topology on estimates of among-site rate variation. *Journal of Molecular Evolution* 42: 308–312.

Sullivan, J., and D. L. Swofford (2001). Should we use model-based methods for phylogenetic inference when we know that assumptions about among-site rate variation and nucleotide substitution process are violated? *Systematic Biology* 50: 723–729.

Swofford, D. L. and W. P. Maddison (1987). Reconstructing ancestral character states under Wagner parsimony. *Mathematical Biosciences* 87: 199–229.

Swofford, D. L., G. J. Olsen, P. J. Waddell, and D. M. Hillis (1996). Phylogenetic Inference *In Molecular Systematics* 2nd ed. D. M. Hillis, C. Moritz, and B. K. Mable (eds.). Sunderland, Massachusetts (USA): Sinauer Associates, Inc., 407–514.

Swofford, D. L. (2002). PAUP*. *Phylogenetic Analysis Using Parsimony (* and other methods). Version 4.0b10.* Sunderland, Massachusetts (USA): Sinauer Associates, Inc.

Tuffley, C., and M. Steel (1997). Links between maximum likelihood and maximum parsimony under a simple model of site substitution. *Bulletin of Mathematical Biology* 59: 581–607.

Yang, Z. (1994). Maximum-likelihood phylogenetic estimation from DNA sequences with variable rates over sites: Approximate methods. *Journal of Molecular Evolution* 39: 306–314.

Phylogenetic analysis using protein sequences

THEORY

Fred R. Opperdoes

8.1 Introduction

Phylogenetic analyses using ribosomal RNA (**rRNA**) sequences, as initiated by Woese and collaborators (Woese and Fox, 1977), suggest that the living world is divided into three domains: **eukaryota, archaebacteria**, and **eubacteria** (Figure 8.1). According to this so-called **universal tree of life** and based on morphological and biochemical evidence, it was originally inferred that the earliest eukaryotic cells would have been archaezoa (i.e., amitochondriate organisms adapted to an anaerobic lifestyle), such as the extant diplomonads (e.g., *giardia*), parabasalia (e.g., *trichomonas*), and the microsporidia (Cavalier-Smith, 1993). Mitochondria then would have been acquired at a later stage from a bacterial endosymbiont belonging to the group of the α-proteobacteria. Also according to this tree, the Euglenozoa, consisting of trypanosomatids, bodonids, and euglenoids, was the first group to have acquired a mitochondrion via endosymbiosis.

However, a major problem associated with this tree of life is that the trunk of early eukaryotic evolution, as well as many of the protist branches, is much longer than any other part of the tree. This can be explained by a sudden increase in size of the rRNA from the 16S bacterial type to the 18S eukaryotic type in order to accommodate additional proteins when the large ribosomal subunit increased in size from 50S to 60S. It is well known that all tree-construction methods are sensitive to the so-called **long-branch attraction** phenomenon (see Section 5.3), placing long branches preferentially together with the **outgroup** toward the bottom of the tree. This led to the suspicion that the part of the rRNA representing protist evolution may not reflect true evolutionary events. Indeed, during the

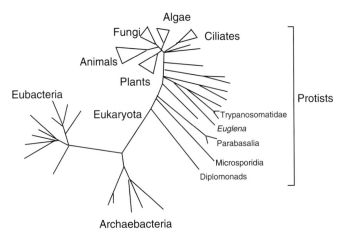

Figure 8.1   Tree of life based on 16S and 18S rRNA sequences. Thus, Archaea may have more similarity to Eubacteria or Eucarya than both of them have to each other, which is in good agreement with the finding that Archaea exhibit a mixture of eucaryal and bacterial traits at the molecular level. This kind of tree often has been called the "universal tree of life." (Modified from Patterson and Sogin, 1992.)

last several years, phylogenetic methods using protein-coding genes have clearly demonstrated that molecular phylogeny based on rRNA often does not delineate phylogenetic relationships between domains or between major lineages of them. Especially in the case of the protists, this renders the use of rRNA sequences highly unsuitable.

In recent years, an increasing number of protein sequences from archaeabacteria, eubacteria, and eukaryotes have become available and it seems that the so-called housekeeping proteins of the eukaryotes have not undergone such a sudden evolutionary drift. Analyses based on these protein sequences suggest an almost simultaneous massive and rapid radiation of protists, algae, fungi, and animals (Keeling and Doolittle, 1996; Philippe and Adoutte, 1998). Much of the new information obtained with these protein sequences also contradicts the original idea that the amitochondriate archaezoa were the first eukaryotes on earth (Germot et al., 1996, 1997; Germot and Philippe, 1999; Philippe and Germot, 2000).

Although rRNAs have been useful in unravelling the phylogenetic relationships of organisms, many questions remain unsolved. Therefore, a valuable alternative for the construction of **phylogenetic trees** is the use of protein-coding sequences; the previous section illustrated that proteins provide interesting answers. In this chapter, it is demonstrated that protein sequences are equally powerful in unravelling the affiliations of organisms, and can do so over long time spans.

8.2 Why protein sequences?

A phylogenetic tree based on phosphoglycerate kinase sequences from all major kingdoms (i.e., animalia, fungi, plantae, protista, eubacteria, and archaebacteria) illustrates the possibility of creating alternative trees of life from protein sequences (Figure 8.2). The following sections develop a number of arguments about why it may be advantageous to use protein-coding rather than nucleic-acid sequences for the construction of phylogenetic trees.

Because it is the DNA that contains all the information to create functional proteins, it is often thought that the DNA should also be used in molecular-evolution studies. However, there are many reasons why it may be more appropriate to use protein sequences for such analyses. The fundamental building blocks of life are proteins. The catalysts of virtually all chemical transformations in the cell are proteins. The functional properties of proteins are determined by the sequence of the 20 amino acids. In many cases, proteins are largely self-folding. For protein-encoding genes, the object on which **natural selection** acts is the protein itself, not the DNA. The underlying DNA sequence reflects this process in combination

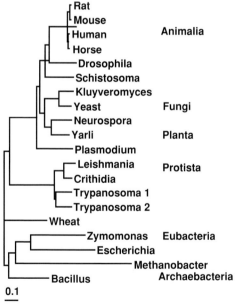

Figure 8.2 The tree was constructed from representative phosphoglycerate kinase protein sequences from all major kingdoms (i.e., animalia, fungi, plantae, protista, eubacteria, and archaebacteria). Note that the wheat sequence represents the chloroplast PGK, which clusters with the bacteria.

with species-specific pressures on DNA sequence, such as the need for thermophiles to have DNA that is resistant to melting or a very high or low GC content. Thus, if function demands that a protein maintain a specific sequence, there still is sufficient room for the DNA sequence to change.

8.2.1 The genetic code

The most important argument for using protein rather than DNA sequences is that the information present in the genes is interpreted only via the intermediacy of the genetic code. Four different nucleotides taken three at a time can result in 64 different possible triplet codes – more than enough to encode 20 amino acids (see Table 1.2). One codon, ATG, representing methionine and also serving as the initiation codon, represents the start of a protein; every other amino acid may be encoded by 1 to 6 different triplet codes. Finally, 3 of the 64 codes, called stop (or termination) codons, specify "end of peptide sequence." Where multiple codons specify the same amino acid, the different codons are used with unequal frequency, depending on the nature of the gene and its level of expression. This distribution of frequency is referred to as *codon usage*, and it varies widely among species.

8.2.2 Codon bias

The amino acids leucine, serine, and arginine are encoded by 6 triplets. Valine, proline, threonine, alanine, and glycine are each encoded by four triplets. Because the base composition of the DNA for organisms may vary, not all organisms have the same codon preference. Yeast, protists, and animals all have different codon preferences; consequently, the same protein sequence in each of these organisms would result in differences in DNA sequence that are more related to codon bias than to evolution. Also, some protists use the codons TAA and TGA to encode glutamin, rather than STOP; in mitochondria, the codon TGA encodes tryptophan, rather than STOP. The inclusion of unique codons in a subset of the sequences tends to make that subset appear more divergent than it really is. Therefore, it may be advantageous to first translate a coding sequence, or *open reading frame* (*ORF*), into its corresponding protein sequence, which results in a peptide sequence in either the one- or three-letter code (see Table 1.1).

8.2.3 Long time horizon

Homologous sequences tend to incorporate mutations, causing them to diverge with time. This makes the two sequences increasingly different from each other as time passes. The chance that a certain position in the DNA incorporates a second mutation, thereby obscuring the first mutational event – or even a back mutation resulting in no observable difference – increases with the total number of mutations incorporated and thus also increases with time. In protein-coding sequences, the first and second positions of each codon are less prone to the incorporation of mutations because this almost always leads to a change in amino acid in the corresponding position of the protein. In most cases, the third position may be mutated without directly affecting the protein. When comparing protein-coding sequences that have diverged for possibly hundreds of millions of years, it is likely that the third position, in the codons have become randomized. By excluding the third position by removing every third nucleotide from the protein-coding sequence (a general technique in phylogenetic analyses), one is actually looking at amino-acid sequences. It may be cumbersome to remove the third positions from the DNA sequences, whereas it is often much easier to simply translate an ORF into its corresponding protein sequence.

8.2.4 Phylogenetic noise reduction

DNA is composed of only four types of unit: A, G, C, and T. If gaps were not allowed, an average of 25% of residues in two randomly chosen aligned sequences would be identical. However, as soon as gaps are allowed, as many as 50% of residues in two randomly chosen aligned sequences can be identical. This situation may obscure any genuine relationship that may exist between two gene sequences, especially when comparing distantly related or rapidly evolving gene sequences. By contrast, the alignment of proteins with their 20 amino acids is less cumbersome. On average, at least 5% of residues in two randomly chosen and aligned sequences would be identical. Even after the introduction of gaps, still only 10 to 15% of residues in two randomly chosen aligned sequences is identical. As a result of the translation of gene sequences into their corresponding protein sequences, the latter are much easier to align. Thus, translation of DNA into 21 different types of codon, 20 amino acids and a terminator allows the signal-to-noise ratio to improve significantly.

8.2.5 Introns and noncoding DNA

When confronted with a DNA sequence, a biologist needs to figure out where the code for a protein starts and stops. This problem is even more difficult because a eukaryotic genome contains significantly more DNA than is needed to encode proteins; the sequence of a random piece of DNA is likely to encode no protein whatsoever. The DNA that encodes proteins is often not continuous, but rather is frequently scattered in separate blocks called *exons* (Figure 8.3). Many of these problems can be reduced by sequencing RNA (via cDNA) rather than DNA itself because the cDNA contains less extraneous material and

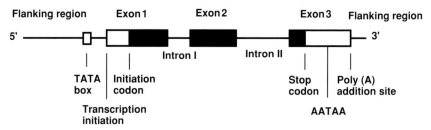

Figure 8.3   Structure of a typical eukaryotic gene with introns and exons.

because the separate exons have been joined in one continuous stretch in the RNA. Hundreds of thousands of these partial sequences are now available in the so-called **expressed sequence tag (EST)** databases being compiled for many organisms.

In general, eukaryotic genes have been fragmented into exons and interspersing **introns.** Due to differences in evolutionary pressure on exons and introns, the rate of incorporation of base substitutions in these two elements of eukaryotic genes may be dramatically different. Therefore, a study of the evolution of a protein using its DNA sequence should only include coding sequences. This requires that in every DNA sequence, all introns are deleted, which may be cumbersome and time-consuming. It may be easier to translate a cDNA into its corresponding protein than to use the genomic DNA sequences.

8.2.6   Multigene families and post-transcriptional editing

Organisms may contain many highly similar genes, whereas only one peptide sequence can be identified (i.e., histones, tubulins, and glyceraldehyde-phosphate dehydrogenase for humans). Using these DNA sequences, it would be difficult to decide which genes are expressed and which are not – and, thus, to decide which genes to include in the analysis.

The DNA sequence does not always translate into amino-acid sequence. The pre-mRNA may require alteration of its coding sequence before it can be translated into a functional protein. This is called **post-transcriptional editing,** in which different mechanisms are known. For example, the RNA editing in the kinetoplastida involves the insertion or deletion of one or more uridines in the pre-mRNA, using guide RNAs as templates. This may lead to frame shifts. Even worse, noncoded initiation codons or entire codons for amino acids may be added or removed during the editing process (Figure 8.4). This may lead to major differences between the DNA and the resulting mature mRNA sequence (Arts and Benne, 1996). In some extreme cases, such as in *Trypanosoma brucei* **mitochondrial DNA,** sometimes more than 50% of a gene is edited; they are called pan-edited genes. In these cases, DNA and mature mRNA are not able to hybridize anymore.

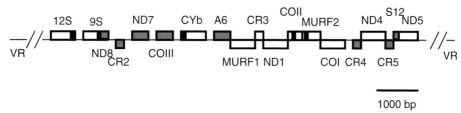

Figure 8.4   Linear map of the 22-kb maxicircle of *T. brucei*. The genes above the line are transcribed from left to right, whereas the genes beneath the line are transcribed from right to left. The 9S and 12S rRNA genes have added uridines at their 3' termini (black boxes). Transcripts from Cyb, COII, and MURF2 have limited internal editing (black boxes). Shaded boxes indicate genes that are extensively or pan-edited. The variable region of the maxicircle is indicated by VR. (Modified from Arts and Benne, 1996.)

8.3   Measurement of sequence divergence in proteins: The PAM

Mutations in the DNA of protein-encoding genes are transmitted to their corresponding proteins. With time, increasingly more mutations are incorporated; consequently, the two descendants of one ancestral protein diverge with time. However, the **observed sequence difference** of two proteins that incorporate mutations is not linear with time, but rather takes the course of a negative exponential (Figure 8.5). This is the result not only of the fact that each position is subject to the incorporation of mutations, but also to back mutations and multiple hits (see Chapter 4). These events increase in number as the **evolutionary distance** between two **homologous proteins** increases, which leads to an underestimation of evolutionary

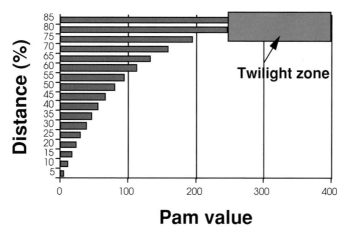

Figure 8.5   When the PAM distance value between two distantly related proteins nears the value 250, it becomes difficult to tell whether the two proteins are homologous or that they are two randomly taken proteins that can be aligned by chance. That case is known as the "twilight zone." (Modified from Doolittle, 1987.)

Table 8.1 Observed versus evolutionary distances expressed in PAM (accepted point mutations per 100 amino acids) between proteins

Observed percentage difference	Evolutionary distance in PAM
1	1
5	5
10	11
15	17
20	23
25	30
30	38
35	47
40	56
45	67
50	80
55	94
60	112
65	133
70	159
75	195
80	246
85	328 <— Twilight zone

As the evolutionary distance increases, the probability of superimposed mutation becomes greater, resulting in a lower observed percent difference. (Adapted from Table 23 in Dayhoff, 1978.)

distances between them. Consequently, the observed percentage of difference between two protein sequences is not proportional to their actual evolutionary difference (Table 8.1) (see Chapter 4). A measure that is proportional to the true *evolutionary distance* between two proteins is the **PAM** value (Dayhoff, 1978), which is the number of **accepted point mutations per 100 amino acids**. Examples of PAM values and their corresponding observed distances are provided in Table 8.1. Two homologous proteins that have a common ancestor and a PAM distance of 250–300 (80–85% distance) or more cannot be distinguished from two randomly chosen and aligned proteins of similar length. Therefore, in general, phylogenetic analyses using protein sequences must be limited to proteins that are less than 80–85% different.

Proteins with functions that are less essential to the organism rapidly evolve. As a result, evolutionary information is quickly erased and these proteins can be used only for the study of closely related organisms. However, housekeeping proteins, such as histones, enzymes of core metabolism, and proteins of the cytoskeleton,

Table 8.2 Protein evolutionary rates

Type of protein	Rate of change (PAMs/100 YA)	Theoretical look-back time (in MYA)
Pseudogenes	400	45
Fibrinopeptides	90	200
Ig lambda chain C	27	670
Somatotropin	25	800
Ribonucleases	21	850
Hemoglobin alpha chain	12	1,500
Acid proteases	8	2,300
Cytochrome c	4	5,000
Adenylate kinase	3.2	6,000
Glyceraldehyde P dehydrogenase	2	9,000
Glutamate dehydrogenase	0.9	18,000

Useful look-back time = 360 PAMs (Adapted from Table 1 in Dayhoff, 1978.)
YA = years ago
MYA = million years ago

evolve slowly and incorporate between 1 to 10 mutations per 100 residues and per 100 million years (Table 8.2). Therefore, it takes considerable time for these proteins to incorporate sufficient substitutions before all evolutionary information has been erased. Because of this slow rate of evolution, housekeeping proteins are excellent tools to trace evolutionary relationships over long periods. For instance, the slow mutation rate of the enzyme glutamate dehydrogenase provides a theoretical look-back window of several times the age of our solar system. Thus, proteins can be excellent tools to closely study the evolutionary relationships of both, as well as distantly related *taxa*.

8.4 Alignment of protein sequences

To study molecular evolution of proteins, the sequences of homologous proteins must be compared; however, this is not easy to do because homologous proteins are never identical, only similar. In addition to having substitutions, there are insertions and deletions in one sequence relative to the other. Moreover, phylogenetic analyses should be carried out on residues with positional homology (see Section 3.15); therefore, it is essential that two or more sequences be properly aligned relative to each other.

Several excellent programs have been developed for the alignment of protein sequences and are available in the public domain, such as ClustalX (see Chapter 3). Alignment programs for two or more sequences can be accessed directly

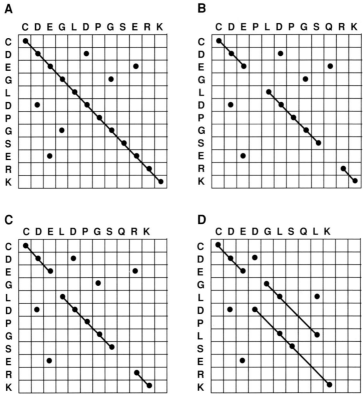

Figure 8.6 Dot matrixes for sets of more or less related sequence pairs. Each identity between two positions leads to a positive score value of 1; a mismatch results in no score. Gaps in the alignment indicated as discontinuities along the diagonal are scored as penalties or negative scores. (A) Two identical sequences, identity score = 12; (B) two sequences with two mismatches, score = 10; (C) two sequences with a gap of one position plus a mismatch, score = 10, gap penalty = −1; and (D) two sequences with both gaps and mismatches. In A, B, and C, there is no ambiguity about how to align the two sequences. In example D, there are two possible ways to align the two sequences. For the upper diagonal, the identity score = 6 with one gap; for the lower diagonal, the score = 7 with two gaps.

on the Internet and multiple alignments created (see the Practice section of this chapter). Most of these programs implement the ***dot-matrix method*** (see Section 3.6.1), in which the two sequences to be aligned are written out as column and row headings of a matrix (Figure 8.6) (see Figures 3.6 and 3.8). Dots are put in the matrix wherever the residues in the two sequences are identical. If the two sequences are identical, dots will be in all the major diagonal cells of the matrix. If the two sequences are different but can be aligned without gaps, dots will be in most of the major diagonal cells. If a gap occurred in one of the sequences, the alignment diagonal will shift vertically or horizontally. Figure 8.7 shows a dot matrix for two

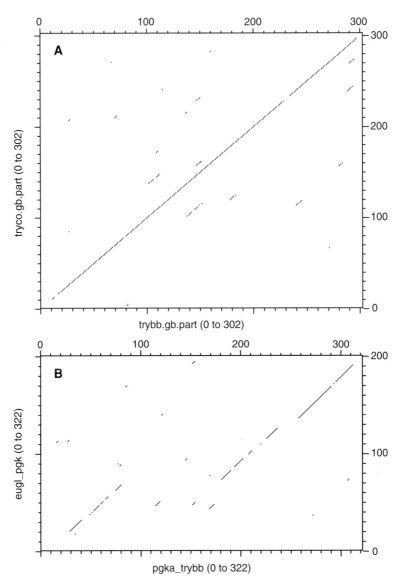

Figure 8.7 (A) Dot matrix of two clearly homologous phosphoglycerate-kinase sequences from *Trypanosoma brucei* and from *T. congolense*. The two sequences are 81% identical. (B) Dot matrix of two less similar (50% identical) phosphoglycerate-kinase sequences from *Euglena* and of *Trypanosoma*. The latter has an 80-amino-acid-long insertion with respect to a partial *Euglena* sequence.

highly similar trypanosome phosphoglycerate kinases and two more distantly related phosphoglycerate-kinase sequences. It is obvious that there is no problem in aligning two sequences as long as they are of similar length and have more than 50% identity. These plots can be obtained from the BioEdit program, which uses amino acids as input data, and by following the procedure in Section 3.10 for nucleotide sequences.

In principle, computer algorithms developed for the alignment of two homologous sequences use the same procedure as the dot-matrix method. Each residue of one sequence is compared with each residue of the other; when there is an identity, a certain value is given to that position in the matrix. Then a diagonal line is drawn connecting the points with the highest score. Horizontal or vertical shifts from the diagonal due to the presence of gaps are penalized (see Chapter 3). The choice of the value for each positive score relative to the gap penalty strongly influences the quality of the resulting alignment. Too low a gap penalty value leads to a situation in which dissimilar regions are not aligned with each other at all, but rather with gap regions, whereas a too high gap penalty leads to alignment of nonhomologous residues or regions. Most programs allow the user to select an appropriate weight matrix to score either identities or similarities of amino acids for adjustment of the gap penalty value and the size of the window that scans the diagonal.

Many different weight matrixes have been developed for use with sequence-alignment programs to reflect observed rules of mutation, including the *identity matrix*, *mutation-cost matrix*, *hydrophobicity matrix*, *log odds matrix*, *PAM250 matrix*, *BLOSUM 62* (Block Substitution) *matrix*, *JTT matrix*, and *Gonnet matrix*. What type of weight matrix chosen depends on the scientific judgment of the user and the data set being analyzed. The identity matrix has only the value of one for identities and zeros for mutations. The mutation-cost matrix scores the minimum number of base changes required to convert one amino acid to another; it contains only the values 0, 1, 2, and 3. The hydrophobicity matrix considers the physicochemical properties of the amino acids, assuming that replacement of one hydrophobic amino acid by another is a more likely event than by a hydrophilic amino acid. Log-odds matrixes use the log-odds ratio S of the probability that two residues, i and j, are aligned by evolutionary descent, and the probability that they are aligned by chance [$S_{ij} = \log(q_{ij}/(p_i \times p_j))$], where [$q_{ij}$] are the frequencies that residue i and j are observed to align in sequences known to be related, and [p_i] and [p_j] are the frequencies of occurrence of residue i and j in the set of sequences. The PAM250 (Dayhoff et al., 1978) and BLOSUM62 (Henikoff and Henikoff, 1992) are log-odds matrixes in which the two probabilities were determined from a limited and a large database, respectively, of aligned homologous protein sequences. The latter two are implemented in the `ClustalX` program. The JTT matrix (Jones et al., 1992) is an update to the PAM matrix. The Gonnet matrix (Gonnet et al., 1992) is a scoring matrix based on the alignment of the entire SWISS-PROT database in which 1.7×10^6 matches were used from sequences differing by

6.4 to 100 PAM; it is implemented in the Darwin program by the same authors. The PAM and BLOSUM62 matrixes are now widely used in alignment
programs.

Several algorithms for the alignment of protein sequences have been developed,
including the better known Pearson-Lipman algorithm (Pearson and Lipman,
1988), used in Pearson's well-known FASTA program; the Needleman-Wunsch
algorithm (1970); the Smith-Waterman algorithm (1981); and the BLAST algorithms (Altschul et al., 1990). They are used in sequence-comparison programs
to search for homologous sequences in large databases and in programs used for
multiple-sequence alignment.

8.4.1 Sequence retrieval and multiple-sequence alignment

To create a multiple alignment of homologous protein sequences, all related sequences from a database have to be collected. Various databases are available;
the SWISS-PROT database (see Chapter 2) is the most reliable source for the
collection of sequences. In this database, each sequence has been checked by
experts and extensively annotated. Moreover, homologous sequences from different organisms have highly similar locus names, which facilitates the recognition and retrieval of all related sequences. The Protein Identification Resource
(PIR) (**http://www-nbrf.georgetown.edu/**) database contains many more protein
sequences; however, they have not all been checked and the database is redundant, which is also the case with the GenPept (i.e., translated GenBank sequences)
and the TrEMBL (i.e., translated EMBL sequences) databases. However, to ensure that no homologous sequences to query protein – even an incomplete sequence or a pseudogene – is missed, these databases should be included in the
search. The Enzyme or EC database (**http://www.expasy.ch/enzyme/**) is a useful tool to simultaneously retrieve all homologous sequences of one specific
enzyme.

8.4.2 Secondary structure-based alignment

PredictProtein   (**http://www.embl-heidelberg.de/predictprotein/predictprotein.
html**) is a program that predicts the secondary structure of a protein, using
information on the mutability of each residue in a multiple-sequence alignment.
It first aligns the query sequence with all homologous sequences available in the
SWISS-PROT database, which is an easy way to collect all homologous sequences
in the database and to create a multiple-sequence alignment that also includes the
query sequence. The Darwin server (**http://cbrg.inf.ethz.ch/Darwin/index.html**)
also aligns the sequence with homologous sequences in the SWISS-PROT database
(see Chapter 2).

The Match-Box algorithm (**http://www.fundp.ac.be/sciences/biologie/bms/matchbox_submit.html**) enables the simultaneous alignment of several protein sequences, in which each aligned position is weighted by a reliability score, using strict statistical criteria. This method circumvents the gap-penalty requirement. Gaps are the result of the alignment and not a governing parameter of the matching procedure. A reliability score is provided below each aligned position. The Match-Box program is particularly suitable for finding and aligning conserved structural motifs within distantly related proteins. It is advisable to try more than one program and to remember that most multiple-sequence alignment programs work best with sequences of similar length.

8.4.3   Prodom, Pfam, and Blocks databases

Another way to create an alignment is to first obtain multiple-sequence alignments from a database to which the user's own sequence can be added. There are three databases of prealigned sequences that share domain structures or homologous blocks of sequences: Prodom, Pfam, and Blocks; are all accessible via the Internet. They have been compiled by comparing and aligning all homologous sequences of the SWISS-PROT database. The Prodom database (**http://www.toulouse.inra.fr/prodom.html**) consists of an automatic compilation of *homologous protein domains*. The Pfam database (**http://www.sanger.ac.uk/Software/Pfam/**) is actually formed in two separate ways. Pfam-A are accurate human-crafted multiple alignments, whereas Pfam-B is an automatic clustering of the rest of SWISS-PROT and TrEMBL. The Blocks database (**http://www.blocks.fhcrc.org/**), compiled using the BLAST algorithm, contains multiply aligned ungapped segments corresponding to the most highly conserved regions of proteins. In general, the compiled sequences in Blocks are shorter than in the other two databases. Prodom, Pfam, and Blocks alignments serve as an excellent basis to start a multiple-sequence alignment and/or phylogeny project for proteins.

8.4.4   Manual adjustment of a protein alignment

An automatically produced multiple-sequences alignment often needs manual adjustment to improve its quality, especially at the position of gaps, which can be achieved by using all the knowledge that is available about a protein. Information about active site residues and elements of secondary structure, such as α-helixes, β-strands, and loops, may be of great help here. When manually adjusting multiple alignments, the user should have knowledge of the physicochemical properties of the 20 amino acids and the rules-of-thumb for the mutability of the various amino acids. In a folded protein, the residues D, R, E, N, and K are preferably mutated to residues of similar properties. Because they are polar or charged, they are mainly found on the surface of the folded protein; moreover, because they play a lesser role in protein-folding, they mutate rather easily. Hydrophobic residues (F, A, M, I, L, Y, V, and W) are preferentially replaced by other hydrophobic residues. They are mainly internal, determine the folding of the protein, and mutate rather slowly. The residues C, H, Q, S, and T are generally indifferent and may be replaced with any other type of residue. The residues (D, R, E, N, K, C, H, Q, S, and T), when conserved throughout the alignment, are likely to be involved in the active site of an enzyme. Therefore, the multiple alignment should be adjusted in such a way as to maintain these residues aligned. Periodicity of charged residues may provide information about the presence of elements of secondary structure, such as α-helixes and β-strands. α-Helixes have a repetition of 3.6 residues per turn. Stretches of more than 12 amino acids with a charged amino acid every 3rd or 4th position in the sequence may indicate the presence of a α-helix. Short stretches with a repetition of charged amino acids every 2nd residue may indicate a α-strand structure. Gaps are almost never found in elements of secondary structure, only in regions with loops. Moreover, the residue P interferes with secondary-structure elements and the residue G allows the polypeptide backbone to make turns. Thus they have a preference for loop regions. Becauses loops easily acquire or lose residues, gaps should always be aligned together with P and G residues. In general, hydrophobicity (or hydropathy) profiles according to Kyte and Doolittle (1982) of two homologous proteins are strikingly similar (Figure 8.8) and also may help in manually adjusting the alignment.

In the case of a crystal structure with at least one enzyme in the alignment, a useful tool for manually aligning the proteins is the NLR-3D database (**http://www-nbrf.georgetown.edu/pirwww/dbinfo/nrl3d.html**), which contains protein sequences plus secondary-structure information about which residues of a protein sequence belong to conserved areas, such as α-helixes and β-strands. This information, if available, is also provided by the SWISS-PROT database.

8.5　Tree-building methods for protein phylogeny

Once a multiple-sequence alignment has been prepared, it may serve further evolutionary analyses. The final goal of such an analysis is to prepare an evolutionary tree describing the relationship of the various sequences with respect to each other. Two main tree-building algorithms can be used for protein sequences: *distance methods*, based on a matrix containing pairwise distance values between all sequences in the alignment, such as *UPGMA*, *Neighbor-Joining* (*NJ*), and *Fitch and Margoliash* (see Chapter 5); and *character-based methods*, which carry out calculations on individual residues of the sequences, such as *maximum likelihood* and *maximum*

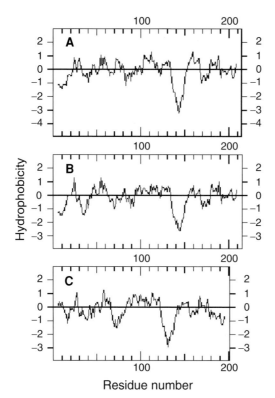

Figure 8.8   In general, hydrophobicity (or hydropathy) profiles according to Kyte and Doolittle (1982) of homologous proteins are strikingly similar and may provide a tool in the alignment of two or more proteins. The three phosphoglycerate-kinase sequences have respectively 81% (*T. brucei* (A) vs. *T. congolense* (B)) and 50% (*E. gracilis* (C) vs. *T. brucei* (A)) identity to each other.

parsimony (see Chapters 6 and 7). In general, character-based methods are much slower for proteins than for nucleotides, because they are central processing unit (CPU) intensive.

Distance-based tree-building algorithms for proteins are identical to the ones described in Chapter 5 for nucleotide sequences. Fitch and Margoliash (Fitch, 1981) and the NJ methods (Saitou and Nei, 1987) both assume an evolutionary model for the transformation of the **observed distances** to evolutionary distances. With proteins, however, pairwise distances have to be calculated using one of the corrected distance matrixes described in Section 8.4. An alternative is to use the approximate formula developed by Kimura (1983) to estimate the **genetic distance** d from the observed distance p:

$$d = -ln(1 - p - 0.2\,p^2) \tag{8.1}$$

where p is the proportion of observed differences between any two homologous sequences. This formula is based on statistical considerations (see Kimura, 1983) and is usually reliable for $p \leq 0.7$ (Nei, 1985). However, because it does not consider which amino acids differ or into which amino acids they change, some information is lost. Genetic distances, according to Equation 8.1 or the PAM matrix, can be calculated with the `Protdist.exe` program implemented in the PHYLIP package. Protein distance estimates can be obtained with the TREE–PUZZLE program as well, which also implements the BLOSUM62 and the JTT matrix.

The maximum-parsimony method does not assume an explicit evolutionary model. This method allows many equally likely trees to be found. Programs available for maximum-parsimony analysis include `Protpars.exe` implemented in PHYLIP (Felsenstein, 1993), and PAUP* (Swofford, 1998).

Maximum-likelihood–based methods evaluate a hypothesis about evolutionary history in terms of the probability that the proposed model and the hypothesized history would give rise to the observed data set (see Chapter 6). The supposition is that a history with a higher probability of reaching the observed state is preferred to one with a lower probability. The method searches for the tree with the highest probability or likelihood. The available programs to analyze protein sequences using maximum likelihood are `ProtML.exe` of the MOLPHY package (Adachi and Hasegawa, 1992) and TREE–PUZZLE (Strimmer and von Haeseler, 1997). The latter program applies a heuristic method and is much faster than the former, but does not necessarily guarantee to find the best tree (see Chapter 7).

Most methods for the inference of phylogeny yield trees that are **unrooted**. Thus, from a tree by itself, it is impossible to tell which of the **operational taxonomic units** (**OTUs**) branched off before all the others. To root a tree, an outgroup should be added to the data set, which is an OTU for which external information (e.g., paleontological information) is available that indicates that the outgroup branched off before all other taxa (see Section 1.4).

An interesting way to root a tree is to exploit an event of **gene duplication** that occurred before speciation that led to the formation of isoenzymes or homologous proteins, which then can be used as an outgroup. Examples are the gene duplications leading to the various vacuolar ATPases present in archaebacteria, eubacteria, and eukaryotes, which allowed the tree of life rooting (Gogarten et al., 1989).

Tree topologies may strongly depend on the following factors: whether distance or parsimony methods were applied; the number of OTUs included in the alignment; the order of the OTUs in the alignment; the selection of an appropriate outgroup; and, finally, the presence of widely varying branch lengths. Therefore, none of the methods may guarantee the one tree with the correct topology. To be aware about the reliability of the topology of the resulting tree, one or all of the following should be done:

1. Apply different tree-building methods to the data set.
2. Vary the parameters used by the different programs, such as the seed value and jumble factor for the order of OTU addition, or change the order of the sequences manually.
3. When in doubt, apply various ***evolutionary models*** for matrix construction.
4. Add or remove one or more OTUs to see the influence on tree topology.
5. Include an outgroup that may serve as a root for the tree.
6. Be aware of taxa with long branches that may be subject to the so-called long-branch attraction effect, which may lead to an anomalous positioning of that taxon.
7. Apply ***bootstrap*** or ***jackknife analyses*** to the data set and prepare a consensus tree of 200–1,000 replicas, depending on size of the data set and on computer power.
8. Remember that with bootstrap analysis, only nodes that occur in more than 90% of the cases are reliable.

Finally, a tree should be considered robust and thus reliable only when widely different methods infer similar or identical tree topologies, and such topologies are supported by good bootstrap values (i.e., more than 95%).

8.6 Some good advice

It is recommended to analyze both DNA and protein data sets. For a group of closely related species or taxa, such as viral proteins or vertebrate enzymes, DNA-based analysis is probably a good method because problems like differences in codon bias or ***saturation*** of the third position of codons can be avoided (see Section 4.10). It is nevertheless strongly recommended to analyze the protein sequence data as well. In case of ambiguity in the alignment of gene sequences, it is recommended

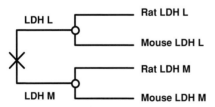

Figure 8.9 The presence of paralogous and orthologous genes within a phylogenetic analysis. ○ indicates speciation and X indicates gene duplication. The mouse isoenzymes L and M are said to be paralogous, whereas the rat and mouse isoenzymes L are orthologous.

to first translate the sequences to their corresponding protein sequences, then align and determine the position of gaps in the DNA sequences according to the protein alignment (see Chapter 3). Multigene families (e.g., genes coding for different but similar isoenzymes) may cause problems and the decision to exclude or include such sequences must be made carefully, which may result in ***paralogous*** sequences in the data set and peculiar-looking phylogenetic trees (Figure 8.9).

PRACTICE

Fred R. Opperdoes

8.7 A phylogenetic analysis of the Leismanial GPD gene carried out via the Internet

This is a practical exercise for the creation of a phylogenetic tree carried out entirely via the World Wide Web. The analysis deals with an ORF coding for the *Leishmania mexicana* NAD-dependent glycerol-3-phosphate dehydrogenase (GPD, EC.1.1.1.8), an enzyme which, in the haemoflagellated protozoan parasite *L. mexicana*, is associated with microbodies (Kohl et al., 1996), and for which the crystal structure was recently solved (Suresh et al., 2000). The following analysis is conducted with no other tools than a computer with access to the World Wide Web and a word-processing program that can cut and paste text files (all files should be saved as "text only" files):

1. Consult the Enzyme database and collect all available protein sequences of a specific enzyme in one step.
2. Search the Brookhaven Protein database for a three-dimensional (3-D) structure.
3. Scan a DNA sequence against GenBank using the Basic Local Alignment Sequence Tool (BLAST) algorithm.
4. Scan a DNA sequence for the presence of ORFs.
5. Translate an ORF into a protein sequence.
6. Do a BLAST homology search using a protein sequence.
7. Create a multiple alignment of homologous protein sequences.
8. Create a publishable alignment figure.
9. Create a phylogenetic tree using the maximum-likelihood method implemented in TREE–PUZZLE.

Before starting the exercise it is necessary to collect general information about the enzyme under study. Connect to the Enzyme, or EC, database, available at the Expasy server in Switzerland (**http://www.expasy.ch/enzyme/**), which contains general information about enzymes, their official names and EC numbers, the reactions they catalyze, and the pathways in which they are involved. It also provides access to all protein sequences available in the SWISS-PROT database. In the Enzyme database, select the enzyme by its EC number 1.1.1.8 to get the requested information, and review all the information available in the database. Also try to find publications in the Medline database about the *Leishmania* enzyme. Retrieve for future use all SWISS-PROT entries referenced on this page using the file transfer protocol (FTP). Name this file "`gpd_sw.pep`".

If a 3-D structure of GPD is already available, it can be found in the Brookhaven Protein Database (PDB) (**http://www.rcsb.org/pdb/**) under the accession codes 1EVY and 1EVZ.

The gene for the *Leishmania* GPD and its flanking nucleotides is available from the GenBank database at the National Center for Biotechnology Information (NCBI) at the National Institutes of Health (NIH) in Bethesda, Maryland (USA), (**http://www.ncbi.nlm.nih.gov**) under the accession number X89739. Retrieve and save the sequence; the ORF also is available from our Web site (**http://www. kuleuven.ac.be/aidslab/phylogenybook.htm**) as "gpd_orf.nuc".

Scan the DNA sequence for the presence of ORFs. Submit the entire nucleotide sequence to the NCBI ORF scanner (**http://www.ncbi.nlm.nih.gov/gorf/gorf.html**) to find all possible coding regions in all 6 ORFs. The longest ORF is found starting at frame +1 from position 643 .. 1, 743, with a length of 1,101 nucleotides or 366 amino acids. The sequence ends with -SKL, a typical microbody targeting sequence.

To determine whether the *Leishmania* ORF has any homology to other nucleic-acid sequences in the GenBank database, perform a BLAST search using the NCBI server (**http://www.ncbi.nlm.nih.gov/BLAST/**) (see Chapter 2). Paste the entire ORF sequence into the BLAST sequence submission window. Use all standard settings and select the complete GenBank nucleotide database (nt). After submitting the sequence, it takes a few minutes for the results.

From the information obtained, it can be concluded that the *Leishmania* ORF codes for a genuine GPD; however, it is intriguing that the highest degree of identity is reported with bacterial rather than eukaryotic sequences.

To improve the sensitivity of the BLAST search, the longest ORF is translated into its corresponding protein sequence. For the translation, access the Protein Machine utility at the European Biotechnology Institute (EBI) (**http://www.ebi.ac. uk/translate/**). Paste the nucleotide sequence comprising the ORF into the sequence window of the Protein Machine and submit it. After a few seconds, the server returns the following output:

```
MSTKQHSAKDELLYLNKAVVFGSGAFGTALAMVLSKKCREVCVW
HMNEEEVRLVNEKRENVLFLKGVQLASNITFTSDVEKAYNGAEI
ILFVIPTQFLRGFFEKSGGNLIAYAKEKQVPVLVCTKGIERSTL
KFPAEIIGEFLPSPLLSVLAGPSFAIEVATGVFTCVSIASADIN
VARRLQRIMSTGDRSFVCWATTDTVGCEVASAVKNVLAIGSGVA
NGLGMGLNARAALIMRGLLEIRDLTAALGGDGSAVFGLAGLGDL
QLTCSSELSRNFTVGKKLGKGLPIEEIQRTSKAVAEGVATADPL
MRLAKQLKVKMPLCHQIYEIVYKKKNPRDALADLLSCGLQDEGL
PPLFKRSASTPSKL
```

A BLAST search of a protein sequence has a much better signal-to-noise ratio than a corresponding search of a nucleic acid. Therefore, use the *Leishmania* protein as query sequence in a protein BLAST search. In this way, it should be possible to find all glycerol-3-phosphate dehydrogenase sequences in the protein data bank. Although the translated GenPep database of GenBank could be used, it is better to search the nonredundant SWISS-PROT database. It is a derived database that has been checked thoroughly by protein scientists for redundancy and accuracy of included sequence information; moreover, this database is extensively annotated (see Chapter 2). Paste the sequence into the BLAST sequence submission window, then select the options BLASTp and the SWISS-PROT database; the BLAST output should be returned within a few seconds. Because of the higher signal-to-noise ratio of a protein search, only glycerol-3-phosphate dehydrogenases are reported at the top of the output.

It is puzzling that in both BLAST searches, bacterial sequences score better than eukaryotic GPD sequences when the eukaryotic protist GPD sequence of *Leishmania* is used as the query sequence. Therefore, a multiple alignment of all available sequences may be better suited to study the evolutionary relationship of the Leishmanial enzyme with other glycerol-3-phosphate dehydrogenases.

The file gpd_sw.pep obtained from the Expasy server by FTP contains all the GPD sequences also reported in the last BLAST search. However, each entry in the file contains a lot of text information not required for the analyses. Moreover, before this file can be used for the next step, the sequences need to be reformatted in FASTA (see Box 2.2). However, there is a useful format-conversion utility on the Web called Readseq, which automatically reformats the sequences. Connect to the Readseq server at the NIH (**http://bimas.dcrt.nih.gov/molbio/readseq/**) and select the FASTA format as the output file. The entire contents of the gpd_sw.pep file is pasted into the sequence window and then submitted. The result, which is immediately returned, can be copied into a text document and saved on disk under the name gpd.fasta. At the top of this file, add the *Leishmania* protein GPD sequence in exactly the same format as for the other sequences; save the file again under the same name. This file will be used as the input file for the construction of a multiple-sequence alignment.

Contact the ClustalW server at Baylor College in Houston, Texas (USA) (**http://cbrg.inf.ethz.ch/Server/MultiAlign.html**) and paste the contents of the gpd.fasta file in FASTA format into the sequence window. Submit the data to the server using all default settings. After a few minutes, the result is returned, consisting of two parts: an alignment in GCG/MSF format and the same alignment in FASTA format. Copy and paste the second alignment into a text file and save it as gpd_aligned.fasta.

To prepare a figure of publishable quality for this alignment, go to the BOXSHADE server at the Swiss EMBL node (**http://www.ch.embnet.org/software/BOX_form.html**). After submitting the sequence alignment in FASTA format, the server allows the user to print the result in various ways, depending on the specific requirements.

Using the `Readseq` server at the NIH, it is better to convert the `ClustalW` alignment to a GCG/MSF format, which is well suited for manual editing in a word processor. The alignment in GCG/MSF format can be saved as `gpd_aligned.msf`.

If any obvious alignment errors are observed, the next step is to manually adjust the alignment with a sequence editor or word-processing program (see Chapter 3). If a 3-D structure is already available, the structural information also can be used to improve the alignment. The edited alignment must be saved as a "text only" file under the name `gpd_corr.msf`, and then pasted into `Readseq` again for reformatting in the PHYLIP format (see Box 2.2). Save the result in a file named `gpd_corr.phy`.

The phylogenetic tree will be constructed using the online TREE–PUZZLE program (see Chapters 4, 5, and 6) available on the French "Institut Pasteur" server (**http://bioweb.pasteur.fr/seqanal/interfaces/Puzzle-simple.html**). Submit the alignment in PHYLIP format to the server (or run TREE–PUZZLE in the local computer). The results will arrive via email, which may take a few minutes because the maximum-likelihood method used by TREE–PUZZLE – although a heuristic method – is computationally demanding (see Chapter 6). All results are returned by email, including the Web address where they can be accessed directly on the server. This URL also allows the preparation and printing of a nicely presented figure of the tree file using the program `Drawtree` (Figure 8.10). Internal branch-point labels containing puzzle frequencies or bootstrap values cannot be shown in `Drawtree`. However, this information is provided in both the `outfile` and the `treefile` of TREE–PUZZLE and can be used by other tree-drawing programs such as `TreeView` (see Chapter 5).

From the topology of the tree, it is immediately obvious that the GPD of the protist *Leishmania* clusters with that of *Trypanosoma brucei* (GPD_TRYBR), another member of the family Trypanosomatidae. Apparently, their GPDs are more related to the bacterial GPDs than to any of their eukaryotic homologues. The clustering of the trypanosomatids with the bacteria is robust with a puzzle frequency of 96%, confirmed by a careful review of the gap positions in the *Leishmania* and *Trypanosoma* sequences in the multiple alignment. Apparently, horizontal gene transfer may have occurred. The availability of more bacterial sequences is necessary before a written explanation of this observation is possible.

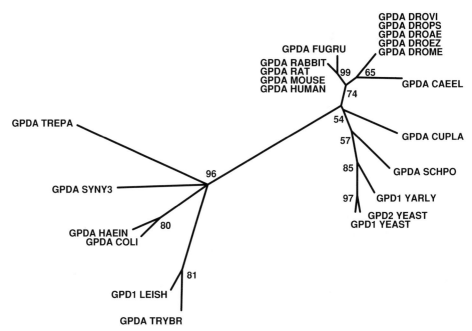

Figure 8.10 Maximum-likelihood tree of glycerol-3-phosphate dehydrogenase sequences available in the SWISS-PROT database. The tree was created using publicly available Web tools. `Tree-Puzzle` frequencies at the branch points were added manually.

8.8 A comparison of the trypanosomatid phylogeny from nucleotide and protein sequences

This exercise deals with the phylogenetic relationship of Trypanosomatidae, parasitic flagellated protists. As long as a limited number of taxa were included in the 18S rRNA tree, members of the genus *Trypanosoma* behaved as a ***paraphyletic*** group of organisms that seemed to have separated from the main tree of trypanosomatid evolution much earlier than other members of the same family (Fernandes et al., 1993). When more *Trypanosoma* rRNA sequences became available and were subsequently added to the tree, the trypanosomes became a ***monophyletic*** group of taxa (Stevens and Gibson, 1999) and joined the other Trypanosomatidae, as had been predicted from protein data (Wiemer et al., 1995; Hannaert et al., 1998). The paraphyly of trypanosomes, as originally reported, apparently was the result of the long-branch attraction phenomenon. When the long branches were transformed into smaller ones by the addition of more taxa, this artifact disappeared. Thus, even within groups of related organisms, differences in the ***evolutionary rate*** of rRNA may strongly influence tree topologies.

In this exercise, the phylogeny of the Trypanosomatidae, as inferred from both the gene sequence and the corresponding protein sequence of the enzyme

Table 8.3 Accession numbers for the
Trypanosomatidae data set

Species	Accession number
Trypanosoma brucei gambiense	AF047499
Trypanosoma congolense	AF047498
Trypanosoma vivax	AF047500
Leishmania major	AF047497
Phytomonas sp.	AF047496
Herpetomonas samuelpessoai	AF047494
Leptomonas seymouri	AF047495
Crithidia fasciculata	AF047493
Trypanosoma brucei	X59955
Trypanosoma cruzi	X52898
Leishmania mexicana	X65226
Trypanoplasma borreli	X74535
Euglena gracilis	L39772

glyceraldehyde-phosphate dehydrogenase, are compared. All nucleotide and protein sequences, either partial cDNAs or complete gene sequences, are available in
the GenBank database (Table 8.3). Retrieve the sequences from the NCBI server at
http://www.ncbi.nlm.nih.gov/ and create two new files entitled `gapdh.pep` and
`gapdh.nuc`. Following the same steps outlined in the preceding section, align
both data sets with the `ClustalW` server or use `ClustalX` or DAMBE in the local
computer (see Chapter 2). Save the edited alignments (gaps excluded) in PHYLIP
format as `gapdh_corr.phy` and `gapdh_nuc_corr.phy` for amino acids and
nucleotides, respectively.

The two files in PHYLIP format with peptide and nucleic-acid sequence alignments are used for phylogenetic analyses using the French phylogeny server at
the Pasteur Institute; all the phylogeny-inference programs can be accessed from
the directory **http://bioweb.pasteur.fr/seqanal/interfaces**. First, a bootstrapped NJ
analysis of the peptide sequences is conducted. In the Web-site directory, select
`protdist.html` and paste the contents of the file `gapdh_corr.phy` into the
sequence window, or browse to locate the file. Choose **K80** as the evolutionary
model for a quick analysis, check the `multiple datasets` box, and take 200
analyses. Then launch the `Protdist` program, which calculates the distances between all pairwise protein sequences and creates a PAM distance matrix. This also
can be accomplished on the local computer with the `Protdist.exe` program in
the PHYLIP package (see Chapter 4 and Box 4.1). The **distance matrix** is used by
the program `Neighbor.exe` (via either the Web server or the local computer) to
create an NJ tree (select the option for analyzing multiple data sets). The 200

different tree files thus obtained are fed into the program `Consense.exe`, which creates a **consensus tree** and calculates the **bootstrap values**. The final tree can be displayed in the usual way by `TreeView` (see Chaper 5); print the tree and compare it to Figure 8.11A.

Create a bootstrapped NJ tree for the `gapdh_nuc_corr.phy` data file by repeating the entire procedure using `DNAdist.exe` to calculate genetic distances. Print the tree and compare it to Figure 8.11B.

T. borreli and *E. gracilis* do not belong to the Trypanosomatidae, but together they belong to the Euglenozoa, to which the Trypanosomatidae also belong. To root the trees, they need to be selected as outgroups in `TreeView`. The two trees obtained by NJ analysis are similar, as are the two ML trees shown in Figure 8.11. When using the same data sets to also conduct a maximum-parsimony analysis, all methods generate essentially the same result. The only difference in the two maximum-likelihood trees in Figure 8.11 is that *H. samuelpessoai* is paraphyletic in the nucleic-acid tree, whereas it forms a monophyletic group with *Phytomonas*

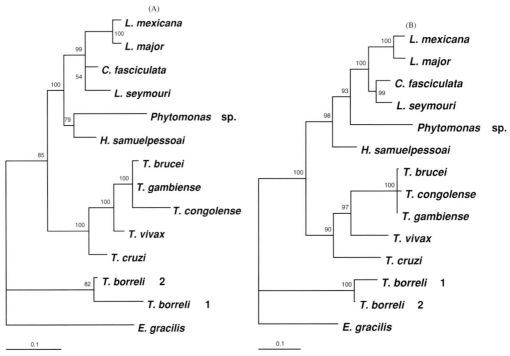

Figure 8.11 Comparison of maximum-likelihood trees of glyceraldehyde-3-phosphate dehydrogenase sequences made from (A) partial protein sequences and (B) partial gene sequences. The tree was created using publicly available Web tools. `Tree-Puzzle` frequencies at the branch points were added manually. The horizontal bar represents 10 mutations per 100 residues.

in the protein tree. However, this part of the tree is not well supported by high bootstrap values. The analyses show that first, protein- and nucleic-acid-based trees have the same or similar results; and second, all Trypanosomatidae form a monophyletic group. Also within this group, the genus *Trypanosoma* is by itself monophyletic. While rRNA trees suffer from unequal rates of evolution resulting in the long-branch attraction artifact, trees based on housekeeping proteins – whether made with protein or nucleic-acid sequences – suffer less or not at all from this problem.

8.9   Implementing different evolutionary models with DAMBE and TREE-PUZZLE

Most of the weight matrixes described in the previous sections (especially Section 8.4) are implemented in DAMBE and TREE-PUZZLE. The details can be found in the TREE-PUZZLE documentation (file `manual.html` in the TREE-PUZZLE doc subfolder and in the DAMBE help menu (Xia, 2000).

REFERENCES

Adachi, J. and M. Hasegawa (1992). Amino-acid substitution of proteins coded for in mitochondrial DNA during mammalian evolution. *Japanese Journal of Genetics*, 67, 187–197.

Altschul, S. F., W. Gish, W. Miller, E. W. Myers, and D. J. Lipman (1990). Basic local alignment search tool. *Journal of Molecular Biology*, 215, 403–410.

Arts, G. J. and R. Benne (1996). Mechanism and evolution of RNA editing in kinetoplastida. *Biochemical Biophysics Acta*, 1307, 39–54.

Cavalier-Smith, T. (1993). Kingdom protozoa and its 18 phyla. *Microbiology Reviews*, 57, 953–94.

Dayhoff, M. O. (ed.) (1978). *Atlas of Protein Sequence and Structure*, Vol. 5, Suppl. 3. Silver Spring, MD: National Biomedical Research Foundation.

Dayhoff, M. O., R. M. Schwartz, and B. C. Orcutt (1978). A model for evolutionary change. In: *Atlas of Protein Sequence and Structure*, vol. 5, suppl. 3, ed. M. O. Dayhoff, pp. 345–352. National Biomedical Research Foundation, Silver Spring, MD.

Doolittle, R. F. (1987). *Of URFs and ORFs*. University Science Books.

Felsenstein, J. (1993). *PHYLIP (Phylogeny Inference Package) Version 3.5c*. Distributed by the author. Seattle, WA: University of Washington, Department of Genetics.

Fernandes, A. P., K. Nelson, and S. M. Beverley (1993). Evolution of nuclear ribosomal RNAs in kinetoplastid protozoa: Perspectives on the age and origins of parasitism. *Proceedings of the National Academy of Sciences of the USA*, 90, 11608–11612.

Fitch, W. M. (1981). A nonsequential method for constructing trees and hierarchical classifications. *Journal of Molecular Evolution*, 18, 30–37.

Germot, A. and H. Philippe (1999). Critical analysis of eukaryotic phylogeny: A case study based on the HSP70 family. *Journal of Eukaryotes Microbiology*, 46, 116–124.

Germot, A., H. Philippe, and H. Le Guyader (1996). Presence of a mitochondrial-type 70-kDa heat shock protein in Trichomonas vaginalis suggests a very early mitochondrial endosymbiosis in eukaryotes. *Proceedings of the National Academy of Sciences of the USA*, 93, 14614–14617.

Germot, A., H. Philippe, and H. Le Guyader (1997). Evidence for loss of mitochondria in Microsporidia from a mitochondrial-type HSP70 in *Nosema locustae*. *Molecular Biochemical Parasitology*, 87, 159–168.

Gogarten, J. P., H. Kibak, P. Dittrich, L. Taiz, E. J. Bowman, B. J. Bowman, M. F. Manolson, R. J. Poole, T. Date, T. Oshima, et al. (1989). Evolution of the vacuolar H+-ATPase: Implications for the origin of eukaryotes. *Proceedings of the National Academy of Sciences of the USA*, 86, 6661–6665.

Gonnet, G. H., M. A. Cohen, and S. A. Benner (1992). Exhaustive matching of the entire protein sequence database. *Science*, 256, 1443–1445.

Hannaert, V., F. R. Opperdoes, and P. A. Michels (1998). Comparison and evolutionary analysis of the glycosomal glyceraldehyde-3-phosphate dehydrogenase from different Kinetoplastida. *Journal of Molecular Evolution*, 47, 728–738.

Henikoff, S. and J. G. Henikoff (1992). Amino-acid substitution matrixes from protein blocks. *Proceedings of the National Academy of Sciences of the USA*, 89, 10915–10919.

Keeling, P. J. and W. F. Doolittle (1996). Alpha-tubulin from early-diverging eukaryotic lineages and the evolution of the tubulin family. *Molecular Biology and Evolution*, 13, 1297–1305.

Kimura, M. (1983). *The Neutral Theory of Molecular Evolution*. Cambridge, UK: Cambridge University Press.

Kohl, L., T. Drmota, C. D. Thi, M. Callens, J. Van Beeumen, F. R. Opperdoes, and P. A. Michels (1996). Cloning and characterization of the NAD-linked glycerol-3-phosphate dehydrogenases of Trypanosoma brucei brucei and Leishmania mexicana mexicana and expression of the trypanosome enzyme in Escherichia coli. *Molecular Biochemical Parasitology*, 76, 159–173.

Kyte, J. and R. F. Doolittle (1982). A simple method for displaying the hydropathic character of a protein. *Journal of Molecular Biology*, 157, 105–132.

Needleman, S. B. and C. D. Wunsch (1970). A general method applicable to the search of similarities in the amino-acid sequences of two proteins. *Journal of Molecular Biology*, 48, 443–453.

Nei, M. (1985). *Molecular Evolutionary Genetics*. New York: Columbia University Press.

Patterson, D. J. and M. L. Sogin (1992). Eukaryotic origins and protistan diversity. In: *The Origin and Evolution of the Cell*, eds. H. Hartmann and K. Matsuno, pp. 13–47. River Edge, NJ: World Scientific Publishing Co.

Pearson, W. R. and D. J. Lipman (1988). Improved tools for biological sequence comparison. *Proceedings of the National Academy of Sciences of the USA*, 85, 2444–2448.

Philippe, H. and A. Adoutte (1998). The molecular phylogeny of eukaryota: Solid facts and uncertainties. In: *Evolutionary Relationships amongst Protozoa*, The Systematics Association Special Volume Series 56, eds. G. H. Coombs, K. Vickerman, M. A. Sleigh, and A. Warren, pp. 25–56. Dordrecht, The Netherlands: Kluwer Academic Publishers.

Philippe, H. and A. Germot (2000). Phylogeny of eukaryotes based on ribosomal RNA: Long-branch attraction and models of sequence evolution. *Molecular Biology and Evolution*, 17, 830–834.

Saitou, N. and M. Nei (1987). The neighbor-joining method: A new method for reconstructing phylogenetic trees. *Molecular Biology and Evolution*, 4, 406–425.

Smith, T. F. and M. S. Waterman (1981). Identification of common molecular subsequences. *Journal of Molecular Biology*, 147, 195–197.

Stevens, J. R. and W. Gibson (1999). The molecular evolution of trypanosomes. *Parasitology Today*, 15, 11432–11437.

Strimmer, K. and A. von Haeseler (1997). Likelihood-mapping: A simple method to visualize phylogenetic content of a sequence alignment. *Proceedings of the National Academy of Sciences of the USA*, 94, 6815–6819.

Suresh, S., S. Turley, F. R. Opperdoes, P. A. Michels, and W. G. Hol (2000). A potential target enzyme for trypanocidal drugs revealed by the crystal structure of NAD-dependent glycerol-3-phosphate dehydrogenase from Leishmania mexicana. *Structure with Folding and Design*, 8, 541–552.

Swofford, D. L. (1998). *PAUP**. *Phylogenetic Analysis Using Parsimony (*and Other Methods), Version 4.* Sunderland, MA: Sinauer Associates, Inc.

Wiemer, E. A., V. Hannaert, P. R. van den IJssel, J. van Roy, F. R. Opperdoes, and P. A. Michels (1995). Molecular analysis of glyceraldehyde-3-phosphate dehydrogenase in *Trypanoplasma borelli*: An evolutionary scenario of subcellular compartmentation in kinetoplastida. *Journal of Molecular Evolution*, 40, 443–454.

Woese, C. R. and G. E. Fox (1977). Phylogenetic structure of the prokaryotic domain: The primary kingdoms. *Proceedings of the National Academy of Sciences of the USA*, 74, 5088–5090.

Xia, X. (2000). *Data Analysis in Molecular Biology and Evolution*. Boston: Kluwer Academic Publishers.

Analysis of nucleotide sequences using TREECON

THEORY

Yves Van de Peer

9.1 Introduction

TREECON is a software package developed primarily for the construction and drawing of *neighbor-joining (NJ) phylogenetic trees* (Saitou and Nei, 1987) on the basis of *evolutionary distances* inferred from nucleic and amino-acid sequences. A unique feature implemented in TREECON is the possibility to measure the *relative substitution rate* of individual sites in a nucleotide-sequence alignment with a method called *substitution rate calibration (SRC)*, which was mainly developed to study *ribosomal RNA* (Van de Peer and De Wachter, 1993; Van de Peer et al., 1996b). The current version of TREECON runs under a standard Windows interface (Van de Peer and De Wachter, 1993; Van de Peer and De Wachter, 1994). Advantages of TREECON for Windows are its device independence, the multitasking environment, and the possibility of displaying large trees containing hundreds of sequences. It has become more user-friendly, in particular because of its drawing interface (Figure 9.1). A limitation of the program is that it only runs under the MS-Windows environment; however, TREECON does run on Macintosh computers that have software to emulate the MS-Windows operating system. More information about TREECON for Windows can be found on the TREECON Web site at http://www.psb.rug.ac.be/bioinformatics/.

9.2 TREECON, distance trees, and among-site rate variation

As discussed in Chapter 5, *distance methods* compute the evolutionary distance for all pairs of *operational taxonomic units (OTUs)*, and a phylogenetic tree is then inferred by considering the relationship between these distance values. Usually, the *genetic distance* is estimated from the *p-distance* by correcting for multiple

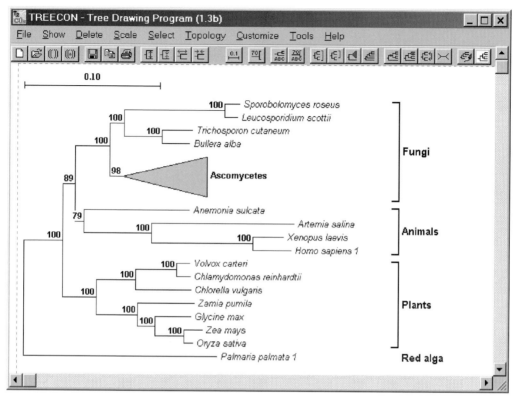

Figure 9.1   Display of the TREECON tree-drawing tool.

mutations (explained in Chapter 4). A number of frequently used equations to convert dissimilarities into distances are implemented in TREECON. As discussed throughout this book, the use of a realistic **substitution model** to estimate distances or likelihoods is important and can have a major influence on the inferred tree topology. Many recent studies have been devoted to the consideration of **among-site rate variation** in molecular markers (see Yang, 1996, for a review; also Felsenstein and Churchill, 1996; Waddell and Steel, 1997; Waddell et al., 1997; Hartmann and Golding, 1998; Sullivan et al., 1999), and **site-to-site substitution rate variation** is now considered to be a major cause of artifacts in phylogeny reconstruction. Olsen (1987) demonstrated that application of the **Jukes and Cantor correction** (1969), which assumes equal **substitution rates** for sites, to sequences composed of sites with unequal rates can lead to artifacts in tree topology caused by the underestimation of the distances. Figure 9.2 explains why distances are underestimated when there are significant differences in substitution rates among sites in the sequence alignment. This is primarily because the majority of substitutions happen at the same sites; that is, the variable positions. Obviously, the more distantly related the sequences, the more pronounced this phenomenon becomes. To correct for among-site rate

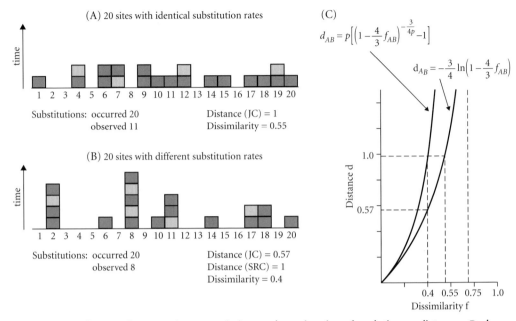

Figure 9.2   Influence of among-site rate variation on the estimation of evolutionary distances. Dark gray squares represent substitutions leading to a different nucleotide than the original one; light gray squares represent those resulting in the original nucleotide. (A) Hypothetical distribution of substitutions in a sequence of 20 nucleotides with equal substitution rates among nucleotide sites. (B) Hypothetical distribution of substitutions in a sequence of 20 nucleotides with different substitution rates among nucleotide sites. (C) Graphic representation of the functions describing the relationship between dissimilarity and evolutionary distance when substitutions are assumed to occur randomly (right curve), and when substitution rates are assumed to differ among sites (left curve).

variation, Olsen (1987) proposed the use of a different evolutionary model that assumed a log-normal distribution of substitution rates over the sequence positions. Jin and Nei (1990) followed a similar approach, but assumed that substitution rates were Γ-distributed (see Section 4.6.1). On the basis of this Γ-*evolutionary model*, which involves a parameter α that describes the extent of the rate variation, they derived several equations to compute the evolutionary distance from the observed sequence dissimilarities (see Section 4.6.1). The estimation of evolutionary distances or likelihoods based on Γ-distributed substitution rates is used extensively and has been implemented in many tree-inference programs.

The main problem with applying Γ-*distances*, however, is the sensitivity of the α estimation to alternative methods, tree topologies, and the number of sequences used (Rzhetsky and Nei, 1994; Sullivan et al., 1995; Yang, 1996; Tourasse and Gouy, 1997; Gu and Zhang, 1997). *Maximum-parsimony* estimates of *nucleotide substitution rates* can be heavily biased (Wakeley, 1993; Yang, 1996; Tourasse and Gouy, 1997), whereas maximum-likelihood estimates may experience computational difficulties because they are too time-consuming (Yang, 1996). A few years ago, a method called SRC was developed for measuring the relative substitution rate of individual sites in a nucleotide-sequence alignment and estimating the α parameter on the basis of a distance approach (Van de Peer and De Wachter, 1993; Van de Peer et al., 1996b). The main advantage of using a distance approach is that nucleotide-variability estimates do not start from a predefined tree topology, and they can be used with numerous sequences. The latter point in particular is important because the more sequences considered, the more accurate the substitution-rate estimates become (Tourasse and Gouy, 1997; Sullivan et al., 1999). Although initially developed for a VAX/VMS environment, the SRC method was implemented in TREECON in 1996 (Van de Peer and De Wachter, 1997a), and has since been successfully applied in several evolutionary studies, mostly based on ribosomal RNA sequences (e.g., Van de Peer et al., 1996a,b, 2000a; Van de Peer and De Wachter, 1997b; Van der Auwera et al., 1998); however, the method is applicable to other nucleotide sequences as well (e.g., Van de Peer et al., 1996c).

Relative nucleotide substitution rates in the SRC method are estimated by observing the frequencies with which sequence pairs differ at *homologous* positions as a function of their evolutionary distance. In practice, this happens as follows (Figure 9.3): for an alignment of n sequences, TREECON computes $n(n-1)/2$ pairwise evolutionary distances d according to the *Jukes and Cantor equation* (see Section 4.4.1). When all pairwise distances have been computed, they are classified in several distance intervals (e.g., four). For each distance interval, the fraction of sequence pairs possessing a different nucleotide is plotted and a curve obeying the following equation:

$$p_i = \frac{3}{4}\left[1 - \exp\left(-\frac{4}{3}v_i d\right)\right] \tag{9.1}$$

which is then fitted to these points by nonlinear regression (see Figure 9.2). Equation 9.1 is the only one fitting the points shown in the plot in Figure 9.2, which are nonlinearly distributed. (The derivation of Equation 9.1 is described in detail in Van de Peer and De Wachter, 1997b.)

This is accomplished for all alignment positions that contain a nucleotide rather than a gap in at least 25% of the sequences. Equation 9.1 expresses the probability p_i that an alignment position i contains a different nucleotide in two sequences, as a function of the evolutionary distance d separating these sequences. The slope of the curve through the origin (see Figure 9.3) yields the specific nucleotide substitution rate v_i for the position under consideration (Van de Peer et al., 1993, 1996b). After estimation of all v_i values, alignment positions are grouped into sets of similar variability and form a spectrum of relative nucleotide substitution rates. This spectrum is shown for an alignment of eukaryotic small subunit ribosomal RNA (SSU rRNA)

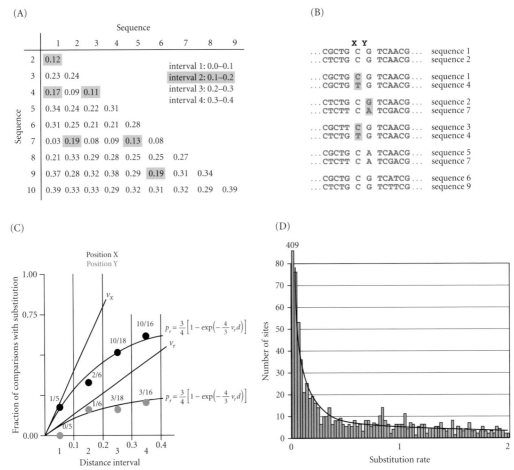

Figure 9.3   Estimation of nucleotide substitution rates in TREECON. (A) Matrix of pairwise evolutionary distances of 10 hypothetical nucleic-acid sequences. After estimation of the evolutionary distances, pairwise distances are classified a number of distance intervals. (B) For all sequence pairs falling within a given distance interval, the fraction accompanied by a nucleotide change in the considered position is computed. (C) Fraction of sequence pairs showing a substitution versus the mean distance of the distance interval. A curve obeying Equation 9.1 (see text) is fitted to the points by nonlinear regression. (D) Substitution rate spectrum obtained from an alignment of 44 eukaryotic small subunit ribosomal RNA sequences (see Figure 9.4).

in Figure 9.3. The rate variation shown by the spectrum can then be modeled by the Γ-evolutionary model (Uzzell and Corbin, 1971; Jin and Nei, 1990; Yang, 1996) which, as already stated, involves a "shape" parameter α that specifies the range of rate variation among sites, with a small α and a large α representing extreme and minor rate variations, respectively (see Sections 4.6.1 and 4.10.1). In TREECON, invariant positions typically are omitted from the analyses, which makes the applied model an **invariant + gamma model**, as described by Gu et al. (1995) and

Waddell and Penny (1996). Ultimately, after estimation of the shape parameter, the final equation used to compute Γ-distances (see Figure 9.2) based on the observed dissimilarities f, and taking into account among-site rate variation is as follows:

$$d = p \left[\left(1 - \frac{4}{3} f \right)^{-\frac{3}{4p}} - 1 \right] \tag{9.2}$$

(Van de Peer et al., 1996b). Equation 9.2 is the same as proposed by Rzhetsky and Nei (1994), with $p = \frac{3}{4} a$.

Note: An empirical distribution of rates like that shown in Figure 9.3D – obtained from an alignment of 44 eukaryotic small subunit ribosomal RNA sequences – can basically be fitted only by a Γ-*distribution function* (it is definitely not a typical distribution) because nearly all distributions are based on among-site rate variation. This is one reason why the Γ-distribution function is used to model rate heterogeneity over sites (Section 4.10.1).

The actual estimation of substitution rates and, therefore, the estimation of the rate parameter p (or α) is somewhat more complicated than described herein. For example, the estimated nucleotide substitution rates described previously are not yet optimal because they are derived on the basis of *Jukes and Cantor distances*. Using this model provides only a first approximation of the relation between dissimilarity and distance because it starts from the rather unrealistic assumption that all nucleotides have the same substitution rate. Therefore, after estimation of all v_i values and finding the optimal shape parameter, evolutionary distances are estimated again using Equation 9.2, after which they are used to reestimate the substitution rate of each alignment position, defining a new and improved spectrum of *evolutionary rates*. This iterative process is repeated several times until the nucleotide substitution rates v_i do not change anymore, at which time the rate spectrum acquires its definitive shape and the final function is known to compute evolutionary distances (Van de Peer et al., 1996b; Van de Peer and De Wachter, 1997a). In general, no more than three iterations are needed for the changes to become imperceptible. A drawback of the SRC method is that (for different reasons) at least 30 sequences with a length of at least 1000 characters should be used to obtain reliable estimates of the nucleotide substitution rates.

9.2.1 Taking into account among-site rate variation: An example

Ribosomal RNA sequences show considerable rate variation (Rzhetsky, 1995; Sullivan et al., 1995; Van de Peer et al., 1996d), and the importance of taking this variation into account when analyzing distantly related sequences was demonstrated previously in numerous studies (e.g., Van de Peer et al., 1996a and b, 2000a). Figure 9.4 illustrates another application of the SRC method, as implemented

(A)

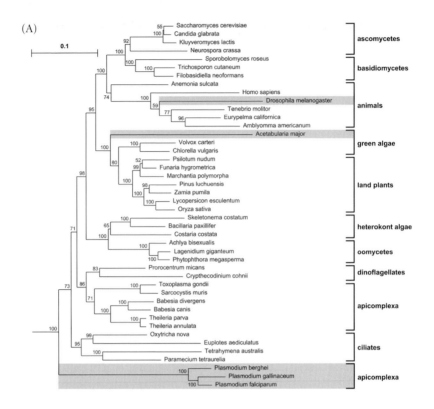

(B)

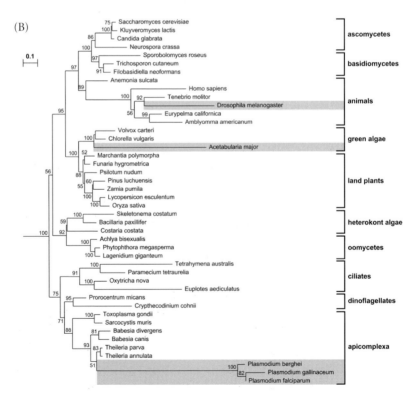

242

in TREECON, and demonstrates again the effect of considering among-site rate variation in ribosomal RNA sequences. The tree in Figure 9.4A shows a ribosomal RNA phylogeny of 44 small subunit ribosomal RNA (SSU rRNA) sequences of different eukaryotic taxa, constructed by NJ on the basis of Jukes and Cantor distances.

Three species were added to the data set for which the SSU rRNA sequences have an increased evolutionary rate, compared to the other sequences: the insect *Drosophila*, the green alga *Acetabularia*, and the apicomplexan genus *Plasmodium*. As shown, the respective positions of these organisms in the tree are somewhat unexpected. First, *Drosophila* does not group with the other insect *Tenebrio*; second, the green alga *Acetabularia* branches off before the divergence of green algae and land plants; and third, *Plasmodium* forms the first branch in the tree and is not clustered with the other apicomplexans.

The tree shown in Figure 9.4B was based on the SRC method. For the particular alignment of the 44 SSU rRNA sequences, substitution rates of individual nucleotides were estimated as explained previously (and in more detail in Van de Peer et al., 1993 and 1996b), leading to the distribution shown in Figure 9.3D, which can be modeled by a function described in Equation 9.2, with $p = 0.58$. This function was then used to convert **observed distances** to evolutionary distances, and an NJ tree was constructed on the basis of these distances. As shown in Figure 9.4B, the position of the three fast-evolving branches has changed considerably and the sequences are now clustered as expected (see Van de Peer and De Wachter, 1997b). The erroneous clustering in Figure 9.4A, which is highly supported by **bootstrap analysis** (Felsenstein, 1985), appears to be artificial and caused by the underestimation of evolutionary distances.

The phenomenon in which long branches are erroneously clustered (the so-called long-branch attraction artifact) in molecular phylogenies has been discussed extensively in recent years (Van de Peer et al., 1996a,b, 2000a; Stiller and Hall, 1999; Philippe and Germot, 2000; Philippe et al., 2000). The reason that long branches are clustered together is quite easy to understand for maximum parsimony, but more difficult to grasp for distance methods. Contrary to parsimony methods, in which covarying **homoplasies** or convergences leading to the same character state in nonrelated branches may mislead parsimony to infer

---

Figure 9.4   (Facing page) Phylogenetic trees of 44 SSU rRNA sequences of different eukaryotic taxa, retrieved from the European SSU rRNA database (Van de Peer et al., 2000b). (A) Neighbor-joining tree based on Jukes and Cantor distances. Bootstrap values above 50% out of 500 resamplings are shown at the internodes. Taxon designations are placed to the right of the corresponding clusters. (B) Neighbor-joining tree of the same set of eukaryotes, but considering among-site rate variation.

(A)

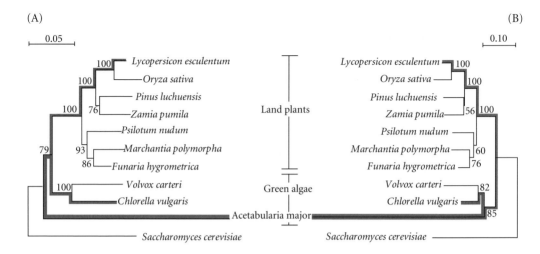

(C)

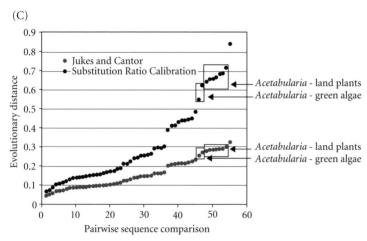

Figure 9.5   Neighbor-joining trees of a subset of the species used in Figure 9.4. The light gray line indicates the evolutionary distance between *Acetabularia* and *Lycopersicon*; the dark gray line
shows the distance between *Acetabularia* and *Chlorella*. (A) Phylogenetic tree based on
Jukes and Cantor distances. (B) Phylogenetic tree based on SRC distances.

the wrong tree (Felsenstein, 1978; Swofford et al., 1996) (see Chapter 7), these misleading parsimonious informative positions are less decisive in distance methods
(Huelsenbeck and Hillis, 1993; Huelsenbeck et al., 1996). As concluded from Figure 9.2, for distance methods, underestimation of evolutionary distances becomes
greater with increasing distances, in particular when among-site rate variation is
present. For small dissimilarities, the underestimation is negligible, which explains
why the tree topologies in Figures 9.4A and 9.4B are the same for closely related
organisms. Conversely, the longest branches in the tree are bound to be the most
underestimated; therefore, distant species seem closer to one another than they

actually are. As a result, long branches are often clustered together, to the exclusion of other sequences for which the evolutionary distance is more accurately estimated.

Alternatively, long branches will be drawn toward the base of the cluster to which they belong or even to the base of the entire tree (Figure 9.4A). It is possible they are pulled close to the *outgroup* for reasons discussed previously; namely, that the distance between the long branch and the outgroup is the most underestimated. Another explanation is that due to the combination of distant relatedness and extensive rate variation – which often results in substitution saturation of the hot-spot regions of the molecule (see Section 4.11) – the number of observed substitutions between the long-branch sequence and any other sequence is similar. Consequently, tree-construction programs often have difficulties elucidating the exact sister relationships between the long-branch sequence and its closest relative(s).

This is shown in Figure 9.5 for a subset of eukaryotic SSU rRNA sequences. The gray dots represent estimated Jukes and Cantor distances computed for all pairs of sequences. The NJ tree obtained on the basis of these distances, illustrated in Figure 9.5A, shows the green alga *Acetabularia* branching off before the divergence between the other green algae and the land plants. The same tree topology is obtained when the *K80 model* (Kimura, 1980) is applied. Maximum parsimony also reconstructs the same topology, grouping *Acetabularia* with *Saccharomyces* in an unrooted tree. The black dots represent genetic distances based on an invariant + gamma model as described previously and implemented in TREECON. The SRC tree topology, illustrated in Figure 9.5B, shows *Acetabularia* now grouped with the other green algae (see Figure 9.4A). Although the dissimilarity values are similar – they do differ slightly because in the SRC method, invariant positions were removed – the obtained distances are quite dissimilar, due to the different shapes of the different functions (see Figure 9.2B). In this case, the effect of using an invariant + gamma distance model is like using a magnifying glass, which now enables discrimination between evolutionary distances estimated from similar but different dissimilarities.

9.3 Conclusions

TREECON for Windows is a software package for constructing and drawing of evolutionary trees based on genetic distances. A strong feature of the program is the implementaion of the SRC method in the estimation of evolutionary distances, taking into account among-site rate variation. This is important because differences in substitution rates among sites in sequence alignments is considered one of the major causes of artifacts in inferred tree topologies.

PRACTICE

Yves Van de Peer

9.4   The TREECON software package

TREECON for Windows, or TREECONW, is written in C (C++ 4.5 and 5.0; Borland International) and runs on computers running Microsoft® Windows® 3.1, Windows® 9x, Windows® 2000, and Windows® NT. TREECONW also runs on Macintosh computers and on those that run UNIX/Linux, if the software is installed to emulate the MS-Windows operating system. TREECONW is simple to use – the program is basically self-explanatory – and only a minimum of prior knowledge about computers is needed. Starting from a simple ASCII text file that contains nucleic-acid or amino-acid sequences listed sequentially, with gaps required for mutual alignment, the user can produce publishable trees in a user-friendly, fast, and straightforward way. TREECON also includes a software module called ForCon (Raes and Van de Peer, 1999) (see Box 2.1) that converts numerous different sequence-alignment formats, such as FASTA, PHYLIP, NEXUS, and MEGA (see Boxes 2.1 and 7.4). A limitation of TREECON is that, although trees can be constructed including thousands of sequences, only one-page trees can be printed. However, trees containing up to 100 sequences can be fit on one page (e.g., see Van de Peer et al., 1996a); larger trees can be copied and pasted into another program for further editing. The program can be downloaded for free from **http://www. psb.rug.ac.be/bioinformatics/**.

In TREECON for Windows, NJ phylogenetic trees can be constructed based on several of the substitution models discussed in the previous chapters, including Γ-distances and distances computed on the basis of transversions corrected only for multiple mutations (Swofford et al., 1996). For amino-acid sequences, distances can be computed according to Zuckerkandl and Pauling (1965) and Dickerson (1971), assuming that the rate of amino-acid substitution at each site follows a **_Poisson distribution_** (Fitch and Margoliash, 1967), according to Equation 8.1 (Kimura, 1983) and to the Tajima and Nei (1984) or Ota and Nei (1994) models. More detailed descriptions of these models are found in other chapters of this book, in the original papers, or in reviews and books such as Swofford et al. (1996) and Nei and Kumar (2000).

9.5   Implementation

The TREECON for Windows software package is actually a compilation of programs that are called by a management program, TREECONW (Figure 9.6). Typically, the

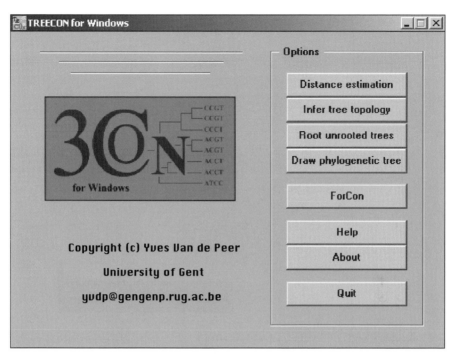

Figure 9.6   Main menu of the TREECON for Windows software package.

different programs are performed one after another, and the output files created by one module serve as input files for the next module. The program MATRIXW, invoked by choosing the option Distance estimation, calculates evolutionary distances for nucleic and amino-acid sequences. TREEW, invoked by the option Infer tree topology, implements tree-construction methods based on distance matrixes, such as NJ (see Chapter 5). The program ROOTW places a root on the tree, by selecting the option Root unrooted trees. Finally, the program DRAWW draws the evolutionary tree, changes its size, and produces drawings suitable for publication.

   When Distance estimation is selected, the program asks for the sequence input file. The TREECON input-file format is similar to the PHYLIP sequential format (see Box 2.2), except that sequence names can contain up to 40 characters (instead of only 10) and that only the length of the sequence alignment must be specified (not the number of sequences). However, TREECON also recognizes the sequential and interleaved file formats used by the PHYLIP package (Felsenstein, 1993), whereas sequence alignments in other formats can be converted to the TREECON or PHYLIP format by the ForCon module (described previously). The number and length of sequences that can be analyzed depends only on the amount of working memory of the computer. When the input file is selected, the program lists all sequences and, if needed, a specific subset of sequences can be selected from which a

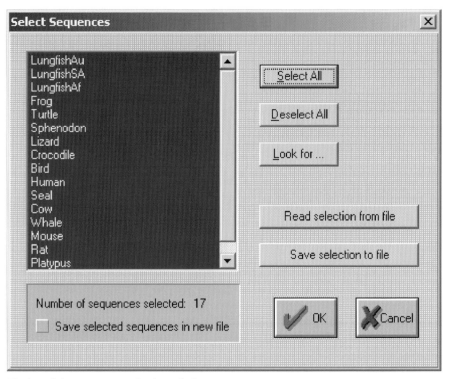

Figure 9.7 Display of the sequence selection window.

tree may be constructed. However, if the sequence names are not listed properly in the sequence list (Figure 9.7), there likely is a format error in the sequence input file.

In the computation of genetic distances, ***indels*** (i.e., insertions and deletions) can be considered or ignored. If they are taken into account, indels are counted separately because the corrections for superimposed events do not hold for insertions and deletions. For example, when indels are considered in TREECON, the evolutionary distance between Sequences A and B according to the Jukes and Cantor model is then estimated as follows:

$$d_{AB} = -\frac{3}{4} \ln\left[1 - \frac{4}{3}\left(\frac{f_S}{f_S + I}\right)\right]\left[1 - \frac{G}{T}\right] + \frac{G}{T} \tag{9.3}$$

where I is the number of identical nucleotides, f_S is the number of positions showing a substitution, and G is the number of gaps in one sequence with respect to the other. T is the sum of I, f_S, and G. The first term of the equation accounts for substitutions and comprises the Jukes and Cantor correction factor for multiple mutations per site (see Section 4.4.1). The second term accounts for deletions and insertions. A row of adjacent gaps is treated as one gap, regardless of its length.

Ambiguities are not taken into account. The following comparison of the imaginary sequences:

```
AGCTGG - - - GCNTA
- - CTAGAAAGACTA
```

would thus give the values $I = 6$, $f_S = 2$, $G = 2$, and $T = 10$. If indels are not taken into account in the estimation of evolutionary distances, then $I = 6$, $f_S = 2$, and $T = 8$.

After the sequences have been read into memory, the user can select the substitution model, decide whether to consider indels or not, and determine whether the data need to be bootstrapped (Figure 9.8). The output of the MATRIXW program is a matrix of distances written on the hard disk. If bootstrap analysis was selected, the program automatically generates the desired number of distance matrixes, based on random resampling from the original sequence alignment (Felsenstein, 1985) (see Section 5.3 and Box 5.3). The tree-inferring program, invoked by the second step in the principal menu, reads these distance matrixes and constructs a tree by *clustering*, the *transformed distance method*, or NJ. These methods are discussed into more detail in Chapter 5; in practice, however, trees are nearly always reconstructed using the NJ algorithm.

Before NJ trees can be displayed on the screen, it is necessary to root the tree. When trees have been inferred by clustering, rooting of the tree should be omitted because clustering methods already produce rooted tree topologies (see Chapter 5). The easiest method to place a root on the NJ tree is to add an additional sequence – an outgroup – to the data set that is *a priori* assumed to lie outside the group of interest (see Section 5.7). However, this may not always be possible. There are different options to root the tree in TREECONW. The first two options allow the user to select the outgroup sequence. In the first option, the root is placed so as to equalize its distance to the outgroup and its average distance to the remaining sequences. However, at times, this criterion cannot be fulfilled because the branch leading to the outgroup sequence is too short. In this case, it is still possible to place the root on the latter branch, forcing the chosen sequence to be the outgroup. If no *a priori* outgroup can be included, a third option places the root at a point where the average distance to the sequences on both sides of the root is the same, thereby scanning the entire tree until such a point can be found. This kind of rooting should be chosen if an outgroup of more than one sequence is used. However, in some trees, several rooting points can be found and the program chooses the first one that it encounters. An alternative approach is to root the tree with one sequence, rerooting it afterward at a particular internal node in the DRAWW program (Figure 9.9). If *bootstrap trees* are constructed (note: remember to select the bootstrap option at every step of the tree-construction process), all the trees must be rooted by the same

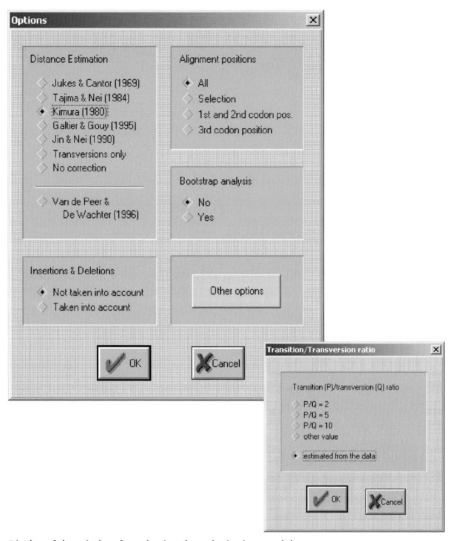

Figure 9.8 Display of the window for selecting the substitution model.

single outgroup sequence. However, if necessary, the tree can be rerooted later in the drawing program.

In the DRAWW program, many options are available to create one-page publishable trees. Trees also can be copied to the clipboard and pasted into other programs. *Taxon* designations can be added in different ways, different fonts can be selected for sequence names and bootstrap values, branches can be swapped, and so on. Trees in TREECON are always drawn so as to minimize the vertical branch lengths, which greatly improves the readability of trees – particularly, for trees that contain hundreds of sequences.

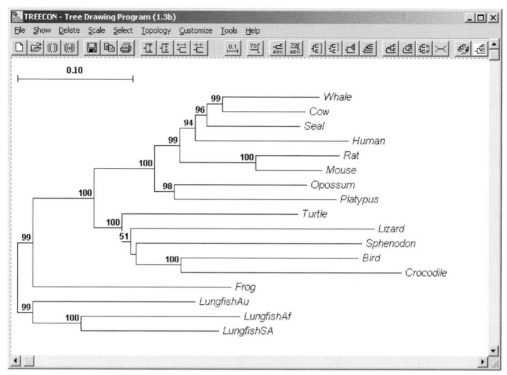

Figure 9.9   Display of the tree-drawing module of TREECON for Windows. The tree shown is based on the mtDNA sequences of 17 different vertebrates. Distances were computed according to the K80 model. The tree was rooted with the three lungfish sequences. The scale on top measures the evolutionary distance in substitutions per nucleotide. Bootstrap values above 50%, out of 500 replications, are indicated.

9.6   Substitution rate calibration

As already discussed in the theoretical part of this chapter, TREECONW contains the SRC method that estimates the amount of substitution rate variation in nucleic-acid sequence alignments and uses this information for estimating the genetic distances. Box 9.1 illustrates schematically how the SRC method works in practice. To use the SRC method, the option substitution rate calibration should be selected in the Distance estimation menu. Estimation of nucleotide substitution rates and the shape parameter that specifies the range of rate variation among sites then proceeds automatically. Afterward, genetic distances can be estimated taking into account site-to-site rate variation using the option Van de Peer and De Wachter (see Figure 9.8). If this option is selected, the program asks for the specific rate-parameter value determined by the SRC method. If the user has chosen to omit invariable sites from the analysis, the SRC method produces a

Box 9.1 Schematic outline of the Substitution Rate Calibration method

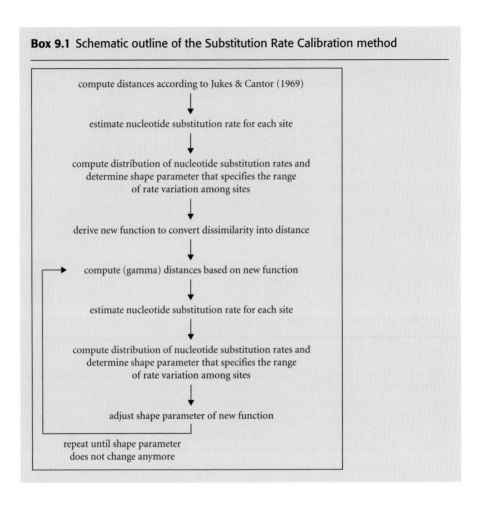

compute distances according to Jukes & Cantor (1969)

↓

estimate nucleotide substitution rate for each site

↓

compute distribution of nucleotide substitution rates and determine shape parameter that specifies the range of rate variation among sites

↓

derive new function to convert dissimilarity into distance

↓

compute (gamma) distances based on new function

↓

estimate nucleotide substitution rate for each site

↓

compute distribution of nucleotide substitution rates and determine shape parameter that specifies the range of rate variation among sites

↓

adjust shape parameter of new function

repeat until shape parameter does not change anymore

new sequence alignment file, in which invariable sites have been removed. This file rather than the original one should be used to construct trees. In the TREECONW user manual, a step-by-step example shows how phylogenetic trees can be inferred on the basis of substitution rate calibrated distances.

REFERENCES

Dickerson, R. E. (1971). The structure of cytochrome C and the rates of molecular evolution. *Journal of Molecular Evolution*, 1, 26–45.

Felsenstein, J. (1978). Cases in which parsimony and compatibility methods will be positively misleading. *Systematic Zoology*, 27, 401–410.

Felsenstein, J. (1985). Confidence limits on phylogenies: An approach using the bootstrap. *Evolution*, 39, 783–791.

Felsenstein, J. (1993). *PHYLIP (Phylogeny Inference Package) Version 3.5c*. Distributed by the author. Seattle, WA: University of Washington, Department of Genetics.

Felsenstein, J. and G. A. Churchill (1996). A hidden Markov model approach to variation among sites in rate of evolution. *Molecular Biology and Evolution* 13, 93–104.

Fitch, W. M. and E. Margoliash (1967). A method for estimating the number of invariant amino acid coding positions in a gene using Cytochrome c as a model case. *Biochemical Genetics*, 1, 65–71.

Gu, X. and J. Zhang (1997). A simple method for estimating the parameter of substitution rates variation among sites. *Molecular Biology and Evolution*, 14, 1106–1113.

Gu, X., Y.-X. Fu, and W-H. Li (1995). Maximum-likelihood estimation of the heterogeneity of substitution rates among nucleotide sites. *Molecular Biology and Evolution*, 12, 546–557.

Hartmann, M. and G. B. Golding (1998). Searching for substitution-rate heterogeneity. *Molecular Phylogenetics and Evolution*, 9, 64–71.

Huelsenbeck, J. P. and D. M. Hillis (1993). Success of phylogenetic methods in the four-taxon case. *Systematic Biology*, 42, 247–264.

Huelsenbeck, J. P., J. J. Bull, and C. W. Cunningham (1996). Combining data in phylogenetic analysis. *Trends in Ecology and Evolution*, 11, 152–157.

Jin, L. and M. Nei (1990). Limitations of the evolutionary parsimony method of phylogenetic analysis. *Molecular Biology and Evolution*, 7, 82–102.

Jukes, T. H. and C. R. Cantor (1969). Evolution of protein molecules. In: *Mammalian Protein Metabolism*, ed. H. H. Munro, pp. 21–132. Academic Press, New York.

Kimura, M. (1980). A simple method for estimating evolutionary rates of base substitutions through comparative studies of nucleotide sequences. *Journal of Molecular Evolution*, 16, 111–120.

Kimura, M. (1983). *The Neutral Theory of Molecular Evolution*. Cambridge: Cambridge University Press.

Nei, M. and S. Kumar (2000). *Molecular Evolution and Phylogenetics*. Oxford: Oxford University Press.

Olsen, G. J. (1987). Earliest phylogenetic branchings: Comparing rRNA-based evolutionary trees inferred with various techniques. *Cold Spring Harbor Symposia on Quantitative Biology*, LII, 825–837.

Ota, T. and M. Nei (1994). Estimation of the number of amino-acid substitutions per site when the substitution rate varies among sites. *Journal of Molecular Evolution*, 38, 642–643.

Philippe, H. and A. Germot (2000). Phylogeny of eukaryotes based on ribosomal RNA: Long-branch attraction and models of sequence evolution. *Molecular Biology and Evolution*, 17, 830–834.

Philippe, H., P. Lopez, H. Brinkmann, K. Budin, A. Germot, J. Laurent, D. Moreira, M. Müller, and H. Le Guyader (2000). Early-branching or fast-evolving eukaryotes? An answer based on slowly evolving positions. *Proceedings of the Royal Society Series B*, 267, 1213–1222.

Raes, J. and Y. Van de Peer (1999). ForCon: A software tool for the conversion of sequence alignments. *Embnet.news*, 6, 10–12.

Rzhetsky, A. (1995). Estimating substitution rates in ribosomal RNA genes. *Genetics*, 141, 771–783.

Rzhetsky, A. and M. Nei (1994). Unbiased estimates of the number of nucleotide substitutions when substitution rate varies among different sites. *Journal of Molecular Evolution*, 38, 295–299.

Saitou, N. and M. Nei (1987). The NJ method: A new method for reconstructing phylogenetic trees. *Molecular Biology and Evolution*, 4, 406–425.

Stiller, J. W. and B. D. Hall (1999). Long-branch attraction and the rDNA model of early eukaryotic evolution. *Molecular Biology and Evolution*, 16, 1270–1279.

Studier, J. A. and K. J. Keppler (1988). A note on the NJ algorithm of Saitou and Nei. *Molecular Biology and Evolution*, 5, 729–731.

Sullivan, J. A., K. E. Holsinger, and C. Simon (1995). Among-site rate variation and phylogenetic analysis of 12S rRNA in Sigmodontine rodents. *Molecular Biology and Evolution*, 12, 988–1001.

Sullivan, J., D. L. Swofford, and G. J. P. Naylor (1999). The effect of taxon sampling on estimating rate heterogeneity parameters of maximum-likelihood models. *Molecular Biology and Evolution*, 16, 1347–1356.

Swofford, D. L., G. J. Olsen, P. J. Waddell, and D. M. Hillis (1996). Phylogenetic inference. In: *Molecular Systematics*, eds. D. M. Hillis, C. Moritz, and B. K. Mable, pp. 407–514. Sunderland, MA, Sinauer Associates.

Tajima, F. and M. Nei (1984). Estimation of evolutionary distance between nucleotide sequences. *Molecular Biology and Evolution*, 1, 269–285.

Tourasse, N. J. and M. Gouy (1997). Evolutionary distances between nucleotide sequences based on the distribution of substitution rates among sites as estimated by parsimony. *Molecular Biology and Evolution*, 14, 287–298.

Uzzell, T. and K. W. Corbin (1971). Fitting discrete probability distributions to evolutionary events. *Science*, 172, 1089–1096.

Van de Peer, Y. and R. De Wachter (1993). TREECON: A software package for the construction and drawing of evolutionary trees. *Computer Applications in the Biosciences*, 9, 177–182.

Van de Peer, Y. and Y. De Wachter (1994). TREECON for Windows: A software package for the construction and drawing of evolutionary trees for the Microsoft Windows environment. *Computer Applications in the Biosciences*, 10, 569–570.

Van de Peer, Y. and R. De Wachter (1997a). Construction of evolutionary distance trees with TREECON for Windows: Accounting for variation in nucleotide substitution rate among sites. *Computer Applications in the Biosciences*, 13, 227–230.

Van de Peer, Y. and R. De Wachter (1997b). Evolutionary relationships among the eukaryotic crown taxa taking into account site-to-site rate variation in 18S rRNA. *Journal of Molecular Evolution*, 45, 619–630.

Van de Peer, Y., S. Rensing, U.-G. Maier, and R. De Wachter (1996a). Substitution rate calibration of small ribosomal subunit RNA identifies chlorarachniophyte endosymbionts as remnants of green algae. *Proceedings of the National Academy of Sciences of the USA*, 93, 7732–7736.

Van de Peer, Y., G. Van der Auwera, and R. De Wachter (1996b). The evolution of stramenopiles and alveolates as derived by "substitution rate calibration" of small ribosomal subunit RNA. *Journal of Molecular Evolution*, 42, 201–210.

Van de Peer, Y., W. Janssens, L. Heyndrickx, K. Fransen, G. Van der Groen, and R. De Wachter (1996c). Phylogenetic analysis of the *env* gene of HIV-1 isolates taking into account individual nucleotide substitution rates. *AIDS*, 10, 1485–1494.

Van de Peer, Y., S. Chapelle, and R. De Wachter (1996d). A quantitative map of nucleotide substitution rates in bacterial ribosomal subunit RNA. *Nucleic Acids Research*, 24, 3381–3391.

Van de Peer, Y., A. Ben Ali, and A. Meyer (2000a). Microsporidia: Accumulating molecular evidence that a group of amitochondriate and suspectedly primitive eukaryotes are just curious fungi. *Gene*, 246, 1–8.

Van de Peer, Y., P. De Rijk, J. Wuyts, T. Winkelmans, and R. De Wachter (2000b). The European small subunit ribosomal RNA database. *Nucleic Acid Research*, 28, 175–176.

Van der Auwera, G., C. J. B. Hofmann, P. De Rijk, and R. De Wachter (1998). The origin of red algae and cryptomonad nucleomorphs: A comparative phylogeny based on small and large subunit rRNA sequences of *Palmaria palmata*, *Gracilaria verrucosa*, and the *Guillardia theta* nucleomorph. *Molecular Phylogenetics and Evolution*, 10, 333–342.

Waddell, P. and D. Penny (1996). Evolutionary trees of apes and humans from DNA sequences. In: *Handbook of Symbolic Evolution*, eds. A. J. Lock and C. R. Peters, pp. 53–73. Clarendon Press, Oxford.

Waddell, P. and M. A. Steel (1997). General time-reversible distances with unequal rates across sites: Mixing Γ and inverse Gaussian distributions with invariant sites. *Molecular Phylogenetics and Evolution*, 8, 398–414.

Waddell, P., D. Penny, and T. Moore (1997). Hadamard conjugations and modeling sequence evolution with unequal rates across sites. *Molecular Phylogenetics and Evolution*, 8, 33–50.

Wakeley, J. (1993). Substitution rate variation among sites in hypervariable region 1 of human mitochondrial DNA. *Journal of Molecular Evolution*, 37, 613–623.

Yang, Z. (1996). Among-site rate variation and its impact on phylogenetic analyses. *Trends in Ecology and Evolution*, 11, 367–372.

Zuckerkandl, E. and L. Pauling (1965). Evolutionary divergence and convergence in proteins. In: *Evolving Genes and Proteins*, eds. V. Bruson and H. J. Vogel, pp. 97–166. Academic Press, New York.

Selecting models of evolution

THEORY

David Posada

10.1 Models of evolution and phylogeny reconstruction

Phylogenetic reconstruction is regarded as a problem of statistical inference. Because statistical inferences cannot be drawn in the absence of a probability model, the use of a model of nucleotide or amino-acid substitution – an *evolutionary model* – becomes necessary when using DNA or amino-acid sequences to estimate phylogenetic relationships among organisms. Evolutionary models are sets of assumptions about the process of nucleotide or amino-acid substitution (see Chapters 4 and 8). They describe the different probabilities of change from one nucleotide or amino acid to another, with the aim of correcting for unseen changes along the phylogeny. Although this chapter focuses on *models of nucleotide substitution*, all the points made herein can be applied directly to models of amino-acid replacement. Comprehensive reviews of models of evolution are offered by Swofford et al. (1996) and Lió and Goldman (1998).

As discussed in the previous chapters, the methods used in molecular phylogeny are based on a number of assumptions about how the evolutionary process works. These assumptions can be implicit, like in *parsimony methods* (see Chapter 7), or explicit, like in *distance* or *maximum-likelihood methods* (see Chapters 5 and 6). The advantage of making a model explicit is that the parameters of the model may be estimated. Distance methods may estimate only from the data of a single parameter of the model – the number of substitutions per site. However, maximum likelihood can estimate all the relevant parameters of the *substitution model*. Parameters estimated via maximum likelihood have desirable statistical properties: as sample sizes get large, they converge to the true parameter value and have the smallest possible variance among all estimates with the same expected value. Most important, as shown in the following sections, maximum likelihood provides a framework in which different evolutionary hypotheses can be statistically tested rigorously and objectively.

10.2   The relevance of models of evolution

It is well established that the use of one evolutionary model or another may change the results of a phylogenetic analysis. When the model assumed is wrong, branch lengths, transition/transversion ratio, and sequence divergence may be underestimated, whereas the strength of rate variation among sites may be overestimated. Simple models tend to suggest that a tree is significantly supported when it cannot be, and tests of evolutionary hypotheses (e.g., the ***molecular clock***) can become conservative. In general, phylogenetic methods may be less accurate (i.e., recover an incorrect tree more often) or inconsistent (i.e., converge to an incorrect tree with increased amounts of data) when the assumed evolutionary model is wrong. Cases in which the use of wrong models increases phylogenetic performance are the exception; they represent a bias toward the true tree due to violated assumptions. Indeed, models are not important just because of their consequences in phylogenetic analysis, but also because the characterization of the evolutionary process at the sequence level is itself a legitimate pursuit.

Evolutionary models are always simplified, and they often make assumptions just to turn a complex problem into a computationally tractable one. A model becomes a powerful tool when, despite its simplified assumptions, it can fit the data and make accurate predictions about the problem at hand. The performance of a method is maximized when its assumptions are satisfied and some indication of the fit of the data to the phylogenetic model is necessary. Unfortunately – and despite their relevance – the unjustified use of evolutionary models is still a common practice in phylogenetic studies. If the model used may influence results of the analysis, it becomes crucial to decide which is the most appropriate model with which to work.

10.3   Selecting models of evolution

In general, more complex models fit the data better than simpler ones. An a priori attractive procedure to select a model of evolution is the arbitrary use of complex, parameter-rich models. However, when using complex models, numerous parameters need to be estimated, which has several disadvantages. First, the analysis becomes computationally difficult, and requires significant time. Second, as more parameters need to be estimated from the same amount of data, more error is included in each estimate. Ideally, it would be advisable to incorporate as much complexity as needed; that is, to choose a model complex enough to explain the data but not so complex that it requires impractical long computations or large data sets to obtain accurate estimates.

The best-fit model of evolution for a particular data set can be selected through statistical testing. The fit to the data of different models can be contrasted through *likelihood ratio tests* (*LRTs*) or *information criteria* to select the best-fit model within a set of possible ones. In addition, the overall adequacy of a particular model to fit the data can be tested using an LRT.

A word of caution is necessary when selecting best-fit models for heterogeneous data; for example, when joining different genes for the phylogenetic analysis or a coding and a noncoding region. Because different genomic regions are subjected to different selective pressures and evolutionary constraints, a single substitution model may not fit well all the data. Although some options exist for the combined analysis of multiple-sequence data (Yang, 1996; Salemi, Desmyter, and Vandamme, 2000a), these are computationally expensive. An alternative solution would be to run separate analyses for each gene or region.

10.4 The likelihood ratio test

In Chapter 6, the *likelihood function* was introduced as the conditional probability of the data (i.e., aligned *homologous* sequences) given the following hypothesis (i.e., a model of substitution with a set of parameters θ – for example, base frequencies or transition/transversion ratio – and the tree τ, including branch lengths):

$$L(\tau, \theta) = \text{Prob}(\text{Data} \,|\, \tau, \theta)$$
$$= \text{Prob}(\text{Aligned sequences} \,|\, \text{tree, model of evolution}) \qquad (10.1)$$

with *maximum-likelihood estimates* (*MLEs*) of τ and θ making the likelihood function as large as possible:

$$\hat{\tau}, \hat{\theta} = \max_{\tau,\theta} L(\tau, \theta) \qquad (10.2)$$

A natural way of comparing two models is to contrast their likelihoods using the LRT statistic:

$$\Delta = 2(\log_e L_1 - \log_e L_0) \qquad (10.3)$$

where L_1 is the maximum likelihood under the more parameter-rich, complex model (i.e., alternative hypothesis) and L_0 is the maximum likelihood under the less parameter-rich, simple model (i.e., null hypothesis). The value of this statistic is always equal to or greater than zero – even if the simple model is the true one – simply because the superfluous parameters in the complex model provide a better explanation of the stochastic variation in the data than the simpler model. When the models compared are nested (i.e., the null hypothesis is a special case of the alternative hypothesis) and the null hypothesis is correct, this statistic is asymptotically distributed as χ^2, with a number of degrees of freedom equal to the difference in number of free parameters between the two models. In other words, the number of degrees of freedom is the number of restrictions on the parameters of the alternative hypothesis required to derive the particular case of the null hypothesis. When the value of the LRT is significant (i.e., <0.05 or <0.01), the conclusion is that the inclusion of additional parameters in the alternative model significantly increases the likelihood of the data and, consequently, the use of the more complex model is favored. Conversely, a difference in the log likelihood close to zero means that the alternative hypothesis does not fit the data significantly better than the null hypothesis (i.e., adding those particular parameters to the null model does not give a better explanation of the data).

That two models are nested means that one model (i.e., null model or constrained model) is equivalent to restrict the possible values that one or more parameters can take in the other model (i.e., alternative, unconstrained, or full model). For example, the **Jukes-Cantor (JC) model** (1969) and the **Felsenstein (F81) model** (1981) are nested. This is because the JC model is a special case of the F81, where the base frequencies are set to be equal (all are 0.25); whereas in the F81 model, these frequencies can be different (e.g., 0.20., 0.60., 0.15, and 0.05). The χ^2 distribution approximation for the LRT statistic may not be appropriate when the null model is equivalent to fixing some parameter at the boundary of its parameter space in the alternative model (Whelan and Goldman, 1999). An example of this situation is the invariable sites test, in which the alternative hypothesis postulates that the proportion of invariable sites could range from 0 to 1. The null hypothesis (i.e., no invariable sites) is a special case of the alternative hypothesis, with the proportion of invariable sites fixed to 0, which is at the boundary of the range of the parameter in the alternative model. In this case, the use of a mixed χ^2 distribution (i.e., 50% χ_0^2 and 50% χ_1^2) is appropriate. Although the difference in likelihoods when comparing current models may be significant and the inaccuracy of the χ^2 approximation may not change results of these tests (Posada, 2001a; Posada and Crandall, 2001a), as more complex and realistic models are developed (in which the differences of likelihoods might be insignificant), the use of a mixed χ^2 distribution may be essential. The use of LRTs for hypothesis testing in phylogeny is reviewed by Huelsenbeck and Crandall (1997) and Huelsenbeck and Rannala (1997).

10.4.1   LRTs and parametric bootstrapping

The χ^2 approximation to assess the significance of the LRT is not appropriate when the two competing hypotheses are not nested, and it may perform poorly when the data include very short sequences relative to the number of parameters to be estimated. In this case, the null distribution of the LRT statistic can be approximated by the *Monte Carlo simulation*. The general strategy is as follows:

1. Select the competing models: one for the null hypothesis H_0 and one for the alternative hypothesis H_1.
2. Estimate the tree and the parameters of the model under the null hypothesis.
3. Use the tree and the estimated parameters to simulate 200–1000 replicate data sets of the same size as the original.
4. For each simulated data set, estimate a tree and calculate its likelihood under the models representing H_0 and H_1 (L_0 and L_1, respectively). Calculate the LRT statistic $\Delta = 2$ ($\log_e L_1 - \log_e L_0$). These simulated Δs form the distribution of the LRT statistic if the null hypothesis was true (i.e., they constitute the null distribution of the LRT statistic).
5. The probability of observing the LRT statistic from the original data set if the null hypothesis is true is the number of simulated Δs bigger than the original Δ, divided by the total number of simulated data sets. If this probability is smaller than a predefined value (usually 0.05), H_0 is rejected.

The main disadvantage of parametric bootstrapping is its computational expensiveness. Because the likelihood calculations must be repeated on each simulated data set, this approach becomes unfeasible when many sequences are considered, even for fast supercomputers. A general discussion on model-fitting through parametric bootstrapping can be found in Goldman (1993a and b). Huelsenbeck et al. (1996) provide an interesting review of the applications of parametric bootstrapping in molecular phylogenetics.

10.4.2 Hierarchical LRTs

Comparing two different nested models through an LRT means testing hypotheses about the data. The hypotheses tested are those represented by the difference in the assumptions among the models compared. Several hypotheses can be tested hierarchically to select the best-fit model for the data set at hand among a set of possible models. It is to our advantage to test one hypothesis at a time: Are the base frequencies equal? Is there a transition/transversion (ti/tv) bias? Are all transition rates equal? Are there invariable sites? Is there rate homogeneity among sites? For example, testing the equal-base-frequencies hypothesis can be done with a LRT comparing JC versus F81, because these models only differ in the fact that F81 allows for unequal base frequencies (i.e., alternative hypothesis), whereas JC assumes equal base frequencies (i.e., null hypothesis). However, the hypothesis also could be evaluated by comparing $JC + \Gamma$ versus $F81 + \Gamma$, or $K80 + I$ versus $HKY + I$, and so forth (see Chapter 4 for more details about the models). Which model comparison is used to compare which hypothesis depends on the starting model of the hierarchy and on the order in which different hypotheses are performed. For example, it could be possible to start with the simple JC or with the most complex GTR + I + Γ. In the same way, a test for equal-base frequencies could be performed first, followed by a test for rate heterogeneity among sites, or vice versa. Many hierarchies of LRTs are possible, and some seem to be more effective in selecting the best-fit model (Posada, 2001a; Posada and Crandall, 2001a). An alternative to the use of a particular hierarchy of LRTs is the use of dynamical LRTs described in the next section. The main steps to perform the hierarchical LRTs are as follows:

1. Estimate a tree from the data (i.e., the **base tree**). This tree has been shown to not have influence in the final model selected as far as it is not a random tree (Posada and Crandall, 2001a). A **neighbor-joining** (**NJ**) tree will be fast and will do fine.
2. Estimate the likelihoods of the candidate models for the given data set and the base tree.
3. Compare the likelihoods of the candidate models through a hierarchy of LRTs (Figure 10.1) to select the best-fit model among the candidates.

The hierarchy of tests can be accomplished easily by using the program MODEL-TEST (Posada and Crandall, 1998).

10.4.3 Dynamical LRTs

An alternative to the use of a predefined hierarchy LRT is to let the data itself determine the order in which the hypotheses are tested. In this case, the hierarchy used does not have to be the same for different data sets. The algorithms suggested proceed as follows:

Algorithm 1 (bottom-up)

1. Start with the simplest model and calculate its likelihood. This is the current model.
2. Calculate the likelihood of the alternative models differing by one assumption and perform the corresponding nested LRTs.
3. If any hypotheses are rejected, the alternative model corresponding to the LRT with smallest associated **P-value** becomes the current model. In the case of several equally smallest p-values, select the alternative model with the best likelihood.
4. Repeat Steps 2 and 3 until the algorithm converges.

Algorithm 2 (top-down)

1. Start with the most complex model and calculate its likelihood. This is the current model.
2. Calculate the likelihood of the null models differing by one assumption and perform the corresponding nested LRTs.

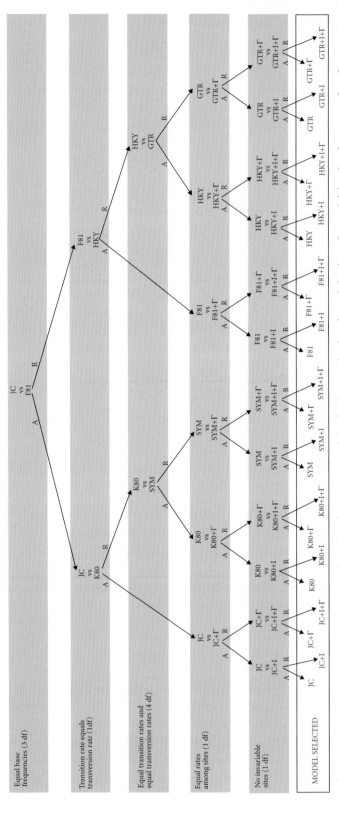

Figure 10.1 A comparison of models of nucleotide substitution. Model-selection methods selected the best-fit model for the data set at hand among 24 possible models. The models of DNA substitution are JC (Jukes and Cantor, 1969), K80 (Kimura, 1980), SYM (Zharkikh, 1994), F81 (Felsenstein, 1981), HKY (Hasegawa et al., 1985), and GTR (Rodríguez et al., 1990). Γ: rate heterogeneity among sites; I: proportion of invariable sites; df: degrees of freedom.

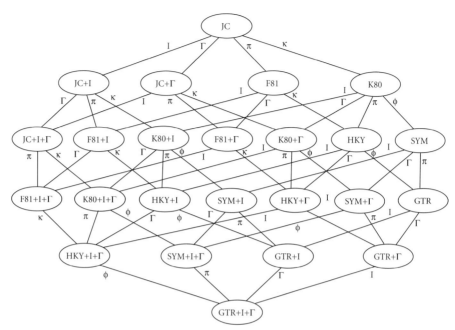

Figure 10.2 Dynamic LRTs. Starting with the simplest (JC) or the most complex model (GTR + I + Γ), LRTs are performed among the current model and the alternative models that maximize the difference in likelihood. π: base frequencies; κ: transition/transversion bias; φ: substitution rates among nucleotides; Γ: rate heterogeneity among sites; I: proportion of invariable sites.

3. If any hypotheses are not rejected, the null model corresponding to the LRT with the biggest associated p-value becomes the current model. In the case of several equally biggest p-values, select the null model with the best likelihood.
4. Repeat Steps 2 and 3 until the algorithm converges.

The alternative paths that the algorithm can generate can be represented graphically (Figure 10.2).

10.5 Information criteria

Whereas the LRTs compare two models at a time, a different approach for model selection is the simultaneous comparison of all competing models. The idea again is to include as much complexity in the model as needed. To do that, the likelihood of each model is penalized by a function of the number of parameters in the model: the more parameters, the bigger the penalty. Two common information criteria are the **Akaike information criterion** (**AIC**) (Akaike, 1974) and the **Bayesian information criterion** (**BIC**) (Schwarz, 1974).

10.5.1   AIC

The AIC is an asymptotically unbiased estimator of the Kullback-Leibler informa-tion quantity (Kullback and Leibler, 1951), which measures the expected distance between the true model and the estimated model. The AIC takes into account not only the goodness of fit, but also the variance of the parameter estimates: the smaller the AIC, the better the fit of the model to the data. An advantage of the AIC is that it also can be used to compare both nested and non-nested models. It is computed as follows:

$$\text{AIC}_i = -2 \log_e L_i + 2 N_i \tag{10.4}$$

where N_i is the number of free parameters in the ith model and L_i is the maximum-likelihood value of the data under the ith model. The AIC calculation is imple-mented in the program MODELTEST.

10.5.2   BIC

The BIC provides an approximate solution to the natural log of the Bayes factor, especially when sample sizes are large and competing hypotheses are nested (Kass and Wasserman, 1994). The Bayes factor measures the relative support that data gives to different models; however, its computation often involves difficult integrals and an approximation becomes convenient. Like the AIC, the BIC can be used to compare nested and nonnested models. Its definition is as follows:

$$\text{BIC}_i = -2 \log_e L_i + N_i \log_e n \tag{10.5}$$

where n is the sample size (sequence length): the smaller the BIC, the better the fit of the model to the data. Because in real data analysis the natural log of n is usually greater than 2, the BIC should tend to choose simpler models than the AIC.

10.6   Fit of a single model to the data

Once a model has been shown to offer a better fit than other models, it is important to assess its general adequacy to the data. To do that, an upper bound to which the likelihood of any model can be compared is needed. This upper bound corresponds to an unconstrained model of evolution, and can be estimated by viewing the sites (i.e., columns) of an alignment as a multinomial sample. The likelihood function under the multinomial distribution for n aligned DNA sequences of length N sites (excluding gapped sites) has the form:

$$L = \prod_{b \in \mathfrak{R}} (p_b)^{n_b} \tag{10.6}$$

where $\mathfrak{R}$ is a set of 4^n possible nucleotide patterns that may be observed at each site, p_b is the probability that any site exhibits the pattern b in $\mathfrak{R}$ given the tree and a substitution model, and n_b is the number of times the pattern b is observed out of the N sites. This comparison provides an idea of how well a particular model explains the observed data. However, this test is very stringent, and most models are usually rejected against the multinomial model. This does not mean that current models are inadequate to provide reasonable estimates, but rather that current models do not provide a perfect description of the underlying evolutionary process. Because a model of evolution is never expected to be correct in every detail, this test is perhaps best used to estimate how far the assumed model deviates from the underlying process that generated the data (Swofford et al., 1996).

Rzhetsky and Nei (1995) also developed several tests using linear invariants for the applicability of a particular model to the data. They tested whether the deviation from the expected invariant would be significant if the evaluated model were true. Although these tests do not require the use of an initial phylogeny and they are independent of evolutionary time, they are model-specific; currently, they can be applied only to a small set of substitution models.

10.7   Testing the molecular clock hypothesis

Between 1962 and 1965, before Kimura postulated the neutral theory of evolution (Kimura, 1968), Zuckerkandl and Pauling published two fundamental papers on the **evolutionary rate** of proteins (Zuckerkandl and Pauling 1962 and 1965). They noticed that the **genetic distance** of two sequences coding for the same protein, but isolated from different species, seems to increase linearly with the divergence time of the two species. Because several proteins showed a similar behavior, Zuckerkandl and Pauling hypothesized that the rate of evolution for any given protein is constant over time. This suggestion implies the existence of a type of molecular clock ticking faster or slower for different genes, but at a more or less constant rate for any given gene among different phylogenetic lineages (Figure 10.3). The hypothesis received almost immediate popularity for several reasons. If a molecular clock exists and the rate of evolution of a gene can be calculated, then this information can easily be used for dating the unknown divergence time between two species just by comparing their DNA or protein sequences. Conversely, if the information about the divergence time between two species (e.g., estimated from fossil data) is known, then the rate of molecular evolution of a given gene can be inferred. Moreover, phylogeny reconstruction is much easier and more accurate under the assumption of a molecular clock (see Chapter 5).

The molecular-clock hypothesis is in perfect agreement with the neutral theory of evolution (Kimura, 1968 and 1983). In fact, the existence of a clock seems to be

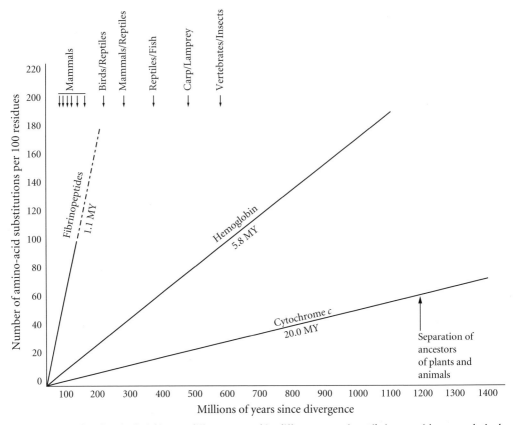

Figure 10.3   Molecular clock ticking at different speed in different proteins. Fibrinopeptides are relatively unconstrained and have a high neutral substitution rate, whereas cytochrome *c* is more constrained and has a lower neutral substitution rate (after Hartl and Clark, 1997).

a major support of the neutral theory against natural selection (see Chapter 1). A detailed discussion of the molecular clock is beyond the scope of this book. Excellent reviews can be found in textbooks of molecular evolution (e.g., Hillis et al., 1996; Li, 1997; Page and Holmes, 1998). The next section focuses more on how to test the clock hypothesis for a group of *taxa* with known phylogenetic relationships.

10.7.1   The relative rate test

According to the molecular-clock hypothesis, two taxa that shared a common ancestor *t* years ago should have accumulated more or less the same number of substitutions during time *t*. In most cases, however, the ancestor is unknown and there is no possibility to directly test the constancy of the evolutionary rate. The problem can be solved by considering an *outgroup*: that is, a more distantly related species (Figure 10.4). Under a perfect molecular clock, d_{AO} – the number of substitutions between taxon A and the outgroup – is expected to be equal to d_{BO} – the number of

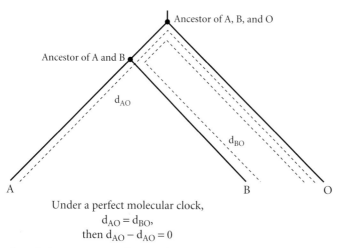

Under a perfect molecular clock,
$$d_{AO} = d_{BO},$$
$$\text{then } d_{AO} - d_{AO} = 0$$

Figure 10.4   The relative rate test. Under a molecular clock, the distance from A to O should be the same as the distance from B to O.

substitutions between taxon B and the outgroup. The relative rate test evaluates the molecular clock hypothesis comparing whether $d_{AO} - d_{BO}$ is significantly different from zero. When this is the case, the sign of the difference indicates which taxon is evolving faster or slower. The relative rate test assumes that the phylogenetic relationships among the taxa are known, which makes the test problematic for taxa, such as the placental mammals with still uncertain phylogeny. In these cases, it would not be a good idea to choose as an outgroup a very distantly related species; a too-distant outgroup means a smaller impact on $d_{AO} - d_{BO}$. In addition, because the more distantly related the outgroup, the higher the probability that multiple substitutions occurred at some sites, the estimation of the genetic distance is less accurate – even employing a sophisticated model of nucleotide substitution (see Chapter 5). A more powerful test for the molecular clock is the LRT.

10.7.2   LRT of the global molecular clock

The phylogeny of a group of taxa is known when the topology and the branch lengths of the phylogenetic tree relating them are known. Of course, whatever the tree topology is, branch lengths can be estimated assuming a constant evolutionary rate along each branch. Clock-like phylogenetic trees are rooted by definition on the longest branch representing the oldest lineage (Figure 10.4). Nonclock-like trees (Figure 10.5A) are unrooted (unless an outgroup is included for rooting the tree; see Chapter 5); in them, a longer branch represents a lineage that evolves faster, which may or may not be an older lineage. Most of the tree-building algorithms, such as the maximum-likelihood, NJ, or *Fitch and Margoliash method*, do not assume a molecular clock; other methods do, such as **UPGMA**.

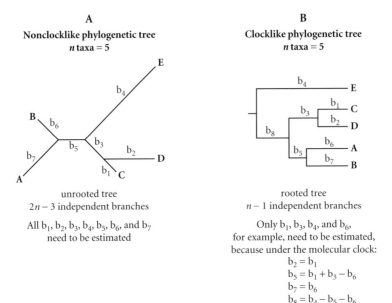

Figure 10.5 Number of free parameters in clock and nonclock trees. Under the free rates model (= nonclock), all the branches need to be estimated ($2n - 3$). Under the molecular clock, only $n - 1$ branches have to be estimated. The difference in the number of parameters among a nonclock and a clock model is $n - 2$.

Maximum-likelihood methods can estimate the branch lengths of a tree by enforcing or not enforcing a molecular clock. In the absence of a molecular clock (the *free-rates model*), $2n - 3$ branch lengths must be inferred for a strictly bifurcating unrooted phylogenetic tree with n taxa (Figure 10.5B). If the molecular clock is enforced, the tree is rooted, and just $n - 1$ branch lengths need to be estimated (see Figure 10.4 and Chapter 1). This should appear obvious considering that under a molecular clock, for any two taxa sharing a common ancestor, only the length of the branch from the ancestor to one of the taxa needs to be estimated, the other one being the same. Statistically speaking, the molecular clock is the null hypothesis (i.e., the rate of evolution is equal for all branches of the tree) and represents a special case of the more general alternative hypothesis that assumes a specific rate for each branch (i.e., free-rates model). Thus, given a tree relating n taxa, the LRT can be used to evaluate whether the taxa have been evolving at the same rate (Felsenstein, 1988). In practice, a model of nucleotide (or amino-acid) substitution is chosen and the branch lengths of the tree with and without enforcing the molecular clock are estimated. To assess the significance of this test, the LRT can be compared with a χ^2 distribution with $(2n - 3) - (n - 1) = n - 2$ degrees of freedom, because the only difference in parameter estimates is in the number of branch lengths that needs to be estimated.

A global molecular clock, ticking at the same rate for all taxa, and a free rate (or nonclock) model, with each taxon evolving at its own rate, are not the only possible scenarios. The clock hypothesis also can be relaxed, allowing a constant rate of evolution within a particular clade but assuming different rates for different clades (i.e., a "local clock" model) (Yoder and Yang 2000). A global molecular clock is a special case of a local molecular clock, which at the same time is a special case of a free-rates model. They can be tested against each other with the LRT; a practical example is discussed later in this chapter. Other relaxation of the molecular clock includes clock models for temporally sampled sequences (i.e., dated tips). These sequences are most frequently from viruses or other fast-evolving pathogens that have been isolated over a range of dates.

An analog to the relative rate test exists within the likelihood framework (Muse and Weir, 1992). Given three nucleotide, amino-acid, or codon sequences and a relevant substitution model, the MLEs can be calculated for the unconstrained three-taxa tree and then for the 3-taxa tree with parameters along two branches constrained to be equal (i.e., the 3rd branch is the outgroup and is estimated independently). A LRT is then performed to determine whether the alternative hypothesis (i.e., all rates are independent) should be accepted or rejected, with the null hypothesis being rates are equal along two given branches.

Molecular-clock calculations based on likelihood methods have been used to date back the origin of viral epidemics, such as the HIV-1 pandemic (Korber et al., 2000; Salemi et al., 2000b), to study substitution dynamics in HIV-1 (Posada and Crandall, 2001b), and to investigate the origin and evolution of the primate T-lymphotropic viruses (PTLVs) (Salemi et al., 1999; Salemi, Desmyter, and Vandamme 2000a). Finally, before applying the LRT for the molecular clock, several precautions need to be taken; specifically, *recombination* has been found to confound this test in such a way that the molecular clock is rejected, when in fact all the lineages are evolving at the same rate (Schierup and Hein, 2000). However, this difficulty can be overcome by using relative ratio tests (Posada, 2001b).

PRACTICE

David Posada

10.8 The model-selection procedure

The different model-selection strategies described in the theory section depend on the estimation of likelihood scores, which can be accomplished in programs like PAUP* (Swofford, 1998), PAML (Yang, 1997), PAL (Drummond and Strimmer, 2001), or HYPHY (Muse and Kosakovsky, 2000). This section demonstrates how to use PAUP* (see Chapter 7) for selecting models of nucleotide substitution and PAML (Box 10.1) for selecting models of amino-acid replacement.

After the likelihood values of the different candidate models are calculated, the model-selection strategies can be applied easily by hand. In the case of nucleotide substitution models, a user-friendly program called MODELTEST (Posada and Crandall, 1998) facilitates this task. The main steps in the model-selection procedure are as follows:

1. Estimate a tree.
2. Calculate the maximum likelihood of the candidate models, given the data and the tree. This provides the MLEs for the parameters of the model.
3. Compare the likelihood of these models using LRTs or information criteria (i.e., AIC or BIC) to select the best-fit model for the data.

Once a model has been selected, it may be interesting to estimate the parameters of the model (e.g., base frequencies, substitution rates, rate variation) while estimating genetic distances or searching for the best phylogeny, given the model and the data. The user might also want to perform an LRT of the molecular clock using this best-fit model. In fact, the LRT of the molecular clock might be viewed as further model testing; that is, considering the molecular clock as just another parameter that might be added to the model. The first step in the model-selection procedure is the estimation of a tree. In the Theory section of this chapter, this tree was called the base tree. The name comes from the fact that the tree is used only to estimate parameters and likelihoods of different models, rather than being considered the final estimate of the phylogenetic relationships among the taxa under investigation. In fact, it has been shown that as long as this tree is a reasonable estimate of the phylogeny (i.e., a maximum-parsimony or NJ tree; never use a random tree!), the parameter estimates and the model selected will be appropriate. An initial tree can be easily estimated in standard phylogenetic programs like PAUP* or PHYLIP. Next, the maximum likelihood for each model, given the base tree and the data, needs to be calculated. In practice, the likelihood and the free parameters of the models

Box 10.1 The PAML package

PAML (Phylogenetic Analysis by Maximum Likelihood) is a freeware software package for phylogenetic analysis of nucleotide and amino-acid sequences using maximum likelihood. Self-extracting archives for MacOs, Windows, and UNIX are available from **http://abacus.gene.ucl.ac.uk/ software/paml.html**. The self-extracting archive creates a PAML directory containing several executable applications (extension .exe in Windows or application icons in MacOs), the compiled files (extension .c, placed in the subdirectory src), an extensive documentation (in the doc subdirectory), and several files with example data sets. Each PAML executable also has a corresponding *control file*, with the same name but the extension .ctl, which needs to be edited with a text editor before running the module. For example, the program baseml.exe has a control file called baseml.ctl, which can be opened with any text editor and looks like the following:

```
      seqfile = hivALN.phy * sequence data file name
      outfile = hivALN.out       * main result file
     treefile = hivALN.tre   * tree structure filename

        noisy = 3   * 0,1,2,3: how much rubbish on the screen
      verbose = 0   * 1: detailed output,  0: concise output
      runmode = 0   * 0: user tree;  1:  semi-automatic;  2: automatic
                    * 3: StepwiseAddition;  (4,5) :PerturbationNNI

        model = 4   * 0:JC69, 1:K80, 2:F81, 3:F84, 4:HKY85, 5:TN93,
                        6:REV, 7:UNREST
        Mgene = 0   * 0:rates, 1:separate; 2:diff pi, 3:diff kapa,
                        4:all diff

    fix_kappa = 0   * 0: estimate kappa; 1: fix kappa at value below
        kappa = 5 * initial or fixed kappa

    fix_alpha = 0    * 0: estimate alpha; 1: fix alpha at value below
        alpha = 0.3  * initial or fixed alpha, 0:infinity (constant
                        rate)

      Malpha = 0   * 1: different alpha's for genes, 0: one alpha
        ncatG = 8   * # of categories in the dG, AdG, or nparK models
                        of rates

        clock = 0   * 0:no clock, 1:clock; 2:local clock; 3:TipDate
        nhomo = 0   * 0 & 1: homogeneous, 2: kappa for branches, 3:
                        N1, 4: N2

        getSE = 0   * 0: don't want them, 1: want S.E.s of estimates
 RateAncestor = 1   * (0,1,2): rates (alpha>0) or ancestral states
```

Box 10.1 (continued)

```
*    Small_Diff = 4e-7
*      cleandata = 0    * remove sites with ambiguity data (1:yes, 0:no)?
*          ndata = 1

*          icode = 0    * (with RateAncestor=1. try "GC" in
                              data,model=4,Mgene=4)
           method = 0   * 0: simultaneous; 1: one branch at a time
```

Each executable has a similar control file. The software modules included in PAML usually require an alignment and a tree topology as input. Users have to edit the control file corresponding to the application they want to employ. This editing consists of adding the name of the sequence input file (next to the = sign of the control variable: seqfile= hivALN.phy in the previous example), adding the name of the file containing one or more phylogenetic trees for the data set under investigation (next to the = sign of the control variable: treefile= hivALN.tre in the previous example), and specifying a name for the output file where results of the computation will be written (outfile = hivALN.out). Other control variables are used to choose among different types of analysis. For example, baseml.exe can estimate maximum-likelihood parameters of a number of **nucleotide substitution models** (see Chapter 4), given a set of aligned sequences and a tree. The control variable of baseml.ctl that needs to be edited in order to choose a model is, in fact, model. In the previous example, by assigning model = 4, the HKY85 substitution model is chosen (see Section 4.6); most of the other control variables are self-explanatory as well. After editing and saving the control file, the corresponding application (.exe extension) can be executed by simply double-clicking on its icon both in MacOS and Windows. Detailed documentation included in the PAML package (doc subdirectory) should be read before using the software.

PAML software modules

The PAML software modules discussed throughout this book are summarized here. Information about the other modules can be found in the PAML documentation.

PAML software module	Input files	Output
baseml.exe	aligned *nt* sequences, phylogenetic tree The tree also can be estimated by baseml.exe choosing runmode = 2 (or 3 or 4) in the control file	ML estimates of different *nt* substitution models
codeml.exe (see also Section 8.9)	aligned *nt* coding (or amino-acid) sequences, phylogenetic tree The tree also can be estimated by codeml.exe choosing runmode = 2 (or 3 or 4) in the control file.	ML estimates of different amino-acid and nucleotide coding substitution models

| yn00.exe | aligned *nt* coding sequences | Analysis of synonymous and nonsynonymous replacements in coding sequences with the YN98 method (see Box 11.1 and Chapter 11) |

PAML input files format

The PAML format is a "relaxed" PHYLIP format (see Box 2.1). Taxa names can be longer than 10 characters and must have at least two blank spaces before starting with the actual sequence. Input trees can be in the usual Newick format (see Figure 5.4). More details can be found in the PAML documentation (in the doc subfolder of the PAML folder).

being compared can be obtained with programs such as PAUP* or PAML. Once the likelihoods of the different models have been obtained, it is straightforward to apply the LRTs or the AIC procedures. This can be done manually with pencil and paper (and maybe a calculator). Moreover, in the case of the LRTs, a chi-square table is also needed to obtain the p-values. If the number of models compared is high – say, 24 or more models – the model-selection procedure can be tedious. The program MODELTEST (Posada and Crandall, 1998) was designed to help in this task.

10.9   The program MODELTEST

MODELTEST is a simple program written in ANSI C and compiled for the Power Macintosh and Windows 95/98/NT using Metrowerks CodeWarrior and for Sun machines using GCC. The MODELTEST package is available for free and can be downloaded from the Web page at **http://bioag.byu.edu/zoology/crandall_lab/ modeltest.htm**. MODELTEST is designed to compare the likelihood of different nested models of DNA substitution and select the best-fit model for the data set at hand.

The input of MODELTEST is a text file containing a matrix of the log-likelihood scores, corresponding to each one of the 24 nucleotide substitution models shown in Figure 10.1, for a specific data set. Such an input file can be generated by executing a particular block of PAUP* commands (Box 10.2), which are written in the modelblock file included in the MODELTEST package. To test different evolutionary models for a given nucleotide data set, first the sequence input file (in NEXUS format) must be executed in PAUP* (see Chapter 7). Then, the modelblock file can be executed with the data in memory. These commands will make PAUP* estimate an NJ tree, calculate the likelihood and parameters of

---

Box 10.2 PAUP* command files

Chapter 7 discusses how to use the PAUP* program by entering commands/options through the command-line interface. Instead of typing all the commands in the command line one by one, separated by a semicolon (see Chapter 7), the user can save them in a text-only document within a so-called PAUP command block, beginning with the keywords Begin PAUP; and ending with the keyword END; (do not forget the semicolon!). For example, a command block could look like the following:

```
BEGIN PAUP;

Set criterion=distance ;

Dset Distance=JC ;

NJ ;

Lset Rates=gamma Shape=Estimate TRatio=Estimate;

Lscore ;

END ;
```

This file could be saved with the .nex extension and successively executed in PAUP*. Such command files, or batch files, are directly executable in PAUP* through the Open . . . item in the File menu. The advantage is that PAUP* users can write their own scripts to perform complex phylogenetic searches and save them for further analyses. Moreover, such scripts often can be modified easily to perform the same or a similar analysis on different data sets.

---

the 24 different models, and save the scores to a file called model.scores, which will be the input file for MODELTEST.

The output of MODELTEST consists of a description of the hierarchical LRT and AIC strategies. For hierarchical LRTs, the particular LRTs performed and their associated p-values are listed, and the model selected with the corresponding parameter estimates (actually calculated by PAUP*) is described. The program also indicates the AIC values and describes the model selected (the one with the smallest AIC) with the corresponding parameter estimates. The output of MODELTEST also provides a block of commands in NEXUS format, which can be executed in PAUP* with the sequence data in memory to automatically implement the selected model. This is useful if the user wants to implement the selected model in PAUP* for further analysis (e.g., to perform an LRT of the molecular clock or to estimate a phylogenetic tree using the best-fit model).

In summary, testing nucleotide substitution models with MODELTEST consists of the following steps:

1. Open the data file and execute it in PAUP*.
2. Execute the command file modelblock3 located in the MODELTEST folder. PAUP* estimates an NJ tree and the likelihood and parameter values for several

models. The task can take from several minutes to several hours, depending on the number of taxa and the computer speed. Once finished, a file called `model.scores` will appear in the same directory as the `modelblock` file.

3. Execute MODELTEST with the file `model.scores`, output from the previous step, as input file. The Mac version of the program has a command-line interface asking the user to select an input file and choose a name for the output file. The PC version requires `model.scores` to be in the same directory where `modeltest.exe` is (this directory, called `Modeltest`, is created during installation of the program). When executing the program, an MS-DOS window appears. To implement the computation, type `modeltest.exe < model.scores > outfile` and press `enter`. The program will save the `outfile` with results in the same directory.

10.10   Implementing the LRT of the molecular clock using PAUP*

Once a substitution model has been selected by MODELTEST, the LRT of the molecular clock can be performed using the current likelihood of the model and a new likelihood can be calculated enforcing a molecular clock on the tree. As discussed in the previous section, the execution of `modelblock3` makes PAUP* infer a simple NJ tree with Jukes and Cantor distances, and uses the tree to estimate likelihood and parameters of the other evolutionary models as well. Therefore, it is possible to evaluate the clock hypothesis by calculating the likelihood of the rooted version of this tree enforcing a molecular clock. The likelihood of such a tree can be compared in an LRT with the likelihood obtained for the correspondent nonclock model, which can be found in the MODELTEST output file. The calculation can be implemented in PAUP* as follows:

1. Infer the NJ tree in PAUP* by executing the **PAUP command block** (see Section 7.8) as follows:

    ```
    BEGIN PAUP;

    DSet distance=JC objective=ME base=equal rates=equal
    pinv=0 subst=all negbrlen=setzero;

    NJ showtree=no breakties=random;

    END;
    ```

    This is precisely the first command block of `modelblock3`. It computes a simple NJ tree with distances estimated with the Jukes and Cantor model.

2. The tree has to be rooted to implement the clock parametrization. This can be achieved with the `root` command, either by choosing an outgroup, if available, or by midpoint rooting.

3. As discussed in the previous section, the output file of MODELTEST contains a command block specifying the parameters of the selected model. Add the PAUP*

command `clock=yes` to the end of the `Lset` block, before the semicolon, and save the entire command block in a separate document as text-only using any text editor. Eventually, the command block will be something like the following:

```
BEGIN PAUP;

Lset Base=(0.4159 0.2281 0.1269) Nst=6 Rmat=(1.0000
2.8596 1.0000 1.0000 5.7951) Rates=gamma Shape=0.6806
Pinvar=0.1698 clock=yes;

END;
```

This PAUP* command block, for example, provides the likelihood settings for the TN $+ \Gamma + I$ model. It specifies the *relative rate parameters* of the distance matrix, the shape parameter of the Γ-*distribution* ($\alpha = 0.6806$ in this case), and the proportion of invariable sites (Pinvar $= 0.1698$), which have all been estimated by PAUP* when executing `modelblock3` (see Chapter 7) using the same NJ tree in memory.

4. Execute the command block in PAUP* (see Chapter 7). The program estimates the log likelihood of the model under the molecular clock, with L_0 representing the probability of the null hypothesis. The log likelihood of the model not enforcing the clock, L_1, is the log likelihood of the selected model written in the output file of MODELTEST.

5. The LRT can now be done manually. Calculate $\Delta = 2^* (L_1 - L_0)$. Because both values L_0 and L_1 are negative, but being that L_1 is bigger than L_0, the Δ value should be positive. The number of degrees of freedom will be the number of taxa -2 (see Section 10.7). The corresponding p-value can be found in a chi-square table. Alternatively, MODELTEST can be used to implement the LRT. Execute the program with the option -c in the argument line (see documentation for different operating systems). Input $|L_0|$ (the absolute value of L_0), $|L_1|$ (the absolute value of L_1), and the number of degrees of freedom (number of taxa -2).

The p-value is interpreted as the probability of observing the obtained LRT statistic (Δ) if the taxa are evolving according to a molecular clock. In other words, if this value is smaller than 0.05 (or 0.01, if a less conservative test is preferred), the molecular clock hypothesis is rejected. When the p-value is marginally significant (close to 0.10–0.01), a more strict way of performing the LRT test would be to use a maximum-likelihood tree. In such a case, first estimate the ML tree with the best-fitting model – which also gives the likelihood of the model without assuming a clock – and then estimate the likelihood of the same model enforcing the clock on the tree.

10.11   Selecting the best-fit model in the example data sets

The first two example data sets were analyzed as described previously using MODELTEST and PAUP*. The candidate models compared were JC, JC + I,

Table 10.1A Hierarchical LRT of models of molecular evolution for the mtDNA data

Null hypothesis	Models compared	−ln Likelihoods	LRT $2(\ln L_1 - \ln L_0)$	df	P-value
Equal base frequencies	H_0: JC69 H_1: F81	−ln L_0: 23646 −ln L_1: 23436	420	3	<0.000001
Equal ti/tv rates	H_0: F81 H_1: HKY	−ln L_0: 23436 −ln L_1: 23009	854	1	<0.000001
Equal ti and equal tv rates	H_0: HKY H_1: GTR	−ln L_0: 23009 −ln L_1: 22677	664	4	<0.000001
Equal rates among sites	H_0: GTR H_1: GTR + Γ	−ln L_0: 22677 −ln L_1: 21156	3042	1	<0.000001
Proportion of invariable sites	H_0: GTR + Γ H_1: GTR + Γ + I	−ln L_0: 21156 −ln L_1: 21148	16	1	0.000063
Molecular clock	H_0: GTR + Γ + I/clock H_1: GTR + Γ + I	−ln L_0: 21192 −ln L_1: 21148	88	15	<0.000001

JC + I + Γ, K80, K80 + I, K80 + I + Γ, SYM, SYM + I, SYM + I + Γ, F81, F81 + I, F81 + I + Γ, HKY, HKY + I, HKY + I + Γ, GTR, GTR + I, and GTR + I + Γ, where I means there is a significant proportion of invariable sites, and Γ means a gamma distribution is being used to account for rate variation among sites.

10.11.1 Vertebrate mtDNA

The first data set is an alignment of mitochondrial sequences from several vertebrates. The model selected by the hierarchical LRTs, dynamical LRTs, AIC, and BIC was the GTR + I + Γ model (Table 10.1A), which is the most complex of the candidate models. In this model, base frequencies are unequal, the six possible types of substitutions among the different bases occur at different rates, there is a significant proportion of invariable sites (p-inv), and some sites evolve faster than others because α, the shape parameter of the gamma distribution, is smaller than 1 (Table 10.1B). The estimates obtained characterize the molecular evolution of this gene: A is the most frequent nucleotide, the most common substitution is between C and T, around 16% of the sites are invariable, and there is medium-rate heterogeneity. As explained in the Theory section, for an idea of how much this model describes the data, it is possible to compare its likelihood to the likelihood of the multinomial model. The log of the likelihood rather than the likelihood itself is

Table 10.1B Estimated parameters for the GTR + Γ + I model

nt Frequencies					Relative substitution rates (G ↔ T = 1.0)						
−lnL	π_A	π_C	π_G	π_T	A ↔ C	A ↔ G	A ↔ T	C ↔ G	C ↔ T	p-inv	α
21148	0.38	0.22	0.18	0.22	3.41	5.10	3.51	0.44	14.99	0.16	0.73

Table 10.2 Hierarchical LRT of models of molecular evolution for the HIV *env* data

Null hypothesis	Models compared	−ln Likelihoods	LRT $2(\ln L_1 - \ln L_0)$	df	P-value
Equal base	H_0: JC69	−ln L_0: 22100	366	3	<0.000001
frequencies	H_1: F81	−ln L_1: 21917			
Equal ti/tv	H_0: F81	−ln L_0: 21917	748	1	<0.000001
rates	H_1: HKY	−ln L_1: 21543			
Equal ti and	H_0: HKY	−ln L_0: 21543	154	4	<0.000001
equal tv rates	H_1: GTR	−ln L_1: 21466			
Equal rates	H_0: GTR	−ln L_0: 21466	1638	1	<0.000001
among sites	H_1: GTR + Γ	−ln L_1: 20647			
Proportion of	H_0: GTR + Γ	−ln L_0: 20647	10	1	0.001565
invariable sites	H_1: GTR + Γ + I	−ln L_1: 20642			
Molecular	H_0: GTR + Γ + Ic	−ln L_0: 20653	22	12	0.037520
clock	H_1: GTR + Γ + I	−ln L_1: 20642			

compared, because likelihoods are often too small to be handled appropriately by standard computers.

To estimate the likelihood of the multinomial model in PAUP*, it is necessary to eliminate positions with ambiguities from the data. The log likelihoods of the GTR + I + Γ model and the multinomial model are then, respectively, −20810 and −11099, which indicates that the selected model does not explain much of the data.

The LRT of the molecular clock can be performed as described previously. The log likelihood of the GTR + I + Γ model under the molecular clock is −21192, which is significantly smaller than the likelihood without assuming a clock (see Table 10.1). Consequently, the global molecular clock hypothesis is rejected for this data set.

10.11.2 HIV envelope gene

The second data set is the *env* gene from HIV and SIV. Again, the model selected by the hierarchical LRTs, dynamical LRTs, AIC, and BIC was the GTR + I + Γ model (Table 10.2). A is the most frequent nucleotide, the most common substitution is between A and G, approximately 13% of the sites have not changed, and there is medium-rate heterogeneity.

Again, to estimate the likelihood of the multinomial model in PAUP*, ambiguities were eliminated from the data. After doing that, the log likelihoods of the GTR + I + Γ model and the multinomial model are, respectively, −15080 and −11361, which indicates that the selected model reasonably explains the data.

The log likelihood of the GTR + I + Γ model under the molecular clock is −20653, which is significantly smaller than the likelihood without assuming a clock (see Table 10.2). However, the p-value is close to being nonsignificant;

Table 10.3 AIC values for different models of amino-acid replacement in the enzyme glycerol-3-phosphate dehydrogenase in bacteria

Model[1]	$-\ln L$	α^2	Free parameters	AIC
Poisson	7704	$\propto$	0	15408
Proportional	7533	$\propto$	19	15104
Empirical				
Jones	7202	$\propto$	0	14404
Dayhoff	7246	$\propto$	0	14492
WAG	7117	$\propto$	0	14234
Empirical + F				
Jones	7208	$\propto$	19	14454
Dayhoff	7246	$\propto$	19	14530
WAG	7110	$\propto$	19	14258
REVAA_0	7205	$\propto$	93	14596
Poisson + Γ	7650	2.5	1	15302
Proportional + Γ	7476	2.3	20	14992
Empirical + Γ				
Jones	7094	1.66	1	14190
Dayhoff	7125	1.56	1	14252
WAG	7043	2.13	1	**14088**
Empirical + F + Γ				
Jones	7099	1.66	20	14238
Dayhoff	7124	1.54	20	14288
WAG	7037	2.13	20	14114
REVAA_0 + Γ	7076	0.004	94	14340

Note: Likelihood values were estimated in PAML 3.0b (Yang, 1997). The model with smallest AIC value is in boldface.

[1] Poisson (Zuckerkandl and Pauling, 1965), Proportional (Hasegawa and Fujiwara, 1993), Jones (Jones et al., 1992), Dayhoff (Dayhoff et al., 1978; Kishino et al., 1990), WAG (Whelan and Goldman, in press), REVAA_0 (Yang et al., 1998); + F: including amino-acid frequencies observed form the data; + Γ: including rate variation as desribed by the gamma distribution.
[2] α is the shape parameter of the gamma distribution.

therefore, the conclusion is not definitive. A larger and more representative HIV data set would be needed to address the issue.

10.11.3 G3PDH protein

The third data set is an amino-acid alignment of the enzyme glycerol-3-phosphate dehydrogenase in bacteria, protozoa, and animals. Because not all models compared (see Chapter 8) are nested, the AIC criterion was used in this case. The model with the best AIC values was the empirical model with the WAG amino-acid replacement matrix (Whelan and Goldman, 2001) with rate variation among sites ($WAG + \Gamma$). The estimated value of the shape parameter of the gamma distribution was 2.13, which indicates that there is moderate rate variation among sites. To estimate the likelihood of the multinomial model in PAML, ambiguities were eliminated from the data. After removing ambiguous positions, the log likelihoods of the WAG model and the multinomial model are, respectively, -5523 and -1574, which indicates that the selected model inadequately explains the data. The log likelihood of the WAG $+ \Gamma$ model under the molecular clock is -7119, which is significantly smaller than the likelihood without assuming a clock (Table 10.3). Consequently, the molecular clock hypothesis should be rejected.

REFERENCES

Akaike, H. (1974). A new look at the statistical model identification. *IEEE Transactions on Automatic Control*, 19, 716–723.

Dayhoff, M. O., R. M. Schwartz, and B. C. Orcutt (1978). A model of evolutionary change in proteins. In: *Atlas of Protein Sequence and Structure*, ed. M. O. Dayhoff, pp. 345–352. Washington, DC.

Drummond, A. and K. Strimmer (2001). PAL: An object-oriented programming library for molecular evolution and phylogenetics. *Bioinformatics*, 17, 662–663.

Felsenstein, J. (1981). Evolutionary trees from DNA sequences: A maximum-likelihood approach. *Journal of Molecular Evolution*, 17, 368–376.

Felsenstein, J. (1988). Phylogenies from molecular sequences: Inference and reliability. *Annual Reviews in Genetics*, 22, 521–565.

Goldman, N. (1993a). Simple diagnostic statistical test of models of DNA substitution. *Journal of Molecular Evolution*, 37, 650–661.

Goldman, N. (1993b). Statistical tests of models of DNA substitution. *Journal of Molecular Evolution*, 36, 182–198.

Hartl, D. L. and A. G. Clark (1997). *Principles of Population Genetics*. Sunderland, MA: Sinauer Associates, Inc.

Hasegawa, M. and M. Fujiwara (1993). Relative efficiencies of the maximum-likelihood, maximum-parsimony, and neighbor-joining methods for estimating protein phylogeny. *Molecular Phylogenetics and Evolution*, 2, 1–5.

Hasegawa, M., K. Kishino, and T. Yano (1985). Dating the human-ape splitting by a molecular clock of mitochondrial DNA. *Journal of Molecular Evolution*, 22, 160–174.

Hillis, D. M., C. Moritz, and B. K. Mable (1996). *Molecular Systematics*. Sunderland, MA: Sinauer Associates, p. 655.

Huelsenbeck, J. P. and K. A. Crandall (1997). Phylogeny estimation and hypothesis testing using maximum likelihood. *Annual Reviews in Ecological Systems*, 28, 437–466.

Huelsenbeck, J. P. and B. Rannala (1997). Phylogenetic methods come of age: Testing hypothesis in an evolutionary context. *Science*, 276, 227–232.

Huelsenbeck, J. P., D. M. Hillis, and R. Jones (1996). Parametric bootstrapping in molecular phylogenetics: Applications and perfomance. In: *Molecular Zoology: Advances, Strategies, and Protocols*, eds. J. D. Ferraris and S. R. Palumbi, pp. 19–45. New York: Wiley-Liss.

Jones, D. T., W. R. Taylor, and J. M. Thornton (1992). The rapid generation of mutation data matrixes from protein sequences. *Computer Applications in the Biosciences*, 8, 275–282.

Jukes, T. H. and C. R. Cantor (1969). Evolution of protein molecules. In: ed. H. M. Munro, *Mammalian Protein Metabolism*, pp. 21–132. New York: Academic Press.

Kass. R. E. and L. Wasserman (1994). *A Reference Bayesian Test for Nested Hypotheses and Its Relationship to the Schwarz Criterion*. Pittsburgh, PA: Carnegie Mellon University, Department of Statistics, p. 16.

Kimura, M. (1968). Evolutionary rate at the molecular level. *Nature*, 217, 624–626.

Kimura, M. (1980). A simple method for estimating evolutionary rate of base substitutions through comparative studies of nucleotide sequences. *Journal of Molecular Evolution*, 16, 111–120.

Kimura, M. (1983). *The Neutral Theory of Molecular Evolution*. Cambridge: Cambridge University Press.

Kishino, H., T. Miyata, and M. Hasegawa (1990). Maximum likelihood inferences of protein phylogeny and the origin of chloroplasts. *Journal of Molecular Evolution*, 31, 151–160.

Korber, B., M. Muldoon, J. Theiler, F. Gao, R. Gupta, A. Lapedes, B. H. Hahn, S. Wolinsky, and T. Bhattarcharya (2000). Timing the ancestor of the HIV-1 pandemic strains. *Science*, 288, 1789–1796.

Kullback, S. and R. A. Leibler (1951). On information and sufficiency. *Annals of Mathematical Statistics*, 22, 79–86.

Li, W.-H. (1997). *Molecular Evolution*. Sunderland, MA: Sinauer Associates.

Liò, P. and N. Goldman (1998). Models of molecular evolution and phylogeny. *Genome Research*, 8, 1233–1244.

Muse, S. V. and S. L. Kosakovsky (2000). *HYPHY: Hypothesis testing using phylogenies*. Raleigh, NC: North Carolina State University, Department of Statistics, Program in Statistical Genetics.

Muse, S. V. and B. S. Weir (1992). Testing for equality of evolutionary rates. *Genetics*, 132, 269–276.

Page, R. D. M. and E. C. Holmes (1998). *Molecular Evolution: A Phylogenetic Approach*. Abingdon, UK: Blackwell Science.

Posada, D. (2001a). The effect of branch-length variation on the selection of models of molecular evolution. *Journal of Molecular Evolution*, 52, 434–444.

Posada, D. (2001b). Unveiling the molecular clock in the presence of recombination. *Molecular Biology and Evolution*, 18, 1976–1978.

Posada, D. and K. A. Crandall (1998). Modeltest: Testing the model of DNA substitution. *Bioinformatics*, 14, 817–818.

Posada, D. and K. A. Crandall (2001a). Selecting the best-fit model of nucleotide substitution. *Systematics Biology*, 50, 580–601.

Posada, D. and K. A. Crandall (2001b). Selecting models of nucleotide substitution: An application to the human immunodeficiency virus 1 (HIV-1). *Molecular Biology and Evolution*, 18, 897–906.

Rodríguez, F., J. F. Oliver, A. Marín, and J. R. Medina (1990). The general stochastic model of nucleotide substitution. *Journal of Theoretical Biology*, 142, 485–501.

Rzhetsky, A. and M. Nei (1995). Tests of applicability of several substitution models for DNA sequence data. *Molecular Biology and Evolution*, 12, 131–151.

Salemi, M., M. J. Lewis, J. F. Egan, W. W. Hall, J. Desmyter, and A.-M. Vandamme (1999). Different population dynamics and evolutionary rates of human T-cell lymphotropic virus type II (HTLV-II) in injecting drug users compared to in endemically infected Amerindian and Pygmy tribes. *Proceedings of the National Academy of Sciences of the USA*, 96, 13253–13259.

Salemi, M., J. Desmyter, and A.-M. Vandamme (2000a). Evolutionary history of human and simian T-lymphotropic viruses (HTLV/STLVs) revealed by analyses of full genome sequences. *Molecular Biology and Evolution*, 3, 374–386.

Salemi, M., K. Strimmer, W. W. Hall, M. Duffy, E. Delaporte, S. Mboup, M. Peeters, and A.-M. Vandamme (2000b). Dating the common ancestor of SIVcpz and HIV-1 group M and the origin of HIV-1 subtypes using a new method to uncover clocklike molecular evolution. *The FASEB Journal*, 2001:15:276–278.

Schierup, M. H. and J. Hein (2000). Recombination and the molecular clock. *Molecular Biology and Evolution*, 17, 1578–1579.

Schwarz, G. (1974). Estimating the dimension of a model. *Annals of Statistics*, 6, 461–464.

Swofford, D. L. (1998). *PAUP* Phylogenetic analysis using parsimony and other methods.* Sunderland, MA: Sinauer Associates.

Swofford, D. L., G. J. Olsen, P. J. Waddell, and D. M. Hillis (1996). Phylogenetic inference. In: *Molecular Systematics*, eds. D. M. Hillis, C. Moritz, and B. K. Mable, pp. 407–514. Sinauer Associates, Sunderland, MA.

Whelan, S. and N. Goldman (1999). Distributions of statistics used for the comparison of models of sequence evolution in phylogenetics. *Molecular Biology and Evolution*, 16, 1292–1299.

Whelan, S. and N. Goldman (2001). A general empirical model of protein evolution derived from multiple protein families using a maximum likelihood approach. *Molecular Biology and Evolution*, 18, 691–699.

Yang, Z. (1996). Maximum-likelihood models for combined analyses of multiple-sequence data. *Journal of Molecular Evolution*, 42, 587–596.

Yang, Z. (1997). PAML: A program package for phylogenetic analysis by maximum likelihood. *Computer Applications in the Biosciences*, 13, 555–556.

Yang, Z., R. Nielsen, and H. Masami (1998). Models of amino-acid substitution and applications to mitochondrial protein evolution. *Molecular Biology and Evolution*, 15, 1600–1611.

Yoder A. D. and Z. Yang (2000). Estimation of primate speciation dates using local molecular clocks. *Molecular Biology and Evolution*, 17, 1081–1090.

Zharkikh, A. (1994). Estimation of evolutionary distances between nucleotide sequences. *Journal of Molecular Evolution*, 39, 315–329.

Zuckerkandl, E. and L. Pauling (1962). Molecular disease, evolution, and genetic heterogeneity. In: *Horizons in Biochemistry*, eds. M. Kasha and B. Pullman, pp. 189–225. Academic Press, New York.

Zuckerkandl, E. and L. Pauling (1965). Evolutionary divergence and convergence in proteins. In: *Evolving Genes and Proteins*, eds. V. Bryson and H. J. Vogel, pp. 97–166. Academic Press, New York.

Analysis of coding sequences

THEORY

Yoshiyuki Suzuki and Takashi Gojobori

11.1 Introduction

Mutants in a population are produced by mutations in the genomes of individuals. The mutation may be a substitution mutation, insertion, deletion, and so forth. Among the mutations, the substitution mutation is one of the most frequent and important factors for producing mutants.

Once the substitution mutations arise, their relative frequencies in the population will change over following generations. Nucleotide substitutions accumulate at a rate that is mainly determined by random genetic drift and natural selection. Roughly speaking, if the relative fitness of a particular mutant (w_m) is similar to the average in the population ($\bar{w}$), the ***rate of nucleotide substitutions*** (r_n; i.e., the rate at which nucleotide substitutions are fixed in the population) is expected to be close to the neutral ***mutation rate*** (r_m; i.e., the rate at which new mutations arise). However, if w_m is higher than $\bar{w}$, r_n is expected to be higher than r_m; if w_m is lower than $\bar{w}$, r_n should be lower than r_m. The former evolutionary mechanism is called ***positive selection***, the latter is called ***negative selection***.

Generally in any codon table, some amino acids are coded by more than one codon (see Table 11.1). Due to this degeneracy, nucleotide substitutions in the protein-coding region can be classified as ***synonymous*** and ***nonsynonymous substitutions***. A synonymous substitution is a nucleotide substitution that does not change the coding amino acid; a nonsynonymous substitution is a nucleotide substitution that changes the encoded amino acid. In this chapter, ***natural selection*** is assumed to operate mainly at the amino-acid sequence level, but only slightly (if at all) at the nucleotide sequence level. This is because most of the important biological functions in the organisms seem to be performed mainly by proteins rather than ***deoxyribonucleic acid*** (***DNA***). Then, the ***rate of synonymous substitutions*** (r_S) may

Table 11.1 Standard codon table[a]

Codon	Amino acid[b]	Codon	Amino acid	Codon	Amino acid	Codon	Amino acid
UUU	Phe	UCU	Ser	UAU	Tyr	UGU	Cys
UUC	Phe	UCC	Ser	UAC	Tyr	UGC	Cys
UUA	Leu	UCA	Ser	UAA	Ter	UGA	Ter
UUG	Leu	UCG	Ser	UAG	Ter	UGG	Trp
CUU	Leu	CCU	Pro	CAU	His	CGU	Arg
CUC	Leu	CCC	Pro	CAC	His	CGC	Arg
CUA	Leu	CCA	Pro	CAA	Gln	CGA	Arg
CUG	Leu	CCG	Pro	CAG	Gln	CGG	Arg
AUU	Ile	ACU	Thr	AAU	Asn	AGU	Ser
AUC	Ile	ACC	Thr	AAC	Asn	AGC	Ser
AUA	Ile	ACA	Thr	AAA	Lys	AGA	Arg
AUG	Met	ACG	Thr	AAG	Lys	AGG	Arg
GUU	Val	GCU	Ala	GAU	Asp	GGU	Gly
GUC	Val	GCC	Ala	GAC	Asp	GGC	Gly
GUA	Val	GCA	Ala	GAA	Glu	GGA	Gly
GUG	Val	GCG	Ala	GAG	Glu	GGG	Gly

[a] Nondegenerate, two-fold degenerate, and four-fold degenerate sites defined by the LWL85 and PBL93 methods are colored with blue, green, and red, respectively.

[b] Amino acids are indicated by three-letter codes; Ala: alanine; Arg: arginine: Asn: aspargine; Asp: aspartate; Cys: cysteine; Gln: glutamine; Glu: glutamate; Gly: glycine; His: histidine; Ile: isoleucine; Leu: leucine; Lys: lysine; Met: methionine; Phe: phenylalanine; Pro: proline; Ser: serine; Thr: threonine; Trp: tryptophan; Tyr: tyrosine; Val: valine; Ter: termination codon.

be more or less similar to r_m, whereas the **rate of nonsynonymous substitutions** (r_N) may vary according to the type and the strength of natural selection. If positive selection takes place, r_N is expected to be faster than r_S; if negative selection operates, then r_N will be slower than r_S. Therefore, natural selection operating at the amino-acid sequence level can be detected by comparing r_N with r_S, where $r_N > r_S$ is an indicator of positive selection and $r_N < r_S$ is an indicator of negative selection (Hughes and Nei, 1988 and 1989).

Both r_S and r_N can be estimated by comparing two nucleotide sequences. First, the number of synonymous substitutions per synonymous site (d_S) and nonsynonymous substitutions per nonsynonymous site (d_N) between two sequences is calculated, and d_S and d_N are divided by twice their divergence time ($2t$); that is, $r_S = d_S/2t$ and $r_N = d_N/2t$. However, t is usually not known for sequences under consideration. Even in such a situation, natural selection can still be detected by comparing d_S and d_N rather than r_S and r_N.

Several methods have been proposed for estimating d_S and d_N between a pair of nucleotide sequences, and can be classified in the following three categories:

(1) *mutation fraction method*, (2) *degenerate site method*, and (3) *codon model method*. The mutation fraction method consists of the methods of Perler et al. (1980), Miyata and Yasunaga (1980), Nei and Gojobori (1986), Kondo et al. (1993), Ina (1995), Zhang et al. (1998), and Yang and Nielsen (2000). The degenerate site method includes the methods of Li et al. (1985), Pamilo and Bianchi (1993) and Li (1993), Comeron (1995), and Moriyama and Powell (1997). The codon model method consists of the methods of Muse and Gaut (1994), Goldman and Yang (1994), Muse (1996), and Yang and Nielsen (1998).

Methods for estimating d_S and d_N at single codon sites throughout the evolution of many nucleotide sequences also have been developed and can be used for detecting natural selection at single amino-acid sites (Nielsen and Yang, 1998; Suzuki, 1999; Suzuki and Gojobori, 1999).

The following sections briefly summarize the algorithms for some of these methods, and demonstrate how they can be used in the actual data analysis using example data. Readers who are interested in this field may also consult Ina (1996) and Nei and Kumar (2000).

11.2 Mutation fraction methods

Mutation fraction methods estimate d_S and d_N between a pair of nucleotide sequences by the following steps: computation of the numbers of synonymous (S_S) and nonsynonymous (S_N) sites per sequence, computation of the numbers of synonymous (C_S) and nonsynonymous (C_N) differences per sequence, and correction of multiple substitutions.

11.2.1 Method of Nei and Gojobori (NG86 method)

To compute S_S and S_N, first classify the nucleotide positions of 61 sense codons into synonymous and nonsynonymous sites. A synonymous site is a nucleotide site in which all possible substitution mutations are synonymous; a nonsynonymous site is a nucleotide site in which all possible substitution mutations are nonsynonymous. For example, the third position of CTT (Leu) is a synonymous site because all possible substitution mutations T → C, T → A, and T → G are synonymous, whereas the first and the second positions of CTT are nonsynonymous sites.

However, some positions of particular codons are not clearly classified as synonymous or nonsynonymous sites. For example, at the third position of TTT (Phe), T → C is synonymous but T → A and T → G are nonsynonymous. In this case, a nucleotide site is divided into fractional numbers of synonymous and nonsynonymous sites, according to the fractions of *synonymous* and *nonsynonymous mutations* at that site (Kondo et al., 1993; Ina, 1995). For example, assuming λ_{ij} as r_m from nucleotide i to j ($j \neq i$), the number of synonymous sites at the third position of TTT is given by $\lambda_{TC}/(\lambda_{TC} + \lambda_{TA} + \lambda_{TG})$. Note that if some substitution mutations produce termination codons, they are ignored in the computation. For example, C $\rightarrow$ A and C $\rightarrow$ G at the third position of TAC (Tyr) produce termination codons. Therefore, the number of synonymous sites at this position is computed as $\lambda_{CT}/\lambda_{CT} = 1$.

The number of synonymous sites $(s_{S(i)})$ for codon i is defined as the sum of the number of synonymous sites at three positions; the number of nonsynonymous sites $(s_{N(i)})$ is given by $3 - s_{S(i)}$. For example, $s_{S(TTT)}$ is $0 + 0 + \lambda_{TC}/(\lambda_{TC} + \lambda_{TA} + \lambda_{TG})$ because all substitution mutations at the first and second positions are nonsynonymous, and $s_{N(TTT)}$ is $3 - s_{S(TTT)}$. In the NG86 method, λ_{ij} is assumed to be the same for any combination of i and j $(j \neq i)$; thus, $s_{S(TTT)} = 1/3$ and $s_{N(TTT)} = 8/3$.

S_S and S_N are computed as the sum of $s_{S(i)}$ and $s_{N(i)}$ at all codon sites, respectively. To estimate d_S and d_N between two sequences, first S_S and S_N for each of the two sequences are computed, then the average numbers for the two sequences are used for further analysis.

The nucleotide differences between two sequences also are classified into synonymous and nonsynonymous differences, by comparing two sequences codon by codon. If two codons at a **homologous** site are different at only one position, the difference is unambiguously classified as a **synonymous** or **nonsynonymous difference**. For example, the difference between TTT (Phe) and TTC (Phe) is synonymous; the difference between CCC (Pro) and CAC (His) is nonsynonymous.

If two codons are different at two positions, it is assumed that these differences are not obtained simultaneously but rather successively, because the probability of the former should be negligibly smaller than the latter. Then there are two possible pathways to obtain the nucleotide differences. For example, the possible pathways between GAC (Asp) and GGA (Gly) are as follows:

Pathway 1: GAC (Asp)-->GGC (Gly)-->GGA (Gly)
Pathway 2: GAC (Asp)-->GAA (Glu)-->GGA (Gly)

Pathway 1 includes one synonymous and one nonsynonymous difference; Pathway 2 includes two nonsynonymous differences. Because it is unknown which pathway has actually taken place, the numbers of synonymous $(c_{S(i, j)})$ and nonsynonymous $(c_{N(i, j)})$ differences between codons i and j are computed as the unweighted averages of those numbers for all pathways. In this example, $c_{S(GAC,GGA)} = (1 + 0)/2 = 1/2$ and $c_{N(GAC,GGA)} = (1 + 2)/2 = 3/2$.

If a termination codon appears in some pathways, it is excluded from the analysis because the termination codon should have destroyed the function of the protein, and the reversion from such a defective protein seems unlikely. For example, the possible pathways between TAT (Tyr) and CAA (Gln) are as follows:

Pathway 1: TAT (Tyr)-->CAT (His)-->CAA (Gln)
Pathway 2: TAT (Tyr)-->TAA (Ter)-->CAA (Gln)

Because Pathway 2 includes a termination codon, only Pathway 1 is considered for computing $c_{S(TAT,CAA)} = 0$ and $c_{N(TAT,CAA)} = 2$.

Similarly, if three positions are different between codons i and j, $c_{S(i,j)}$ and $c_{N(i,j)}$ are computed by taking into account six possible pathways. C_S and C_N are computed as the sum of $c_{S(i,j)}$ and $c_{N(i,j)}$ for all codon sites, respectively.

The proportion of synonymous differences per synonymous site (p_S) and that of nonsynonymous differences per nonsynonymous site (p_N) are estimated by

$$\hat{p}_S = \frac{C_S}{S_S}, \quad \hat{p}_N = \frac{C_N}{S_N} \tag{11.1a,b}$$

with variances

$$V(\hat{p}_S) = \sum_{i,j} \frac{\left\{ c_{S(i,j)} - \hat{p}_S \left(s_{S(i)} + s_{S(j)} \right)/2 \right\}^2}{S_S^2},$$

$$V(\hat{p}_N) = \sum_{i,j} \frac{\left\{ c_{N(i,j)} - \hat{p}_N \left(s_{N(i)} + s_{N(j)} \right)/2 \right\}^2}{S_N^2} \tag{11.2a,b}$$

where $\sum_{i,j}$ is the summation over all codon sites and i and j are codons at a given codon site (Ota and Nei, 1994). The hat on some of the symbols means they are statistical estimates.

To estimate d_S and d_N, multiple substitutions are corrected by the method of Jukes and Cantor (1969) (JC69 method) (see Chapter 4) as

$$\hat{d}_S = -\frac{3}{4} \ln \left(1 - \frac{4}{3} \hat{p}_S \right), \quad \hat{d}_N = -\frac{3}{4} \ln \left(1 - \frac{4}{3} \hat{p}_N \right) \tag{11.3a,b}$$

with variances

$$V(\hat{d}_S) = \frac{V(\hat{p}_S)}{\left(1 - \frac{4}{3} \hat{p}_S \right)^2}, \quad V(\hat{d}_N) = \frac{V(\hat{p}_N)}{\left(1 - \frac{4}{3} \hat{p}_N \right)^2} \tag{11.4a,b}$$

(Ota and Nei 1994) (see Chapter 4).

11.2.2   Method of Zhang et al. (ZRN98 method)

In the NG86 method, r_m from nucleotide i to j ($j \neq i$) is assumed to be the same for any combination of nucleotides i and j for computing S_S and S_N. However, in reality, the rate of *transitional mutations* (α_m) is often higher than that of *transversional mutations* (β_m) (see Chapter 4). In this situation, the NG86 method may underestimate S_S and overestimate S_N because transitional mutations cause synonymous mutations more frequently than transversional mutations (see Table 11.1). For example, for $\alpha_m/\beta_m = 1$ is $s_{S(AAA)} = 1/3$; whereas for $\alpha_m/\beta_m = 5$ is $s_{S(AAA)} = 5/7 \, (>1/3)$ (see Equation 11.1; the same reasoning applies). As a result, p_S and d_S may be overestimated and p_N and d_N may be underestimated by using the NG86 method.

To correct the difference between α_m and β_m, the ZRN98 method assumes the Kimura model (1980) (K80 model; see Chapter 4) for r_m in the computation of S_S and S_N. In general, $s_{S(i)}$ and $s_{N(i)}$ for any codon i are described as the function of α_m/β_m (Kondo et al., 1993; Ina, 1995). For example, $s_{S(AAA)}$ is given by

$$\lambda_{AG}/(\lambda_{AT} + \lambda_{AC} + \lambda_{AG}) = \alpha_m/(\alpha_m + 2\beta_m) = (\alpha_m/\beta_m)/(\alpha_m/\beta_m + 2).$$

Therefore, if α_m/β_m for the nucleotide sequences under consideration is known, S_S and S_N can be computed. In the ZRN98 method, α_m/β_m is estimated by using an appropriate method. C_S, C_N, p_S, p_N, d_S, and d_N are computed the same way as in the NG86 method.

11.2.3 Method of Ina (I95 method)

In the NG86 and ZRN98 methods, r_n from nucleotide i to j $(j \neq i)$ is assumed to be the same for any combination of nucleotides i and j, and the JC69 method is used for correcting multiple substitutions at **synonymous** and **nonsynonymous** **sites.** However, when the rate of **transitional substitutions** (α_n) is higher than that of **transversional substitutions** (β_n), the JC69 method is known to underestimate the number of nucleotide substitutions (see Chapter 4). Therefore, the NG86 and ZRN99 methods also may underestimate d_S and d_N.

In the I95 method, the K80 model is assumed not only for computing S_S and S_N, but also for correcting multiple substitutions. S_S and S_N are computed the same way as in the ZRN98 method. C_S and C_N are computed the same way as in the NG86 method, but the numbers of **synonymous transitional** $(c_{STs(i,j)})$ and **transversional** $(c_{STv(i,j)})$ **differences**, and those of **nonsynonymous transitional** $(c_{NTs(i,j)})$ and **transversional** $(c_{NTv(i,j)})$ **differences** between codons i and j are computed separately. For example, the difference between TTT (Phe) and TTC (Phe) is one synonymous transition, and that between CCC (Pro) and CAC (His) is one nonsynonymous transversion.

When two or three positions are different between two codons, all possible pathways for obtaining the differences are considered for classifying the nucleotide differences, as in the case for the NG86 method. For example, the possible pathways between CTG (Leu) and ATA (Leu) are as follows:

Pathway 1: CTG (Leu)-->ATG (Met)-->ATA (Ile)
Pathway 2: CTG (Leu)-->CTA (Leu)-->ATA (Ile)

Pathway 1 includes one nonsynonymous transition and one nonsynony-mous transversion, and Pathway 2 contains one synonymous transition and one nonsynonymous transversion. Therefore, $c_{STs(CTG,ATA)} = (0+1)/2 = 1/2$, $c_{STv(CTG,ATA)} = (0+0)/2 = 0$, $c_{NTs(CTG,ATA)} = (1+0)/2 = 1/2$, and $c_{NTv(CTG,ATA)} = (1+1)/2 = 1$. The numbers of synonymous transitional (C_{STs}) and transversional (C_{STv}) differences and those of nonsynonymous transitional (C_{NTs}) and transversional (C_{NTv}) differences per sequence are computed as the sum of $c_{STs(i,j)}$, $c_{STv(i,j)}$, $c_{NTs(i,j)}$, and $c_{NTv(i,j)}$ for all codon sites, respectively.

The proportions of transitional (P_S) and transversional (Q_S) differences per synonymous site and those of transitional (P_N) and transversional (Q_N) differences per nonsynonymous site are estimated by

$$\hat{P}_i = \frac{C_{iTs}}{S_i}, \quad \hat{Q}_i = \frac{C_{iTv}}{S_i} \tag{11.5a,b}$$

with variances

$$V\left(\hat{P}_i\right) = \frac{\hat{P}_i\left(1-\hat{P}_i\right)}{S_i}, \quad V\left(\hat{Q}_i\right) = \frac{\hat{Q}_i\left(1-\hat{Q}_i\right)}{S_i} \tag{11.6a,b}$$

where i is S or N. Both d_S and d_N are estimated by using the K80 method as

$$\hat{d}_i = \frac{1}{2}\ln\frac{1}{1-2\hat{P}_i-\hat{Q}_i} + \frac{1}{4}\ln\frac{1}{1-2\hat{Q}_i} \tag{11.7}$$

with variance

$$V\left(\hat{d}_i\right) = \frac{1}{\hat{S}_i}\left\{a_i^2\hat{P}_i + b_i^2\hat{Q}_i - \left(a_i\hat{P}_i + b_i\hat{Q}_i\right)^2\right\} \tag{11.8}$$

where

$$a_i = \frac{1}{1-2\hat{P}_i-\hat{Q}_i}, \quad b_i = \frac{1}{2}\left(\frac{1}{1-2\hat{P}_i-\hat{Q}_i} + \frac{1}{1-2\hat{Q}_i}\right) \tag{11.9a,b}$$

As in the ZRN98 method, α_m/β_m is required for computing S_S and S_N in the I95 method. To estimate α_m/β_m, two methods (1 and 2) are proposed in the I95 method. In Method 1, the numbers of transitional (d_{3Ts}) and transversional (d_{3Tv}) substitutions at the third position of codons are estimated by using the K80 method, and $2\hat{d}_{3Ts}/\hat{d}_{3Tv}$ is used as the estimate for α_m/β_m. Method 2 estimates the numbers of synonymous transitional (d_{STs}) and transversional (d_{STv}) substitutions, and $2\hat{d}_{STs}/\hat{d}_{STv}$ is used for inferring α_m/β_m. Here, however, because most of the synonymous substitutions at the **two-fold** and **three-fold degenerate** sites are transitional substitutions, $2\hat{d}_{STs}/\hat{d}_{STv}$ should be multiplied by $f_4/(f_2 + f_3 + f_4)$ for estimating α_m/β_m. The frequencies of two-fold degenerate, three-fold degenerate, and **four-fold degenerate** sites are f_2, f_3, and f_4, respectively (see the next section and Section 1.1 for the definitions of these sites. Both d_{STs} and d_{STv} are estimated by using the K80 method, as follows:

$$\hat{d}_{STs} = \frac{1}{2}\ln\frac{1}{1 - 2\hat{P}_S - \hat{Q}_S} - \frac{1}{4}\ln\frac{1}{1 - 2\hat{Q}_S}, \quad \hat{d}_{STv} = \frac{1}{2}\ln\frac{1}{1 - 2\hat{Q}_N}$$

$$(11.10a,b)$$

However, in these equations, α_m/β_m is required for estimating P_S and Q_S. Then, the iteration method is used for estimating α_m/β_m. Namely, α_m/β_m is replaced by $(2\hat{d}_{STs}/\hat{d}_{STv}) \cdot \{f_4/(f_2 + f_3 + f_4)\}$ in each iteration cycle until a stable estimate for α_m/β_m is obtained.

Because α_m/β_m is estimated by $2\hat{d}_{3Ts}/\hat{d}_{3Tv}$ in I95 Method 1, this method is not applicable when there is no transitional or transversional substitution at the third positions of codons between two sequences. Similarly, I95 Method 2 is not applicable when there is no transitional or transversional substitution at the synonymous sites between two sequences.

11.3 Degenerate site methods

Degenerate site methods first classify the nucleotide sites into **nondegenerate**, two-fold degenerate, and four-fold degenerate sites, where none, one, and three of the possible substitution mutations are synonymous, respectively (see Section 1.1). For example, the first position of CGG (Arg) is a two-fold degenerate site, because $C \rightarrow A$ is synonymous but $C \rightarrow T$ and $C \rightarrow G$ are nonsynonymous. Similarly, the second and third positions of CGG are nondegenerate and four-fold degenerate sites, respectively. The third positions of ATT (Ile), ATC (Ile), and ATA (Ile) are classified as the two-fold degenerate sites for computational convenience, although they are actually the three-fold degenerate sites (where two of the possible substitution mutations are synonymous). When a pair of nucleotide sequences is compared for estimating d_S and d_N, the numbers of nondegenerate (n_0), two-fold degenerate (n_2), and four-fold degenerate (n_4) sites are computed for each of two sequences, and the average numbers for two sequences are used for further analysis.

11.3.1　Method of Li et al. (LWL85 method)

The nucleotide differences between two sequences are counted separately at the nondegenerate, two-fold degenerate, and four-fold degenerate sites. In addition, transitions and transversions are counted separately at each degenerate site. For example, the difference between ACT (Thr) and ACG (Thr) is classified as a transversion at the four-fold degenerate site because a transversion is observed at the third position, which is the four-fold degenerate in both codons. Similarly, the difference between CAA (Gln) and CAG (Gln) is a transition at the two-fold degenerate site.

Occasionally, however, the homologous positions between two codons do not belong to the same degeneracy group. For example, the first position of TTA (Leu) is a two-fold degenerate site, whereas that of ATA (Ile) is a nondegenerate site. In this situation, one nucleotide difference is divided into two fractional numbers with equal weights. In this example, half the difference is at the two-fold degenerate site and half is at the nondegenerate site.

Moreover, for computational convenience in the LWL85 method, any synonymous difference is regarded as a transition and any nonsynonymous difference is regarded as a transversion at a two-fold degenerate site. Thus, the differences between CGA (Arg) and AGA (Arg) and between CGG (Arg) and AGG (Arg) are regarded as transitions, although they are actually transversions. Similarly, the differences between ATT (Ile) and ATA (Ile) and between ATC (Ile) and ATA (Ile) are regarded as transitions.

When two positions are different between two codons, all possible pathways for obtaining the differences are considered for classifying the nucleotide differences. For example, the possible pathways between GAG (Glu) and GCA (Ala) are as follows:

Pathway 1: GAG (Glu)-->GCG (Ala)-->GCA (Ala)
Pathway 2: GAG (Glu)-->GAA (Glu)-->GCA (Ala)

In Pathway 1, there is one transversion at the nondegenerate site and one transition at the four-fold degenerate site. In Pathway 2, there is one transition at the two-fold degenerate site and one transversion at the nondegenerate site. In the LWL85 method, different pathways are assumed to have occurred with different probabilities. If w_1 and w_2 are the probabilities for the occurrences of Pathways 1 and 2, respectively, the average numbers of nucleotide differences between GAG and GCA are computed as $w_1 \times 1 + w_2 \times 1 = 1$ transversion at the nondegenerate site, $w_1 \times 1 = w_1$ transition at the four-fold degenerate site, and $w_2 \times 1 = w_2$ transition at the two-fold degenerate site. If some pathways include termination codons, they are excluded from the analysis, as in the mutation fraction method. The weights for different pathways between two codons are determined by using the relative likelihood $(l_{(i,j)})$ of substitution between codons i and j, which are different at only one position; $l_{(i,j)}$ is predetermined empirically (for the actual values of $l_{(i,j)}$, see Li et al., 1985). In this example, the relative likelihood for the occurrences of Pathways 1 and 2 are computed as $l_{(GAG,GCG)} \cdot l_{(GCG,GCA)}$ and $l_{(GAG,GAA)} \cdot l_{(GAA,GCA)}$, respectively, and w_1 and w_2 are determined as

$$w_1 = l_{(GAG,GCG)} \cdot l_{(GCG,GCA)} / \left(l_{(GAG,GCG)} \cdot l_{(GCG,GCA)} \right.$$

$$\left. + l_{(GAG,GAA)} \cdot l_{(GAA,GCA)} \right) \quad (11.12a)$$

$$w_2 = l_{(GAG,GAA)} \cdot l_{(GAA,GCA)} / \left(l_{(GAG,GCG)} \cdot l_{(GCG,GCA)} \right.$$

$$\left. + l_{(GAG,GAA)} \cdot l_{(GAA,GCA)} \right) \quad (11.12b)$$

The numbers of transitional (C_{0Ts}) and transversional (C_{0Tv}) differences at the nondegenerate site, those of transitional (C_{2Ts}) and transversional (C_{2Tv}) differences at the two-fold degenerate site, and those of transitional (C_{4Ts}) and transversional (C_{4Tv}) differences at the four-fold degenerate site per sequence are computed as the sums of those numbers for all codon sites.

The proportions of transitional $(\hat{P}_i)$ and transversional $(\hat{Q}_i)$ differences per i-fold degenerate site are estimated by the following equations:

$$\hat{P}_i = \frac{C_{Ts}}{n_i}, \quad \hat{Q}_i = \frac{C_{Tv}}{n_i} \qquad (11.13a,b)$$

with variances

$$V\left(\hat{P}_i\right) = \frac{\hat{P}_i\left(1 - \hat{P}_i\right)}{n_i}, \quad V\left(\hat{Q}_i\right) = \frac{\hat{Q}_i\left(1 - \hat{Q}_i\right)}{n_i} \qquad (11.14a,b)$$

The numbers of transitional (A_i) and transversional (B_i) substitutions per i-fold degenerate site are estimated by using the K80 method as

$$\hat{A}_i = \frac{1}{2}\ln\frac{1}{1 - 2\hat{P}_i - \hat{Q}_i} - \frac{1}{4}\ln\frac{1}{1 - 2\hat{Q}_i}, \quad \hat{B}_i = \frac{1}{2}\ln\frac{1}{1 - 2\hat{Q}_i} \qquad (11.15a,b)$$

with variances

$$V\left(\hat{A}_i\right) = \frac{1}{n_i}\left\{a_i^2\hat{P}_i + b_i^2\hat{Q}_i - \left(a_i\hat{P}_i + b_i\hat{Q}_i\right)^2\right\},$$

$$V\left(\hat{B}_i\right) = \frac{1}{n_i}c_i^2\hat{Q}_i\left(1 - \hat{Q}_i\right) \qquad (11.16a)$$

where

$$a_i = \frac{1}{1 - 2\hat{P}_i - \hat{Q}_i}, \quad b_i = \frac{1}{2}\left(\frac{1}{1 - 2\hat{P}_i - \hat{Q}_i} - \frac{1}{1 - 2\hat{Q}_i}\right),$$

$$c_i = \frac{1}{1 - 2\hat{Q}_i} \qquad\qquad (11.16b,c,d)$$

The total number (d_i) of nucleotide substitutions per i-fold degenerate site is estimated by

$$\hat{d}_i = \hat{A}_i + \hat{B}_i = \frac{1}{2}\ln\frac{1}{1 - 2\hat{P}_i - \hat{Q}_i} + \frac{1}{4}\ln\frac{1}{1 - 2\hat{Q}_i} \qquad (11.17)$$

with variance

$$V\left(\hat{d}_i\right) = \frac{1}{n_i}\left\{a_i^2\,\hat{P}_i + e_i^2\,\hat{Q}_i - \left(a_i\,\hat{P}_i + e_i\,\hat{Q}_i\right)^2\right\} \qquad (11.18a)$$

where

$$e_i = \frac{1}{2}\left(\frac{1}{1 - 2\hat{P}_i - \hat{Q}_i} + \frac{1}{1 - 2\hat{Q}_i}\right) \qquad (11.18b)$$

A four-fold degenerate site is equivalent to a synonymous site because all possible substitution mutations are synonymous. In addition, a nondegenerate site corresponds to a nonsynonymous site because none of the possible substitution mutations are synonymous. If r_m from nucleotide i to j ($j \neq i$) is supposed to be the same for any combination of nucleotides i and j, one-third of the two-fold degenerate sites can be classified as synonymous sites and two-third as nonsynonymous sites. As a result, S_S and S_N can be computed as $S_S = n_4 + n_2/3$ and $S_N = n_0 + 2n_2/3$.

Nucleotide substitution at a four-fold degenerate site is always synonymous; at a nondegenerate site it is always nonsynonymous. Moreover, in the LWL85 method, transitional and transversional substitutions at two-fold degenerate sites are synonymous and nonsynonymous substitutions, respectively. As a result, the number of synonymous substitutions per all synonymous sites in the sequence is computed as $n_4\,\hat{A}_4 + n_4\,\hat{B}_4 + n_2\,\hat{A}_2$; the number of nonsynonymous substitutions is computed as $n_0\,\hat{A}_0 + n_0\,\hat{B}_0 + n_2\,\hat{B}_2$.

Therefore, d_S and d_N are estimated by the following equation:

$$\hat{d}_S = \frac{n_4\,\hat{A}_4 + n_4\,\hat{B}_4 + n_2\,\hat{A}_2}{n_4 + n_2/3}, \quad \hat{d}_N = \frac{n_0\,\hat{A}_0 + n_0\,\hat{B}_0 + n_2\,\hat{B}_2}{n_0 + 2n_2/3} \qquad (11.19a,b)$$

with variances

$$V\left(\hat{d}_S\right) = \frac{9\left\{n_4^2 V\left(\hat{d}_4\right) + n_2^2 V\left(\hat{A}_2\right)\right\}}{(3n_4 + n_2)^2}, \quad V\left(\hat{d}_N\right) = \frac{9\left\{n_0^2 V\left(\hat{d}_0\right) + n_2^2 V\left(\hat{B}_2\right)\right\}}{(3n_0 + 2n_2)^2}$$

$$(11.20a,b)$$

11.3.2 Method of Pamilo and Bianchi, and Li (PBL93 method)

In the LWL85 method, r_m from nucleotide i to j ($j \neq i$) is assumed to be the same for any combination of nucleotides i and j for computing S_S and S_N. However, this assumption may cause an underestimation of S_S and an overestimation of S_N when $\alpha_m > \beta_m$, as discussed previously. To avoid this problem, the PBL93 method does not compute S_S and S_N. Instead, d_S is estimated as the sum of the numbers of transitional (d_{STs}) and transversional (d_{STv}) substitutions per synonymous site, and d_N is estimated as the sum of the numbers of transitional (d_{NTs}) and transversional (d_{NTv}) substitutions per nonsynonymous site. Note that d_{STs}, d_{STv}, d_{NTs}, and d_{NTv} are estimated by

$$\hat{d}_{STs} = \left(n_4 \hat{A}_4 + n_2 \hat{A}_2\right)/(n_4 + n_2), \quad \hat{d}_{STv} = \hat{B}_4, \quad \hat{d}_{NTs} = \hat{A}_0, \quad \text{and}$$

$$\hat{d}_{NTv} = \left(n_0 \hat{B}_0 + n_2 \hat{B}_2\right)/(n_0 + n_2)$$

Therefore,

$$\hat{d}_S = \frac{n_4 \hat{A}_4 + n_2 \hat{A}_2}{n_4 + n_2} + \hat{B}_4, \quad \hat{d}_N = \hat{A}_0 + \frac{n_0 \hat{B}_0 + n_2 \hat{B}_2}{n_0 + n_2} \qquad (11.21a,b)$$

with variances

$$V\left(\hat{d}_S\right) = \frac{n_4^2 V\left(\hat{A}_4\right) + n_2^2 V\left(\hat{A}_2\right)}{(n_4 + n_2)^2} + V\left(\hat{B}_4\right) - \frac{c_4 \hat{Q}_4 \left\{2a_4 \hat{P}_4 - b_4\left(1 - \hat{Q}_4\right)\right\}}{n_4 + n_2}$$

$$(11.22a)$$

$$V\left(\hat{d}_N\right) = V\left(\hat{A}_0\right) + \frac{n_0^2 V\left(\hat{B}_0\right) + n_2^2 V\left(\hat{B}_2\right)}{(n_0 + n_2)^2} - \frac{c_0 \hat{Q}_0 \left\{2a_0 \hat{P}_0 - b_0\left(1 - \hat{Q}_0\right)\right\}}{n_0 + n_2}$$

$$(11.22b)$$

where a_i, b_i, and c_i ($i = 0, 4$) are as indicated in the previous section.

11.4 Codon model methods

Codon model methods first assume a substitution model among 61 sense codons with the parameters for r_S and r_N (Gojobori, 1983). Then, the maximum-likelihood method is used for obtaining estimates of r_S and r_N, which are used for estimating d_S and d_N.

11.4.1 Method of Muse (M96 method)

In the M96 method, the 61×61 substitution rate matrix $Q = \{q_{ij}\}$ among sense codons is assumed so that the substitution rate q_{ij} from codon i to j ($j \neq i$) is $\alpha \pi_{n_{ij}}$ for synonymous substitutions and $\beta \pi_{n_{ij}}$ for nonsynonymous substitutions, where α and β are the relative rate of synonymous and nonsynonymous substitutions, respectively. Here, $\pi_{n_{ij}}$ stands for the frequency of nucleotide n_{ij}, and n_{ij} is the nucleotide l if the substitution from codon i to j ($j \neq i$) requires the substitution from nucleotide k to l. Consider, for example, that TTT is i, TTC is j, and $\pi_{n_{ij}}$ is the frequency of C. The frequencies of nucleotides are estimated from the data. q_{ij} is assumed to be 0 when more than one position is different between codons i and j. q_{ii} is set as $q_{ii} = -\sum_j q_{ij}$ for any i, so that the sum of any changes (q_{ij}) and no changes (q_{ii}) equals 0 for any i.

The **transition probability matrix** among 61 sense codons during time period t, $P(t) = \{p_{ij}(t)\}$ can be computed as $P(t) = e^{Qt}$ (see Section 4.4). Then, the probability ($f_{i,j}(t)$) of observing codons i and j at a homologous site of two sequences is given by $f_{i,j}(t) = \Pi_i p_{ij}(t)$, where Π_i is the equilibrium frequency of codon i. Π_i is given by $\Pi_i = \pi_k \pi_l \pi_m / (1 - \pi_T \pi_A \pi_A - \pi_T \pi_G \pi_A - \pi_T \pi_A \pi_G)$, if i consists of nucleotides k, l, and m. Likelihood ($L(t, \alpha, \beta)$) and log likelihood ($l(t, \alpha, \beta)$) of observing two sequences under consideration are given by

$$L(t, \alpha, \beta) = \prod_{i,j} \{f_{i,j}(t)\}^{n_{i,j}}, \quad l(t, \alpha, \beta) = \sum_{i,j} n_{i,j} \ln f_{i,j}(t) \qquad (11.23a,b)$$

where $n_{i,j}$ is the number of codon sites at which codons i and j are observed in two sequences. By maximizing the latter function numerically, the maximum-likelihood estimates (MLEs) of αt and βt are obtained.

Both d_S and d_N are estimated by $d_S = E(c_S)/s_S$ and $d_N = E(c_N)/s_N$. Here, $E(c_S)$ and $E(c_N)$ indicate the expected numbers of **synonymous** and **nonsynonymous substitutions per codon**, respectively, and s_S and s_N indicate the effective numbers of **synonymous** and **nonsynonymous sites per codon**, respectively. $E(c_S)$ and $E(c_N)$ are computed as $E(c_S) = \hat{t} \sum_i \sum_{j \neq i} \Pi_i \hat{s}_{ij}$ and $E(c_N) = \hat{t} \sum_i \sum_{j \neq i} \Pi_i \hat{n}_{ij}$, where $\hat{s}_{ij} = \hat{q}_{ij}$ when codons i and j code for the same amino acid and $\hat{s}_{ij} = 0$ otherwise; and $\hat{n}_{ij} = \hat{q}_{ij}$ when codons i and j code for the different amino acids and $\hat{n}_{ij} = 0$ otherwise. In the M96 method, a synonymous site is defined as the nucleotide site where all of three possible substitution mutations are synonymous. This definition is similar to that in the mutation fraction and the degenerate site methods. However, the M96 method does not classify each nucleotide site in the codons into synonymous or nonsynonymous sites, or fractional numbers of them. Instead, s_S and s_N are simply given by $s_S = \sum_i \Pi_i c_{S(i)}/3$ and $s_N = \sum_i \Pi_i c_{N(i)}/3$, where $c_{S(i)}$ and $c_{N(i)}$ indicate the numbers of possible synonymous and nonsynonymous mutations from codon i, respectively. For example, $c_{S(GGG)} = 3$ and $c_{N(GGG)} = 6$.

As a result, d_S and d_N are estimated by

$$\hat{d}_S = \frac{\hat{t}\sum_i\sum_{j\neq i}\Pi_i\hat{s}_{ij}}{\sum_i\Pi_i c_{S(i)}/3}, \quad \hat{d}_N = \frac{\hat{t}\sum_i\sum_{j\neq i}\Pi_i\hat{n}_{ij}}{\sum_i\Pi_i c_{N(i)}/3} \tag{11.24a,b}$$

11.4.2 Method of Yang and Nielsen (YN98 method)

In the M96 method, r_m from nucleotide i to j ($j \neq i$) is assumed to be the same for any combination of nucleotides i and j, which is unrealistic. Therefore, in the YN98 method, the difference between α_m and β_m is incorporated into the codon substitution matrix $Q = \{q_{ij}\}$; q_{ij} is assumed to be Π_j for synonymous transversional substitutions, $\kappa\Pi_j$ for synonymous transitional substitutions, $\omega\Pi_j$ for nonsynonymous transversional substitutions, and $\omega\kappa\Pi_j$ for nonsynonymous transitional substitutions. Here, κ indicates α_m/β_m and ω indicates r_N/r_S.

Q is scaled so that $\sum_i\Pi_i\sum_{j\neq i}q_{ij} = 1$, and the log-likelihood function is formulated as in the case for the M96 method. t, κ, and ω are estimated numerically by maximizing the log-likelihood function. Then, because Q has been scaled so that $\sum_i\Pi_i\sum_{j\neq i}q_{ij} = 1$, $\hat{t}$ is regarded as the number of nucleotide substitutions per codon between two sequences. Moreover, $\hat{\rho}_S = \sum_i\sum_j\Pi_i\hat{s}_{ij}$ and $\hat{\rho}_N = \sum_i\sum_j\Pi_i\hat{n}_{ij}$ are regarded as the proportions of synonymous and nonsynonymous substitutions, respectively. Thus, the numbers of synonymous and nonsynonymous substitutions per codon are given by $\hat{t}\hat{\rho}_S$ and $\hat{t}\hat{\rho}_N$, respectively.

In the previous discussion, by setting $\omega = 1$ and estimating t and κ, $\hat{\rho}_S$ and $\hat{\rho}_N$ can be regarded as the proportions of synonymous and nonsynonymous sites, respectively. If ρ_S^* and ρ_N^* are the estimators of ρ_S and ρ_N under the constraint of $\omega = 1$, the numbers of synonymous and nonsynonymous sites per codon is represented as $3\hat{\rho}_S^*$ and $3\hat{\rho}_N^*$, respectively.

Therefore, d_S and d_N can be estimated by

$$\hat{d}_S = \frac{\hat{t}\hat{\rho}_S}{3\hat{\rho}_S^*}, \quad \hat{d}_N = \frac{\hat{t}\hat{\rho}_N}{3\hat{\rho}_N^*} \tag{11.25a,b}$$

11.5 Methods for estimating d_S and d_N at single codon sites

More recently, methods have been developed for estimating d_S and d_N at single codon sites throughout the evolution of many nucleotide sequences. These methods make use of the **phylogenetic tree** relating the sequences under investigation to estimate d_S and d_N, as explained in the following sections.

11.5.1 Method of Suzuki and Gojobori (SG99 method)

First, for each codon site, the ancestral codon is inferred at each node of the phylogenetic tree. So far, the **maximum-parsimony** (**MP**) **method** (Fitch, 1971; Hartigan, 1973) and the **Bayesian method** (Yang et al., 1995; Zhang and Nei, 1997) are available for inferring ancestral codons, and both methods can be used for the present purpose. Only one ancestral codon is inferred at each node in the phylogenetic tree with the Bayesian method. However, with the MP method, more than one ancestral codon may be inferred at some nodes (see Chapter 7). In this case, the same probability is assumed for the existence of each codon.

Second, the average numbers of synonymous (s'_S) and nonsynonymous (s'_N) sites per codon throughout the phylogenetic tree are computed for each codon site. For that purpose, first the average numbers of synonymous $(b'_{S(i)})$ and nonsynonymous $(b'_{N(i)})$ sites for each branch i in the phylogenetic tree are computed. $b'_{S(i)}$ and $b'_{N(i)}$ are computed by using codons j and k at the ends of branch i. If j and k are either the same or different at only one position, $b'_{S(i)}$ and $b'_{N(i)}$ are computed simply as $b'_{S(i)} = (s_{S(j)} + s_{S(k)})/2$ and $b'_{N(i)} = (s_{N(j)} + s_{N(k)})/2$, where $s_{S(j)}$ and $s_{S(k)}$ are computed the same way as in the I95 method, assuming the appropriate mutation matrix among nucleotides (Suzuki, 1999). If j and k are different at two positions, two possible pathways for obtaining the differences are considered for computing $b'_{S(i)}$ and $b'_{N(i)}$. The average numbers of synonymous and nonsynonymous sites are computed for each pathway, and those numbers are averaged over all pathways with equal weights. For example, assuming ACG (Thr) for j and AAT (Asn) for k, the possible pathways are:

Pathway 1: ACG (Thr)-->AAG (Lys)-->AAT (Asn)
Pathway 2: ACG (Thr)-->ACT (Thr)-->AAT (Asn)

The average number of synonymous sites for Pathway 1 is computed as $(s_{S(ACG)} + s_{S(AAG)} + s_{S(AAT)})/3$, and that for Pathway 2 is $(s_{S(ACG)} + s_{S(ACT)} + s_{S(AAT)})/3$. $b'_{S(i)}$ is given by $b'_{S(i)} = \{(s_{S(ACG)} + s_{S(AAG)} + s_{S(AAT)})/3 + (s_{S(ACG)} + s_{S(ACT)} + s_{S(AAT)})/3\}/2$. $b'_{N(i)}$ is computed in a similar way. If the termination codon appears in some pathways, they are excluded from the analysis; $b'_{S(i)}$ and $b'_{N(i)}$ between two codons that are different at three positions are computed similarly.

Both s'_S and s'_N are computed as the weighted averages of $b'_{S(i)}$ and $b'_{N(i)}$ over all branches in the phylogenetic tree, respectively. The weights are assumed to be proportional to the branch length; thus, $s'_S = \sum_i b'_{S(i)} \cdot l_i/l_t$ and $s'_N = \sum_i b'_{N(i)} \cdot l_i/l_t$, where l_i is the length of branch i and l_t is the total branch length in the phylogenetic tree. This is because the length of the branch reflects its contribution to the overall evolution time.

Third, the total numbers of **synonymous** (c'_S) and **nonsynonymous** (c'_N) **differences per codon** throughout the phylogenetic tree are counted for each codon site. The numbers of synonymous and nonsynonymous differences are counted for each branch by comparing the two codons at its ends, and those numbers are summed over all branches in the phylogenetic tree to obtain c'_S and c'_N. The numbers of synonymous and nonsynonymous differences between two codons are computed the same way as in the NG86 and ZRN98 methods.

Finally, d_S and d_N are estimated by

$$ d_S = \frac{c'_S}{s'_S}, \quad d_N = \frac{c'_N}{s'_N} \tag{11.26a,b} $$

11.6   Test of neutrality for two sequences

Several statistical tests were developed to investigate whether two homologous sequences are evolving according to positive selection or **neutrality**; the most common are summarized in the following sections.

11.6.1   Z test

If d_S and d_N are estimated by either the mutation-fraction (see Section 11.2) or the degenerate-site method (see Section 11.3), the null hypothesis of **neutral evolution** at the amino-acid sequence level $E\left(d_N\right) = E\left(d_S\right)$ can be tested by using the Z test statistic, as follows:

$$ Z = \frac{\hat{d}_N - \hat{d}_S}{\sqrt{V\left(\hat{d}_N\right) + V\left(\hat{d}_S\right)}} \tag{11.27} $$

This test assumes that Z follows the standard normal distribution. A significantly larger value of $\hat{d}_N$ than $\hat{d}_S$ is an indicator of positive selection, whereas the inverse is an indicator of negative selection.

11.6.2   Likelihood ratio test (LRT)

In the codon model method, a **likelihood ratio test** (**LRT**) (see Chapter 10) can be used for testing neutrality of amino-acid substitutions. An **evolutionary model** with a restriction of $r_N = r_S$ is compared with a model without such a restriction. In the M96 method, Model 1 may assume $\alpha = \beta$, whereas Model 2 may estimate α

and β separately. The maximum log-likelihood values from Model 1 (l_1) and Model 2 (l_2) are compared by using the likelihood ratio ($2(l_2 - l_1)$). The distribution of this value is approximated by a χ^2 distribution with one degree of freedom, under the null hypothesis of $\alpha = \beta$ (see Section 10.4). Therefore, if $2(l_2 - l_1)$ is larger than the χ^2 value for a given significance level, $\alpha < \beta$ indicates positive selection; $\alpha > \beta$ indicates negative selection.

Similarly, in the YN98 method, it may be set $\omega = 1$ in Model 1, whereas no assumption may be made for ω in Model 2. The null hypothesis of $\omega = 1$ can be tested by comparing $2(l_2 - l_1)$ with a χ^2 distribution with one degree of freedom. If $2(l_2 - l_1)$ is larger than the χ^2 value for a given significance level, $\omega > 1$ indicates positive selection and $\omega < 1$ indicates negative selection.

11.6.3 Window analysis

Different amino-acid sites may have different functions in a protein; therefore, natural selection is not likely to operate uniformly on all amino-acid sites. It may be difficult to detect natural selection operating in only a limited region of a protein by comparing d_S and d_N when they are estimated as the average numbers over all codon sites in a protein-coding region. Window analysis (Clark and Kao, 1991; Ina et al., 1994; Endo et al., 1996) was developed to detect natural selection operating on a limited region of a protein.

Window analysis starts with setting a window of usually 20 consecutive codon sites at the 5′-terminus of an alignment for two protein-coding sequences under consideration. The window slides along the alignment by one codon site for each step toward the 3′-terminus. At each step, d_S and d_N are estimated within the window and are plotted in two dimensions, in which the abscissa indicates the codon site and the ordinate indicates both d_S and d_N separately. Although no statistical test has been developed for testing the difference between d_S and d_N within a window, this approach may provide an approximate idea about natural selection operating on the protein.

11.7 Test of neutrality at single codon sites

As discussed in Section 11.6.3, window analysis can provide an approximate idea about whether natural selection operates within different regions of a protein. However, it is still difficult to detect natural selection operating at single amino-acid sites. For that purpose, it is no longer useful to compare two nucleotide sequences. Instead, a number of sequences have to be compared to detect natural selection operating at single amino-acid sites.

11.7.1 Method of Nielsen and Yang (1998) (NY98 method)

The NY98 method is an extension of the YN98 method. In this method, two models are first compared for testing the existence of positively selected amino-acid sites in the protein under consideration. In Model 1, codon sites are assumed to be classified into Categories 1 and 2, in which $\omega = 0$ and $\omega = 1$, respectively. The proportions of Categories 1 and 2 are described as p_1 and p_2, respectively. In Model 2, an additional Category 3 is considered, in which ω is regarded as a free parameter to be estimated. The proportion of Category 3 is described as p_3.

It is assumed that the phylogenetic tree has already been inferred for the sequences by using some appropriate method. Then, the probability ($f(x_h)$) of observing data x_h at codon site h under Model 1 or 2 is given by $f(x_h) = \sum_k p_k f(x_h | k)$, where k indicates the category number and $f(x_h | k)$ is the conditional probability of observing x_h under the assumption that codon site h belongs to category k (for the computation of $f(x_h | k)$, see Felsenstein, 1981). The log likelihood for the entire sequence data is $\sum_h \ln\{f(x_h)\}$.

Parameters are estimated by numerically maximizing log-likelihood values for each model. Assume that l_1 and l_2 are the maximum log-likelihood values for Models 1 and 2, respectively. Because Model 2 has two more free parameters (ω for Category 3 and p_3) than Model 1, the log-likelihood ratio $2(l_2 - l_1)$ is assumed to be χ^2 distributed with two degrees of freedom. If $2(l_2 - l_1)$ is larger than the χ^2 value for a given significance level and $\hat{\omega}$ for Category 3 is larger than 1, positively selected codon sites may exist in the nucleotide sequence under consideration.

If the existence of positively selected codon sites is inferred by this test, they can be identified by computing the posterior probability ($p(k = 3 | x_h)$) that a certain codon site belongs to Category 3. The prior probability of that event is assumed as $\hat{p}_3$. Then, $p(k = 3 | x_h)$ is

$$p(k = 3 | x_h) = \frac{\hat{p}_3 \cdot f(x_h | k = 3)}{\sum_{k=1}^{3} \hat{p}_k \cdot f(x_h | k)} \tag{11.28}$$

If $p(k = 3 | x_h)$ is larger than a given confidence level at a certain codon site, that site is inferred as positively selected.

11.7.2 SG99 method

In the SG99 method, natural selection operating at single codon sites can be detected by computing the binomial probability (p) of obtaining the observed or more biased numbers of c'_S and c'_N for each codon site. The probabilities for the occurrences of synonymous and nonsynonymous substitutions are assumed as $s'_S/3$ and $s'_N/3$, respectively. If p is smaller than a given significance level, $(c'_N/s'_N) > (c'_S/s'_S)$ indicates positive selection, whereas $(c'_N/s'_N) < (c'_S/s'_S)$ indicates negative selection.

PRACTICE

Yoshiyuki Suzuki and Takashi Gojobori

11.8 Software for analyzing coding sequences

Most of the methods described previously to estimate d_S and d_N and test neutrality between two nucleotide sequences are implemented in the software package MEGA2 (NG86, ZRN98, LWL85, and PBL93), and in the program PAML (YN98), both of which are freeware. MEGA2 is available at **http://www.megasoftware.net**. Only the DOS and Windows versions exist for MEGA2, but the program can be installed on Macintosh under Virtual PC. MEGA2 infers phylogenetic trees according to the *minimum-evolution* or the maximum-parsimony criterion (see Chapters 5 and 7), and it performs a number of statistical tests on nucleotide and amino-acid sequences. Input files must be in the MEGA format described in Box 2.1. PAML is available at **http://abacus.gene.ucl.ac.uk/software/PAML.html** (see Box 10.1). The format of the input files and the basic instructions to execute the applications are briefly summarized in Box 10.1.

11.9 Estimation of d_S and d_N in an HCV data set

HCV is a causative agent of chronic hepatitis, which may progress to cirrhosis and hepatocellular carcinoma (Alter et al., 1992). HCV is an enveloped, nonsegmented, single-stranded, and positive-sense RNA virus (Choo et al., 1989), and is the sole member of the genus *Hepacivirus* in the family *Flaviviridae* (Murphy et al., 1995). The genome of HCV is approximately 9.5 kilobases long, encoding a single polyprotein of approximately 3000 amino acids (Kato et al., 1990). The polyprotein is co- and post-translationally cleaved into core protein (C), envelope glycoprotein 1 (E1), E2, p7, nonstructural protein 2 (NS2), NS3, NS4A, NS4B, NS5A, and NS5B, in the order from its N- to C-terminus, by the cellular signalase and the viral proteinases (Grakoui et al., 1993).

In what follows, d_S and d_N are estimated between two HCV subtype 1b strains, HCV-JS (Tanaka et al., 1995) and HCV-JT (Tanaka et al., 1992), which can be downloaded from any of the nucleotide databases discussed in Chapter 2. The entire coding region is divided into eight regions (C, E1, E2, NS2, NS3, NS4, NS5A, and NS5B). p7 and E2 are combined as E2, and NS4A and NS4B are combined as NS4, since both p7 (63 amino acids) and NS4A (54 amino acids) are of a length unlikely to yield meaningful d_S and d_N values. The numbers of codon sites used for C, E1, E2, NS2, NS3, NS4, NS5A, and NS5B are 191, 192, 426, 217, 631, 315, 447, and 591, respectively. The input files in MEGA or PAML

Box 11.1 The control file for YN00

```
    seqfile = hivALN.phy * sequence data file name
    outfile = hivALN.yn           * main result file
    verbose = 1 * 1: detailed output (list sequences), 0:
                 concise output
      icode = 0 * 0:universal code; 1:mammalian mt; 2-10:
                 see below
  weighting = 0 * weighting pathways between codons (0/1)?
commonf3x4 = 0 * use one set of codon freqs for all pairs
                 (0/1)?
```

```
* Genetic codes: 0:universal, 1:mammalian mt., 2:yeast mt.,
     3:mold mt.,
* 4: invertebrate mt., 5: ciliate nuclear, 6: echinoderm mt.,
* 7: euplotid mt., 8: alternative yeast nu. 9: ascidian mt.,
* 10: blepharisma nu.
* These codes correspond to transl_table 1 to 11 of GENEBANK.
```

The file YN00.ctl, shown here, specifies the sequence input file (hivALN.phy, in this case) and the name of the output file (hivALN.yn). It is also possible to choose among different genetic codes (icode option), whether to give different weights to different codon substitution pathways (option weighting = 1), and whether to use equal codon frequencies (option commonf3x4 = 0) or a table of empirically estimated codon frequencies for each of the 61 coding triplets (option commonf3x4 = 1).

After editing and saving the YN00.ctl file (any text editor can be used; see Box 10.1), the program is executed by double-clicking on the YN00.exe icon in the PAML folder (remember to place the input file in the same folder!).

---

format can be prepared with any text editor, according to Boxes 10.1 and 2.1, respectively.

11.9.1 Estimation of d_S and d_N with NG86, ZRN98, LWL85, and PBL93 methods (MEGA2)

1. Double-click on the MEGA2 icon to start the program, select Click me to activate a data file in the main window, and open the input file. MEGA2 can activate only data files in MEGA format, but the most common formats like PHYLIP, FASTA, or Clustal (see Box 2.2) can be converted to MEGA by selecting Convert to MEGA Format ... from the File menu.
2. Select Choose model ... from the Distances menu. In the new window, select the Models button and the model of choice from the Syn-Nonsynonymous submenu.
3. Choose Compute Pairwise ... from the Distances menu.

At the end of the computation, the user is asked to enter the name of an output file. The final result is also displayed in the MEGA main window. The output is self-explanatory.

11.9.2 Estimation of d_S and d_N with YN98 method (PAML)

Using PAML is more complicated. The input file has to be placed in the PAML folder. Each PAML software module has a corresponding ***control file***, with the same name but the extension .ctl, which needs to be edited with a text editor before running the module (see Box 10.1). The control file for the YN98 method (implemented in the executable YN00.exe in the PAML folder) is shown in Box 11.1. Open the file YN00.ctl with any text editor, type in the names of the input and output files, and close it after saving (remember to save the changes as text only!). YN00.exe can now be run by double-clicking on its icon. Results of the computation are written as simple text in the specified output file (see Box 10.1 for the PAML format).

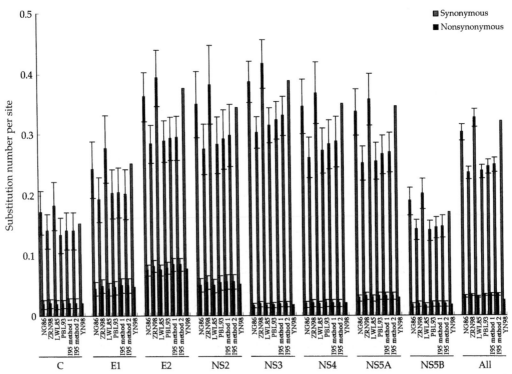

Figure 11.1 Both d_S and d_N estimated for each coding region, as well as the entire coding region, of HCV-JS and HCV-JT by using the NG86, ZRN98, LWL85, PBL93, I95, and YN98 methods. Blue and red rectangles indicate d_S and d_N, respectively, and the standard errors are indicated by black bars. (See color section.)

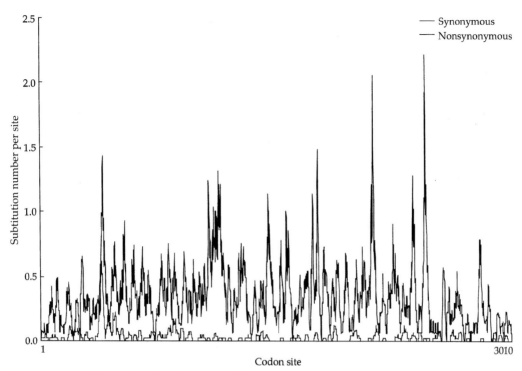

Figure 11.2 Window analysis for the entire coding region of HCV-JS and HCV-JT. The window is set as 20 codon sites long, and it slides by one codon site for each step. Within the window at each step d_S and d_N are estimated and plotted in two dimensions, in which the abscissa indicates the codon site and the ordinate indicates d_S and d_N. (See color section.)

Figure 11.1 illustrates d_S and d_N estimated for each coding region of HCV-JS and HCV-JT by using the NG86, ZRN98, LWL85, PBL93, I95 (MEGA2), and YN98 (PAML) methods. For all coding regions, the NG86 and LWL85 methods give larger $\hat{d}_S$ values than the ZRN98, PBL93, and I95 methods. In contrast, $\hat{d}_N$ values are smaller for the NG86 and LWL85 methods than for the ZRN98, PBL93, and I95 methods; this is probably because the former methods do not consider the difference between α_m and β_m in the computation of S_S and S_N. On average, α_m/β_m is estimated to be 10.7 to 18.1 for HCV-JS and HCV-JT, which may have caused the underestimation of S_S and overestimation of S_N in the NG86 and LWL85 methods. The YN98 method gives $\hat{d}_S$ and $\hat{d}_N$ values similar to the NG86 and LWL85 methods in this case, although it considers the difference between α_m and β_m.

11.9.3 Comparing different estimates of d_S and d_N

From Figure 11.1, it is clear that $\hat{d}_S$ is larger than $\hat{d}_N$ for all coding regions, probably because of the functional constraint operating at the amino-acid sequence level. There is no evidence for positive selection as long as $\hat{d}_S$ and $\hat{d}_N$ are compared for

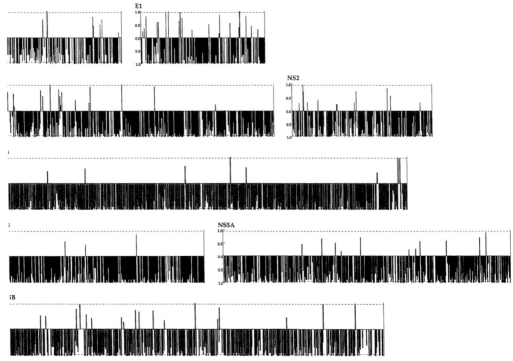

Figure 11.3 Detection of natural selection at single-codon sites for each coding region of HCV subtype
1b by using the SG99 method. The binomial probability (p) of obtaining the observed or
more biased numbers of c'_S and c'_N is computed for each codon site, and $1 - p$ is plotted in
the ordinate against the codon site in the abscissa. At the codon site where $d_N \geq d_S$, $1 - p$ is
indicated above the abscissa; where $d_N < d_S$, $1 - p$ is indicated below the abscissa. Dotted
lines indicate the 5% significance level.

each coding region. Note that $\hat{d}_S$ and $\hat{d}_N$ are different among coding regions: $\hat{d}_S$ is
smaller for C and NS5B than for other regions. It has been reported that the coding
regions for C and NS5B form secondary structures in the genomic RNA (Ina et al.,
1994; Smith and Simmonds, 1997). Therefore, functional constraints may operate
at the nucleotide-sequence level for these regions. $\hat{d}_N$ is larger for E2 than for other
regions. E2 is known to be the major target of the host's immune response and
contains hypervariable region 1 (HVR1) (Hijikata et al., 1991; Weiner et al., 1991).
Therefore, functional constraints operating at the amino-acid sequence level may
be weaker than other coding regions, or positive selection may operate on some
amino-acid sites in E2. However, these possibilities cannot be distinguished with
the present analysis.

11.10   An example of window analysis

For the window analysis of d_S and d_N, the programs WINDOWS and WINA are
available from **ftp://ftp.nig.ac.jp/pub/unix/windows** and **ftp://ftp.nig.ac.jp/pub/**

Table 11.2 Functions of positively selected amino-acid sites in the entire coding region of HCV subtype 1b

Site[a]	Protein	Function	References
75	C	B-cell epitope	Pereboeva et al. (1998)
231	E1	B-cell epitope	Zibert et al. (1999)
235	E1	B-cell epitope	Zibert et al. (1999)
253	E1	B-cell epitope	Zibert et al. (1999)
345	E1	N.A.[b]	N.A.
387	E2	B-cell epitope	Zibert et al. (1997)
461	E2	B-cell epitope	Zibert et al. (1999)
			Nakano et al. (1999)
574	E2	B-cell epitope	Zibert et al. (1999)
			Nakano et al. (1999)
827	NS2	N.A.	N.A.
1384	NS3	T-cell epitope	Tabatabai et al. (1999)
			Wang and Eckels (1999)
		B-cell epitope	Pereboeva et al. (2000)
1644	NS3	T-cell epitope	Tabatabai et al. (1999)
2719	NS5B	N.A.	N.A.
2968	NS5B	N.A.	N.A.

[a] Amino-acid sites are numbered according to HCV-JS.
[b] Not available.

Bio/wina, respectively. These programs estimate d_S and d_N within a window by using the NG86 method, but they are available only for UNIX and are not described here. However, a window analysis can be implemented through the Los Alamos HIV database (see Chapter 2) using the SNAP (Synonymous/Nonsynonymous Analysis Program) Web site at **http://hiv-web.lanl.gov/seq-db.html**.

Figure 11.2 illustrates the results of a window analysis for the entire coding region of HCV-JS and HCV-JT. The reduction of synonymous substitutions in C and NS5B shown in the previous analysis is not clear due to a large stochastic error that occurred because the window used is relatively short (i.e., 20 codon sites). However, it can be seen that $\hat{d}_N > \hat{d}_S$ in several windows, although the statistical significance is not clear.

11.11   Detection of positive selection at single amino acid sites

For estimating d_S and d_N and testing neutrality at single codon sites, ADAPTSITE is available at **http://mep.bio.psu.edu/adaptivevol.html**. The analysis at single-codon sites for each coding region of HCV subtype 1b by using the SG99 method is shown in Figure 11.3. The correlation with the function of the amino acid is given in Table 11.2.

11.12   Conclusions

Knowing the type and the strength of natural selection operating at the amino-acid sequence level is essential for understanding the evolutionary mechanisms of proteins. Several methods have been developed for estimating d_S and d_N over an entire region and even at single-codon sites for protein-coding sequences. Statistical tests are available for testing the null hypothesis of neutrality to detect the type of natural selection. Moreover, d_N/d_S may indicate the strength of natural selection. As the sequence data are accumulating at an enormous speed in the international nucleotide sequence database (DDBJ/EMBL/GenBank), the number of *open reading frames* of unknown function is also increasing. Even for those unknown proteins, it is possible to infer the type and the strength of natural selection by analyzing d_S and d_N. Such information may be helpful for predicting the functions of those proteins.

REFERENCES

Alter, M. J., H. S. Margolis, K. Krawczynski, F. N. Judson, A. Mares, W. J. Alexander, P. Y. Hu, J. K. Miller, M. A. Gerber, R. E. Sampliner, E. L. Meeks, and M. J. Beach (1992). The natural history of community-acquired hepatitis C in the United States. *The New England Journal of Medicine*, 327, 1899–1905.

Choo, Q.-L., G. Kuo, R. Ralston, A. J. Weiner, L. R. Overby, D. W. Bradley, and M. Houghton (1989). Isolation of a cDNA clone derived from a blood-borne non-A, non-B hepatitis genome. *Science*, 244, 359–362.

Clark, A. G. and T.-H. Kao (1991). Excess nonsynonymous substitution at shared polymorphic sites among self-incompatibility alleles of Solanaceae. *Proceedings of the National Academy of Sciences of the USA*, 88, 9823–9827.

Comeron, J. M. (1995). A method for estimating the numbers of synonymous and nonsynonymous substitutions per site. *Journal of Molecular Evolution*, 41, 1152–1159.

Endo, T., K. Ikeo, and T. Gojobori (1996). Large-scale search for genes on which positive selection may operate. *Molecular Biology and Evolution*, 13, 685–690.

Felsenstein, J. (1981). Evolutionary trees from DNA sequences: A maximum-likelihood approach. *Journal of Molecular Evolution*, 17, 368–376.

Fitch, W. M. (1971). Toward defining the course of evolution: Minimum change for a specific tree topology. *Systematic Zoology*, 20, 406–416.

Gojobori, T. (1983). Codon substitution in evolution and the "saturation" of synonymous changes. *Genetics*, 105, 1011–1027.

Goldman, N. and Z. Yang (1994). A codon-based model of nucleotide substitution for protein-coding DNA sequences. *Molecular Biology and Evolution*, 11, 725–736.

Grakoui, A., C. Wychowski, C. Lin, S. M. Feinstone, and C. M. Rice (1993). Expression and identification of hepatitis C virus polyprotein cleavage products. *Journal of Virology*, 67, 1385–1395.

Hartigan, J. A. (1973). Minimum mutation fits to a given tree. *Biometrics*, 29, 53–65.

Hijikata, M., N. Kato, Y. Ootsuyama, M. Nakagawa, S. Ohkoshi, and K. Shimotohno (1991). Hypervariable regions in the putative glycoprotein of hepatitis C virus. *Biochemical and Biophysical Research Communications*, 175, 220–228.

Hughes, A. L. and M. Nei (1988). Pattern of nucleotide substitution at major histocompatibility complex class I loci reveals overdominant selection. *Nature*, 335, 167–170.

Hughes, A. L. and M. Nei (1989). Nucleotide substitution at major histocompatibility complex class II loci: Evidence for overdominant selection. *Proceedings of the National Academy of Sciences of the USA*, 86, 958–962.

Ina, Y. (1995). New methods for estimating the numbers of synonymous and nonsynonymous substitutions. *Journal of Molecular Evolution*, 40, 190–226.

Ina, Y. (1996). Pattern of synonymous and nonsynonymous substitutions: An indicator of mechanisms of molecular evolution. *Journal of Genetics*, 75, 91–115.

Ina, Y., M. Mizokami, K. Ohba, and T. Gojobori (1994). Reduction of synonymous substitutions in the core protein gene of hepatitis C virus. *Journal of Molecular Evolution*, 38, 50–56.

Jukes, T. H. and C. R. Cantor (1969). Evolution of protein molecules. In: *Mammalian Protein Metabolism*, ed. H. N. Munro, pp. 21–132. Academic Press, New York.

Kato, N., M. Hijikata, Y. Ootsuyama, M. Nakagawa, S. Ohkoshi, T. Sugimura, and K. Shimotohno (1990). Molecular cloning of the human hepatitis C virus genome from Japanese patients with non-A, non-B hepatitis. *Proceedings of the National Academy of Sciences of the USA*, 87, 9524–9528.

Kimura, M. (1980). A simple method for estimating evolutionary rates of base substitutions through comparative studies of nucleotide sequences. *Journal of Molecular Evolution*, 16, 111–120.

Kondo, R., S. Horai, Y. Satta, and N. Takahata (1993). Evolution of hominoid mitochondrial DNA with special reference to the silent substitution rate over the genome. *Journal of Molecular Evolution*, 36, 517–531.

Li, W.-H. (1993). Unbiased estimation of the rates of synonymous and nonsynonymous substitution. *Journal of Molecular Evolution*, 36, 96–99.

Li, W.-H., C.-I. Wu, and C.-C. Luo (1985). A new method for estimating synonymous and nonsynonymous rates of nucleotide substitution considering the relative likelihood of nucleotide and codon changes. *Molecular Biology and Evolution*, 2, 150–174.

Miyata, T. and T. Yasunaga (1980). Molecular evolution of mRNA: A method for estimating evolutionary rates of synonymous and amino-acid substitutions from homologous nucleotide sequences and its application. *Journal of Molecular Evolution*, 16, 23–36.

Moriyama, E. N. and J. R. Powell (1997). Synonymous substitution rates in *Drosophila*: mitochondrial versus nuclear gene. *Journal of Molecular Evolution*, 45, 378–391.

Murphy, F. A., C. M. Fauquet, D. H. L. Bishop, S. A. Ghabrial, A. W. Jarvis, G. P. Martelli, M. A. Mayo, and M. D. Summers (1995). *Virus Taxonomy*. Vienna: Springer-Verlag.

Muse, S. V. (1996). Estimating synonymous and nonsynonymous substitution rates. *Molecular Biology and Evolution*, 13, 105–114.

Muse, S. V. and B. S. Gaut (1994). A likelihood approach for comparing synonymous and nonsynonymous nucleotide substitution rates, with application to the chloroplast genome. *Molecular Biology and Evolution*, 11, 715–724.

Nakano, I., Y. Fukuda, Y. Katano, and T. Hayakawa (1999). Conformational epitopes detected by cross-reactive antibodies to envelope 2-glycoprotein of the hepatitis C virus. *Journal of Infectious Diseases*, 180, 1328–1333.

Nei, M. and T. Gojobori (1986). Simple methods for estimating the numbers of synonymous and nonsynonymous nucleotide substitutions. *Molecular Biology and Evolution*, 3, 418–426.

Nei, M. and S. Kumar (2000). *Molecular Evolution and Phylogenetics*. Oxford, NY: Oxford University Press.

Nielsen, R. and Z. Yang (1998). Likelihood models for detecting positively selected amino-acid sites and applications to the HIV-1 envelope gene. *Genetics*, 148, 929–936.

Ota, T. and M. Nei (1994). Variances and covariances of the number of synonymous and nonsynonymous substitutions per site. *Molecular Biology and Evolution*, 11, 613–619.

Pamilo, P. and O. Bianchi (1993). Evolution of the *Zfx* and *Zfy* genes: Rates and interdependence between the genes. *Molecular Biology and Evolution*, 10, 271–281.

Pereboeva, L. A., A. V. Pereboev, and G. E. Morris (1998). Identification of antigenic sites on three hepatitis C virus proteins using phage-displayed peptide libraries. *Journal of Medical Virology*, 56, 105–111.

Pereboeva, L. A., A. V. Pereboev, L. F. Wang, and G. E. Morris (2000). Hepatitis C epitopes from phage-displayed cDNA libraries and improved diagnosis with a chimeric antigen. *Journal of Medical Virology*, 60, 144–151.

Perler, F., A. Efstratiadis, P. Lomedico, W. Gilbert, R. Kolodner, and J. Dodgson (1980). The evolution of genes: The chicken preproinsulin gene. *Cell*, 20, 555–566.

Smith, D. B. and P. Simmonds (1997). Characteristics of nucleotide substitution in the hepatitis C virus genome: Constraints on sequence change in coding regions at both ends of the genome. *Journal of Molecular Evolution*, 45, 238–246.

Suzuki, Y. (1999). *Molecular Evolution of Pathogenic Viruses*. Dissertation, Department of Genetics, School of Life Science, The Graduate University for Advanced Studies, Hayama, Japan.

Suzuki, Y. and T. Gojobori (1999). A method for detecting positive selection at single amino-acid sites. *Molecular Biology and Evolution*, 16, 1315–1328.

Tabatabai, N. M., T.-H. Bian, C. M. Rice, K. Yoshizawa, J. Gill, and D. D. Eckels (1999). Functionally distinct T-cell epitopes within the hepatitis C virus nonstructural 3 protein. *Human Immunology*, 60, 105–115.

Tanaka, T., N. Kato, M. Nakagawa, Y. Ootsuyama, M. J. Cho, T. Nakazawa, M. Hijikata, Y. Ishimura, and K. Shimotohno (1992). Molecular cloning of hepatitis C virus genome from a single Japanese carrier: Sequence variation within the same individual and among infected individuals. *Virus Research*, 23, 39–53.

Tanaka, T., N. Kato, M.-J. Cho, and K. Shimotohno (1995). A novel sequence found at the 3′ terminus of hepatitis C virus genome. *Biochemical and Biophysical Research Communications*, 215, 744–749.

Thompson, J. D., D. G. Higgins, and T. J. Gibson (1994). CLUSTALW: Improving the sensitivity of progressive multiple-sequence alignment through sequence weighting, position-specific gap penalties, and weight-matrix choice. *Nucleic Acids Research*, 22, 4673–4680.

Wang, H. and D. D. Eckels (1999). Mutations in immunodominant T-cell epitopes derived from the nonstructural 3 protein of hepatitis C virus have the potential for generating escape variants that may have important consequences for T-cell recognition. *Journal of Immunology*, 162, 4177–4183.

Weiner, A. J., M. J. Brauer, J. Rosenblatt, K. H. Richman, J. Tung, K. Crawford, F. Bonino, G. Saracco, Q.-L. Choo, M. Houghton, and J. H. Han (1991). Variable and hypervariable domains are found in the regions of HCV corresponding to the Flavivirus envelope and NS1 proteins and the Pestivirus envelope glycoproteins. *Virology*, 180, 842–848.

Yang, Z. and R. Nielsen (1998). Synonymous and nonsynonymous rate variation in nuclear genes of mammals. *Journal of Molecular Evolution*, 46, 409–418.

Yang, Z. and R. Nielsen (2000). Estimating synonymous and nonsynonymous substitution rates under realistic evolutionary models. *Molecular Biology and Evolution*, 17, 32–43.

Yang, Z., S. Kumar, and M. Nei (1995). A new method of inference of ancestral nucleotide and amino-acid sequences. *Genetics*, 141, 1641–1650.

Zhang, J. and M. Nei (1997). Accuracies of ancestral amino-acid sequences inferred by the parsimony, likelihood, and distance methods. *Journal of Molecular Evolution*, 44, S139–S146.

Zhang, J., H. F. Rosenberg, and M. Nei (1998). Positive Darwinian selection after gene duplication in primate ribonuclease genes. *Proceedings of the National Academy of Sciences of the USA*, 95, 7308–7313.

Zibert, A., W. Kraas, H. Meisel, G. Jung, and M. Roggendorf (1997). Epitope mapping of antibodies directed against hypervariable region 1 in acute self-limiting and chronic infections due to hepatitis C virus. *Journal of Virology*, 71, 4123–4127.

Zibert, A., W. Kraas, R. S. Ross, H. Meisel, S. Lechner, G. Jung, and M. Roggendorf (1999). Immunodominant B-cell domains of hepatitis C virus envelope proteins E1 and E2 identified during early and late time points of infection. *Journal of Hepatology*, 30, 177–184.

SplitsTree: A network-based tool for exploring evolutionary relationships in molecular data

THEORY

Vincent Moulton

12.1 Exploring evolutionary relationships through networks

The standard way to represent evolutionary relationships between a given set of *taxa* is to use a **bifurcating tree**, in which internal nodes represent hypothetical ancestors and leaves are labeled by present-day species (see Chapter 1). Using such a tree presumes that the underlying evolutionary processes are **bifurcating**. However, in instances where this is not the case, it is questionable whether a **bifurcating-leaf**-labeled tree is the best structure to represent the phylogenetic relationships. For example, the phenomenon of explosive evolutionary radiation (e.g., when an AIDS virus infects a healthy person) might best be modeled not by a bifurcating tree, but by a **multifurcating tree** (see Chapter 1). In addition, it may be necessary to label internal nodes by taxa if ancestors and present-day species coexist, as has been observed with fast-evolving viruses.

In certain cases, one might want to allow even more general structures than multifurcating trees to represent evolutionary histories. For example, certain viruses/plants are known to exhibit **recombination/hybridization**, and this process is probably not best represented by a tree. A tree implicitly assumes that once two lineages are created, they subsequently never interact with one another later. However, if it is assumed that such interactions might have occurred, then a simplistic representation might look something like the **network** (or **graph**) presented in Figure 12.1. In this network, the **nodes** (or **vertices**) are the dots labeled by *1, 2,.., 7*, and the **branches** (or **edges**) are the lines connecting the nodes. In this example, *1, 2, 3, 4* can be thought of as potential hypothetical ancestors and *5, 6, 7* as extant taxa. As with a **rooted tree**, this structure indicates that all taxa have evolved from *1*, but that at a certain point in time, *2* and *3* interacted, resulting in *4*. The key difference between networks and trees illustrated by this example is that **cycles**

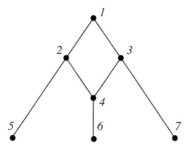

Figure 12.1   A simple network relating three taxa.

are allowed – paths that begin and start at the same node. For example, the nodes labeled *1, 2, 3, 4* form a cycle of length four.

Recombination might be regarded as quite special and, therefore, it might be reasonable to think that networks are useful only for studying certain types of evolution. However, networks also could serve another purpose. They do not implicitly assume a tree-like evolutionary process; therefore, they will not "force" the data onto a tree. Most tree-building programs will output a tree for any input, even when it is evidently not tree-like. Even when the taxa in question did arise from a treelike evolutionary process, it may be that errors in the phylogenetic analysis have led to a data set that is no longer best described by a tree. In this situation, using networks allows the possibility of spotting any deviation from tree-likeness – or at least confirms that a tree is still probably the best structure to use. This is analogous to some tree-building programs, which allow the user to explore the possibility of using various trees for representing the same data. Networks provide a useful complementary technique to building trees, allowing the user to explore and better understand the data, as well as derive the sought-after evolutionary histories. Moreover, networks provide the possibility to visualize the data structure, thereby identifying any underlying patterns, that a tree representation might miss. A good example is the use of networks for the "phylogenetic analysis" of the *Canterbury Tales* (Barbrook et al., 1998).

At this point, it is natural to ask, "What kinds of networks are there and how can they be built?" Various networks have been considered, usually with certain special types of data in mind. For example, **median networks** have been used to represent the evolution of **mtDNA** (Bandelt et al., 1995). However, this chapter focuses on only one **network-based** approach to phylogenetic analysis, which is implemented in the SplitsTree program (Huson, 1998). **Split-graphs**, the networks generated by SplitsTree, are primarily constructed from distances using the mathematics of split-decomposition theory (Bandelt and Dress, 1992a,b and 1993). The remainder of this chapter includes an intuitive introduction to split-decomposition theory and how it is used to construct networks. Other descriptions of the split-decomposition

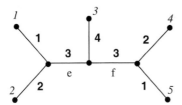

Figure 12.2   A weighted tree relating five taxa.

technique for phylogenetic analysis may be found in Dress, Huson, and Moulton (1996), Page and Holmes (1998), and Swofford et al. (1996).

12.2   An introduction to split-decomposition theory

A cornerstone of split-decomposition theory is the notion of splits, from which split-graphs are constructed. Consider the weighted tree in Figure 12.2, which is labeled by the set of taxa $\{1, 2, 3, 4, 5\}$ with weights on edges denoted in boldface. If any branch of this tree is removed, the result is a natural partition of the data into two nonempty, disjointed sets, each of which represents one of the two resulting trees. For example, removing the edge e results in the partition $A = \{1, 2\}$, $B = \{3, 4, 5\}$. Such a partition is called a split, and it is usually denoted by $S = \{A, B\}$. Any possible pair of splits given by this tree satisfies a special condition; for example, consider the two splits $\{A = \{1, 2\}, B = \{3, 4, 5\}\}$ and $\{C = \{1, 2, 3\}, D = \{4, 5\}\}$ corresponding to the edges e and f, respectively. Then A and D do not intersect (i.e., $A \cap D = \emptyset$), whereas the intersections $A \cap C, B \cap C, B \cap D$ are all nonempty. Using this example, two splits $S = \{A, B\}$ and $T = \{C, D\}$ of a set X are said to be compatible if precisely one of the intersections $A \cap C, A \cap D, B \cap C, B \cap D$ is empty. Otherwise, if all intersections are nonempty, then S and T are incompatible (convince yourself that these are really the only two possibilities!).

In 1971, Buneman presented a groundbreaking paper in which he proved that a collection of splits corresponds to the collection of edges of a **phylogenetic tree** (which is not necessarily a strictly bifurcating tree), if and only if every pair of splits in this collection is compatible. Thus, a search for trees to represent the evolutionary relationships between a given set of taxa is in fact looking for compatible collections of splits of the taxa. However, even for relatively small data sets, there are many possible splits; for example, for 5 taxa, there are 15 possible splits and for a set of n taxa there are $2^{(n-1)} - 1$ splits. Searching for a fully resolved bifurcating tree is equivalent to searching for a collection of $2n - 3$ **compatible splits**. For example, when $n = 15$, there is a collection of 27 compatible splits within 16,383 possible: a very computational-intensive task! One possibility of solving this problem is using the data to search for significant collections of compatible splits (i.e., those that are highly supported by the data). Although many solutions have been proposed, only the elegant one provided by Buneman (1971) is considered here because it also provides a good review for understanding split decomposition.

12.2.1 The Buneman tree

As shown in Chapter 4, there are various ways to estimate the **genetic distance** d on a set of taxa X. Once this has been accomplished, d can be used to select significant splits using the Buneman technique. Given a split $S = \{A, B\}$ of X, some x, y in A and some u, v in B put

$$\beta(xy|uv) = \min\{d(x, u) + d(y, v), d(x, v) + d(y, u)\}$$
$$-(d(x, y) + d(u, v)) \tag{12.1}$$

and define the **Buneman index** β_S of S as $^1\!/_2 \min \beta(xy|uv)$ taken over all x, y in A and u, v in B. For example, in Figure 12.2 the genetic distance d_T between any pair of taxa $\{1, 2, 3, 4, 5\}$ on the weighted tree is defined as the sum of the weights of the edges on the path between them (e.g., $d_T(2,5) = 2 + 3 + 3 + 1 = 9$). Consider the split $S = \{\{1, 2\}, \{3, 4, 5\}\}$. Calculating β for all possible pairs of this split results in $\beta(12|34) = 6$, $\beta(12|35) = 6$, and $\beta(12|45) = 12$. Therefore, it is $^1\!/_2 \beta_S = 6/2 = 3$; that is, exactly the weight of the branch corresponding to S.

The remarkable fact that Buneman noticed was that if the distances d between a set of taxa X is taken, then the collection of splits S, for which $\beta_S > 0$ holds, is compatible and, therefore, corresponds to a tree. Thus, considering β_S as a measure of the significance of the split S, Buneman's method tells which splits to keep (i.e., those with positive Buneman index) and which to discard (i.e., those with nonpositive Buneman index), so that the resulting set of splits corresponds to a tree. This tree, with weight β_S assigned to that edge of the tree corresponding to split S, is called the **Buneman tree** corresponding to genetic distance d.

Any distance tree-building method should satisfy the following properties:

1. The method applied to genetic distances d arising from a labeled, weighted tree T should give back the tree T.
2. The method applied to genetic distances d should depend "continuously" on d; that is, small changes in d should not result in large changes in the topology of the output tree.
3. It should be possible to perform the method efficiently (so the computer can handle it!).
4. The tree output by the method should not depend on the order in which the taxa are input.

Although these seem like reasonable properties, some of the well-known distance-based phylogeny methods do not satisfy them. For example, **UPGMA** does not

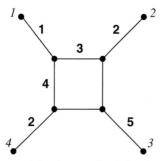

Figure 12.3 A weighted network relating four taxa.

satisfy No. 1 and **neighbor-joining** (*NJ*) does not necessarily satisfy No. 2 or 4 (see Chapter 5) (Moulton and Steel, 1999). The construction of the Buneman tree does indeed satisfy all of the demands; however, as a result, it is quite conservative and usually elects to discard too many splits because only the *minimum* values of $\beta(xy|uv)$ over all x, y in A and u, v in B are taken. Therefore, just one of these values has to be nonnegative in order for S to be rejected. One possibility for addressing this problem might be to take some type of average of the $\beta(xy|uv)$s – paying attention, of course, that a tree is still obtained. Although this idea can be made to work, leading to refined Buneman trees (Moulton and Steel, 1999), the next section focuses on another possibility, which is at the heart of the split-decomposition method.

12.2.2 Split decomposition

The difference between split decomposition and Buneman's approach is a simple but fundamental alteration to the definition of the index β_S of a split S. Given a set X taxa, a related distance matrix, and some split $S = \{A, B\}$ of X, for each x, y in A and u, v in B, the quantity

$$\alpha(xy|uv) = \max\{d(x, u) + d(y, v), d(x, v) + d(y, u)\}$$
$$- (d(x, y) + d(u, v)) \tag{12.2}$$

is defined. This is almost exactly the same formula as the one given for the quantity $\beta(xy|uv)$, except that the maximum rather than a minimum is considered. Proceeding as with the definition of β_S, the **isolation index** α_S of the split S is defined as one half the minimum value of $\alpha(xy|uv)$ taken over all x, y in A and u, v in B (Bandelt and Dress, 1992a). For example, consider the weighted network presented in Figure 12.3, labeled by the taxa $\{1, 2, 3, 4\}$ with weights indicated in boldface. In this network, as with trees, the distance d_N between any pair of taxa is the sum of the weights of the branches on the shortest path between the taxa. There may be several shortest paths, as opposed to a tree in which there is always precisely one;

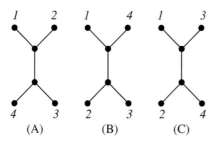

Figure 12.4   Three possible binary trees of four taxa.

for example, $d_N(1, 3) = 1 + 3 + 4 + 5 = 13$. Here, there are two possible ways to go between 1 and 3: either around the top of the central square or around the bottom. Then, $\alpha\ (14|23) = 6$, so that for the split $S = \{\{1, 4\}, \{2, 3\}\}$, it is $\alpha_S = 3$. Similarly, the split $T = \{\{1, 2\}, \{3, 4\}\}$ results in $\alpha_T = 4$. It is no accident that the isolation indexes just computed correspond to the weights on the two sets of parallel edges of the square in this network.

Moreover, this example illustrates several important points. First, the removal of any set of parallel edges results in a division of the network into two networks, each labeled by one part of the split, and the isolation index determines the weight of these edges. This is generally true for split-graphs and mimics the behavior of trees discussed in the previous section. Second, the splits S, T are incompatible (see Section 12.2) and, therefore, cannot correspond to two edges of a tree. Thus, as opposed to the Buneman index, the collection of splits that has positive isolation index (i.e., those splits S for which $\alpha_S > 0$) will no longer necessarily be compatible. Both splits S, T in the example have relatively high isolation indexes, and there is no advantage in retaining more of the significant splits; however, there is a loss of compatibility (i.e., the splits obtained will not necessarily correspond to a tree). Rather than choosing any particular collection of splits, the technique of *spectral analysis* actually assigns a significance to every split and proceeds from there (Lockhart et al., 1995).

Consider the example in Figure 12.3; the isolation index of the split $U = \{\{1, 3\}, \{2, 4\}\}$ is $\alpha_U = 0$. Thus, this split is not one of those in the set $\{1, 2, 3, 4\}$ with a positive isolation index and, consequently, it is discarded. In other words, for the three possible binary tree topologies of the four taxa illustrated in Figure 12.4, the isolation index would keep (A) and (B), but discard (C); the Buneman index would keep (A) and discard (B) and (C). Thus, the network in Figure 12.3 in some sense represents a combination of the two trees (A) and (B), neither being more significant than the other.

Buneman (1971) showed that the collection of all splits S with β_S positive is compatible; however, although the set of splits S with isolation index α_S positive is not necessarily compatible, it still satisfies some relaxation of compatibility (Bandelt

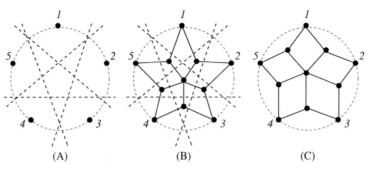

Figure 12.5   (A, B, C) Outer planar networks.

and Dress, 1992b). It is weakly compatible; that is, for every *three* splits $S = \{A, B\}$, $T = \{C, D\}$, $U = \{E, F\}$ in this collection, at least one of the intersections $A \cap C \cap E$, $A \cap D \cap F$, $B \cap C \cap F$, $B \cap D \cap E$ is empty.

Although collections of weakly compatible splits are fairly complicated mathematical objects, the most important consequences of weak compatibility can be summarized as follows (Bandelt and Dress, 1992a): if X has n elements, then the number of splits with positive isolation index is at most $n(n-1)/2$, and, as with the Buneman index, these can be computed efficiently. Moreover, the four desired properties listed previously for tree-building methods also hold for the split-decomposition method.

12.3   From weakly compatible splits to networks

Having computed a collection of weakly compatible splits, each one with its associated isolation index, it is possible to represent it by some weighted network. In general, this can always be achieved using a median network (Dress et al., 1997). These networks have the problem that they may not be, in general, planar (i.e., it may not be possible to draw them in the plane without crossing edges), in contrast to trees, where it is always possible. However, if the collection of splits under consideration is circular, then it can always be represented by a so-called **outer planar network**. These networks are what the program SplitsTree usually computes. Consider the set $\{1, 2, 3, 4, 5\}$ of taxa arranged on a circle, as shown in Figure 12.5. The splits of the taxa may be represented by lines in the plane; for example, the split $\{\{1, 2\}, \{3, 4, 5\}\}$ is represented by one of the dotted lines in Figure 12.5B. A collection of splits of a given set of taxa is circular if the taxa on a circle can be arranged so that every split in the collection can be represented by a line. Weakly compatible collections of splits are not necessarily circular.

Consider the collection of circular splits represented by the lines in Figure 12.5A. If a node is placed in each of the bounded regions created by the lines, and if two nodes in regions separated by exactly one line are connected by an edge, the network illustrated in Figure 12.5B is obtained. This is almost the split-graph representing the original collection of splits: the split-graph, which is shown in Figure 12.5C, can be obtained by adjusting the network in Figure 12.5B to give parallel edges. The removal of any parallel set of edges in the split-graph results in two networks, each labeled by the two parts of the split corresponding to the line passing through these edges. The technique for constructing the split-graph used in this example is based on the mathematical principle of **DeBruijn dualization**; in general, the SplitsTree program computes networks using a related construction (Dress, Huson, and Moulton, 1996). Finally, the split-graph can be weighted by giving the parallel edges corresponding to a split weight equal to the isolation index of that split. The distances between taxa in the resulting **weighted network**, computed using shortest paths, gives a representation d_N of the distance. In general, if the splits are computed with positive isolation index for given distances d on a set of taxa X, and a circular-split system is obtained, then the distances d_N between the taxa given by the network will be an approximation of the original distances d. Therefore, the isolation index tells how to factor out as many splits as possible from the distances d to give the representation d_N of d, leaving some **split-prime residue** $d - d_N$, which is precisely zero when the distance d is represented exactly (because in this case $d = d_N$) and can be disregarded when this is not the case.

To measure how well the distance d_N approximates d, it is typical to compute a simple statistic called the **fit index**, which is defined as $[\sum(d - d_N)(x, y)/\sum d(x, y)]*100\%$, where both sums are taken over all pairs of elements in X. Then, the fit is precisely 100% when all $d = d_N$.

PRACTICE

Vincent Moulton

12.4 The `SplitsTree` program

Several versions of the program `SplitsTree` are available. The most current, `SplitsTree3`, is formatted for Windows95 and UNIX. A `SplitsTree2` version of the program is available for MacOS, and a simple Web-based version of `SplitsTree` is also available at **http://bibiserv.techfak.uni-bielefeld.de/splits/**. Before installing `SplitsTree`, `Tcl/Tk` for Windows95 must be installed. This programming and scripting extension can be downloaded from **ftp://www.scriptics. com/pub/tcl/tcl8_0/** as `TCL805.EXE`, which is the installer file for Version 8.05 of `Tcl/Tk` for Windows. Newer versions of the software (**http://scriptics.com/ software**) have not been tested with `SplitsTree`. To install `SplitsTree`, go to **ftp://ftp.uni-bielefeld.de/pub/math/splits/**, click on the link for the current version, and download `Splits31.zip` (direct link: **ftp://ftp.mathematik.uni-bielefeld.de/pub/math/splits/splitstree3/**). The compressed Zip-archive contains the program files and some documentation. The executable is called `Splits3-1.exe`. Before the program is usable, one bug needs to be fixed: go to the `Tcl/Tk` folder and set the folder properties to show hidden files. Then copy the files `tcl80.dll` and `tk80.dll` into the folder with `Splits3-1.exe`.

12.5 Using `SplitsTree` on the mtDNA data set

In this section, the menu items of the `SplitsTree` program are described using the mtDNA sequences as a sample data set. To use the `SplitsTree` software, data need to be in a variant of the Nexus format (see Chapter 7). The `README.txt` file in the `Splits31.dir` contains all the details for preparing the input file in the format accepted by the program.

Double-click on the `SplitsTree` icon. A window labeled `SplitsTree3` will appear. Go to the `File` menu, select `Open`. Two new windows appear: the `log` window, which will be described later, and the `Open nexus` window. Go to the latter window and open the mtDNA data-set file. A split-graph immediately appears in the document window, which looks like the one shown in Figure 12.6. Inside the window is a scale indicator (labeled 0.1) and at the bottom of the window is information on the input data (e.g., how the graph is computed [-dsplits, hamming]), and the fit index. The fit index has a value of 79.2%, which is quite a reasonable fit; that is, the split-graph represents about 80% of the original distance, whereas 20% – hopefully including any noise – has been discarded. It is not possible to give

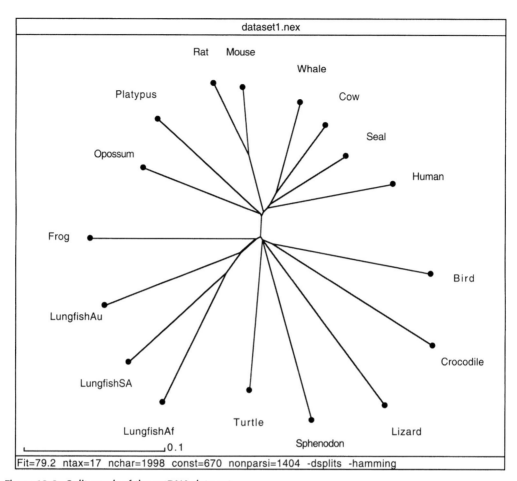

Figure 12.6 Split-graph of the mtDNA data set.

precise values for a good or bad fit; however, based on experience, anything with fit indexes above 80% is quite reasonable; above 90%, split-graphs can be considered robust; and anything much below 60–70% should be treated carefully. The split-graph in Figure 12.6 is quite well resolved and treelike. Thus, in view of the high fit index, it is expected that other ***distance-based methods***, such as NJ, would produce similar trees.

The `SplitsTree` program has many useful options for presenting a split-graph, which can be important for publication purposes. Begin by enlarging the document window as much as possible and clicking in the neighborhood of the graph, not on it. Hold down the button and move the mouse to drag the graph into any position within the window. Moving the cursor to a branch in the graph and clicking on it makes the branch darker than the others. By holding down the mouse button, the branch can move independently from the rest of the graph and be positioned at any angle. Similarly, by clicking on a taxon name, it is possible to move it to any position.

When the graph is positioned with the angles of edges adjusted as desired, go to the Edit menu. The options vertex labels and edge labels allow the labels and edge to be shown so that taxa can be displayed by numbers, names, or both. After **bootstrapping** (explained later in this section), it is possible to use the edge label option to display the **bootstrap values** on each edge. In the Edit menu, the options copy/cut/paste are also present, but they are not yet available for the Windows version of the program. A possible solution to this problem is to obtain the shareware program pstoedit (see Huson's Web link in Section 12.4). This program enables the conversion of a postscript file generated by SplitsTree into a file that can be processed with a drawing program.

Now that the split-graph has been labeled, go to the layout menu. The menu includes various items for modifying the layout of the graph, such as zooming in/out, rotating right/left, and flipping left-right/up-down. The layout menu also contains the cycle item, which forces the order in which the taxa are displayed on the outside of the graph (described in the Theory section using a circle). This option is useful when recomputing split-graphs with other options: it ensures that each graph displays the taxa in the same order, making the graphs easier to compare. Turn on the cycle option, and the cycle window appears in which the cycle order can be chosen. Another option on the layout menu is equal edges, which displays the same split-graph except that all edges are presented with equal length rather than a length proportional to the isolation index. The equal edges option can help discern branching patterns, which may be missed when edges are very short when drawn to scale. Select the to scale option to return to the original graph.

Using the Option menu, it is possible to further revise the split-graph. The taxa option allows the user to remove some taxa so that a new split-graph will be computed with these taxa left out of the calculation. This option is helpful when exploring the effect of a particular subset of taxa on the topology of the graph. Selecting characters from the Option menu allows the removal of characters from the calculation; the user can exclude, for example, codon positions by toggling on or off. Gaps, missing data, nonparsimony sites, and constant sites can be excluded as well by turning them on or off. Distance transformations (e.g., Hamming distances, Jukes and Cantor) are also available in the Option menu. The Nei Miller option computes distances from restriction site data, and the PAM 250 option applies to protein data. Finally, the force triangle inequalities option is a technical option that forces triangle inequalities that do not hold for the data set by adding an appropriate offset to all distances.

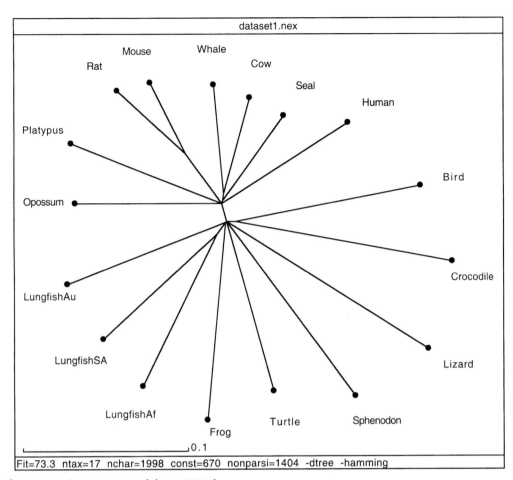

Figure 12.7   Buneman tree of the mtDNA data set.

The `Method` menu generates split-graphs with different algorithms. By default, `SplitsTree` computes the graph with the split-decomposition method, as described in Section 12.2.2. Buneman trees can be computed using the `Buneman tree` option with the formula discussed in Section 12.2.1. Figure 12.7 shows a Buneman tree for the mtDNA data. This tree is less resolved than the split-graph, precisely because of the reasons discussed in Section 12.2.1, and the Buneman tree is more conservative. Other options in the `Method` menu, `P-trees` and `Spectral tree`, are not described here (see Huson, 1998). Bootstrap values can be computed for the split-graph by selecting the `Bootstrap` in the `Option` menu. After `SplitsTree` computes these bootstraps, each edge in the graph is labeled by a value that gives the percentage of computed graphs in which the split corresponding to that edge occurred.

One of the latest additions to `SplitsTree` is the `Fit` menu, which allows the use of various methods for reestimating edge lengths. For example, the `Least squares` item optimizes the edge lengths under a least-squares criterion. This is a helpful alternative to isolation indexes, which can still result in rather conservative estimates of edge length because they are computed by taking a certain minimum.

The `Window` menu contains a list of currently open windows. As previously discussed, when the program is first executed, `SplitsTree` displays the `log` window, which is used to print error messages and/or general information about the current data. Other information is displayed by selecting `syntax` and `show`. Finally, the `command` option creates a window that accepts typed commands and Nexus scripts (Huson, 1998).

Split-graphs can be saved or printed by selecting those options from the `File` menu.

12.6   Using `SplitsTree` on the HIV-1 data set

Consider the HIV sample data set: running `SplitsTree` produces a graph that is almost tree-like, except for a thin box with two pendant taxa U27399 and U43386 (Figure 12.8). The fit, quite high at 88.2%, indicates that the data are very tree-like. It is expected that other tree-building methods (e.g., NJ) would give similar trees for this data set. The graph is well resolved, although the node to which taxa U08443, AF042106, L02317, AF025763 are joined appears to be a small square. Also, resolution at the central node from which the box emanates is poor, being a node of degree six. The fact that there are nodes with degrees higher than three, contrary to what is expected in a strictly bifurcating tree (see Chapter 1), could indicate an explosive star-like evolution of HIV (Dress and Wetzel, 1993). The Buneman tree looks similar to the split-graph, again indicating that the data set is tree-like. The split-graph for the mtDNA data set is even more tree-like (with fit at 88.9%), suggesting that traditional tree-building methods are appropriate for investigating the phylogenetic relationships of these sequences.

In these analyses, distances were computed by simply counting the number of differences between each sequence pair: the Hamming distance, which is the default distance calculation used by `SplitsTree` (see Section 12.5). However, it is possible to compute split-graphs using a ***distance matrix*** based on any nucleotide (or amino-acid) substitution model. For example, a distance matrix calculated with PAUP* and exported in NEXUS format (see Chapter 7) can be used as the input file for `SplitsTree`. In this way, the effect of different substitution models on the resulting split-graphs can be compared.

Figure 12.9 illustrates a split-graph arising from a non-tree-like data set. The split-graph was constructed using HCV data (Allain et al., 2000) from a study on

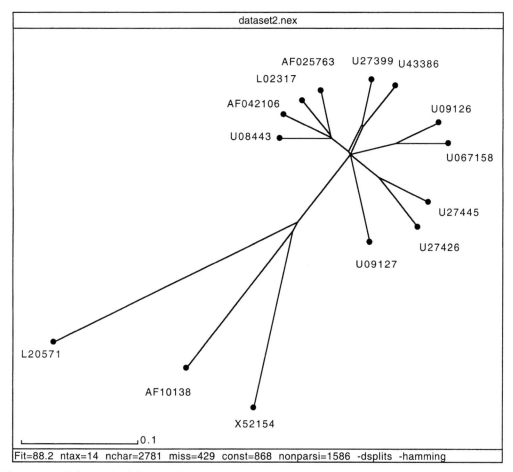

dataset2.nex

Fit=88.2 ntax=14 nchar=2781 miss=429 const=868 nonparsi=1586 -dsplits -hamming

Figure 12.8 Split-graph of the HIV data set.

host immune response to hepatitis C using infected donor-recipient clusters. The
example shows that phylogenetic relationships in this data set are best represented
by a network rather than a tree. The three clusters of taxa in Figure 12.9 are labeled
by the prefixes 603, 163, and 31. The taxa labeled with prefix 603 correspond to viral
sequences taken from a blood donor, and those prefixed by 163 and 31 correspond to
sequences taken from two recipients. The graph is rather box-like, which indicates
incompatibilities in the data that cannot be explained by a simple tree-like evolution
scenario. However, the fit index is 96.3%, indicating that the split-graph accurately
represents the distances. For example, the largest rectangle indicates that the two
splits of the data corresponding to the edges of the rectangle both have high support,
although they are incompatible. Some nodes are multiply labeled; for example, one
vertex is labeled by both 31/7 and 31/13, meaning that the isolation index could find
no support for any split that separated the taxa 31/7 and 31/13. The node labeled

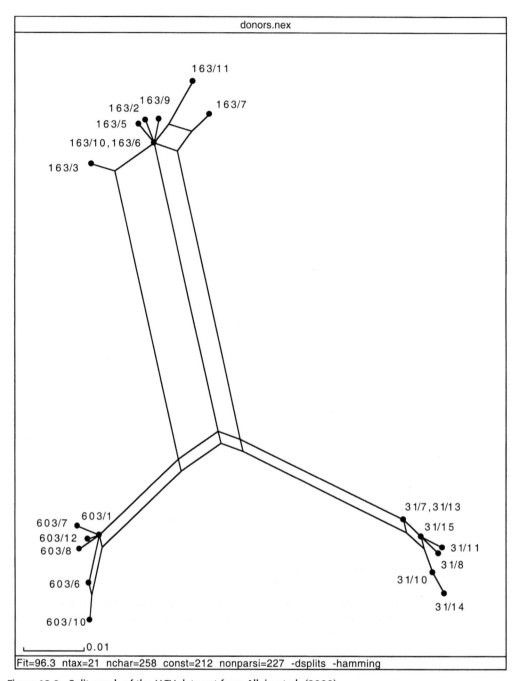

Figure 12.9   Split-graph of the HCV data set from Allain et al. (2000).

31/7, 31/13 is an internal node, not a leaf. Similarly, the node labeled 163/10, 163/6 is an internal node and could be interpreted as the ancestor of the sequences 163/5, 163/2, and 163/9. More examples in which `SplitsTree` is used to analyze viral data can be found in Dopazo et al. (1993) and Plikat, Nieselt-Struwe, and Meyerhans (1997). Additional examples provided in the `SplitsTree` package are recommended for further practice.

     `SplitsTree` does have limitations, the most notable being that for large numbers of sequences, split-graphs tend to become very unresolved. The problem is that to compute isolation indexes, the algorithm takes the minimum over a large number of values for each split, increasing the chance that some value will be found that excludes that split. However, `SplitsTree` is under constant development (Strimmer and Moulton, 2000), and new solutions are continually being sought for extending and improving the split-decomposition technique.

REFERENCES

Allain, J. P., Y. Dong, A.-M. Vandamme, V. Moulton, and M. Salemi (2000). Evolutionary rate and genetic drift of hepatitis C virus are not correlated with host immune response. *Journal of Virology*, 74, 2541–2549.

Bandelt, H.-J. and A. Dress (1992a). A canonical decomposition theory for metrics on a finite set. *Advances in Mathematics*, 92, 47–105.

Bandelt, H.-J. and A. Dress (1992b). Split-decomposition: A new and useful approach to phylogenetic analysis of distance data. *Molecular Phylogenetics and Evolution*, 1, 242–252.

Bandelt, H.-J. and A. Dress (1993). A relational approach to split-decomposition. In: *Information and Classification*, eds. O. Opitz, et al., pp. 123–131. Springer, New York.

Bandelt, H.-J., P. Forster, B. C. Sykes, and M. B. Richards (1995). Mitochondrial portraits of human population using median networks. *Genetics*, 141, 743–753.

Barbrook, A., C. Howe, N. Blake, and P. Robinson (1998). The phylogeny of the Canterbury Tales, *Nature*, 394, 839.

Buneman, P. (1971). The recovery of trees from measures of dissimilarity. In: *Mathematics in the Archeological and Historical Sciences*, eds. F. Hodson et al., pp. 387–395. Edinburgh University Press.

Dopazo, J., A. Dress, and A. von Haeseler (1993). Split-decomposition: A technique to analyze viral evolution, *Proceedings of the National Academy of Sciences of the USA*, 90, 10320–10324.

Dress, A. and R. Wetzel (1993). The human organism – A place to thrive for the immunodeficiency virus. *FSPM-preprint*, University of Bielefeld, Bielefeld, Germany.

Dress, A., D. Huson, and V. Moulton (1996). Analyzing and visualizing sequence and distance data using SplitsTree. *Discrete and Applied Mathematics*, 71, 95–109.

Dress, A., M. Hendy, K. Huber, and V. Moulton (1997). On the number of vertixes and edges in the Buneman graph. *Annals in Combinatorics*, 1, 329–337.

Huson, D. (1998). SplitsTree: A program for analyzing and visualizing evolutionary data. *Bioinformatics*, 14, 68–73.

Lockhart, P., D. Penny, and A. Meyer (1995). Tetsing the phylogeny of swordtail fishes using split-decomposition and spectral analysis. *Journal of Molecular Evolution*, 41, 666–674.

Moulton, V. and M. Steel (1999). Retractions of finite distance functions onto tree metrics. *Discrete Applied Mathematics*, 91, 215–233.

Pages, R. D. M. and E. C. Holmes (1998). *Molecular Evolution: A Phylogenetic Approach*. Oxford: Blackwell Science.

Plikat, P., K. Nieselt-Struwe, and A. Meyerhans (1997). Genetic drift can dominate short-term HIV-1 nef quasispecies evolution in vitro. *Journal of Virology*, 71, 4233–4240.

Strimmer, K. and V. Moulton (2000). Likelihood analysis of phylogenetic networks using directed graphical models. *Molecular Biology and Evolution*, 17, 875–881.

Swofford, D. L., G. J. Olsen, P. J. Waddell, and D. M. Hillis (1996). Phylogenetic Inference. In: *Molecular Systematics*, 2nd ed., eds. D. M. Hillis, C. Moritz, and B. K. Mable, pp. 490–493. Sinauer Associates, Sunderland, MA.

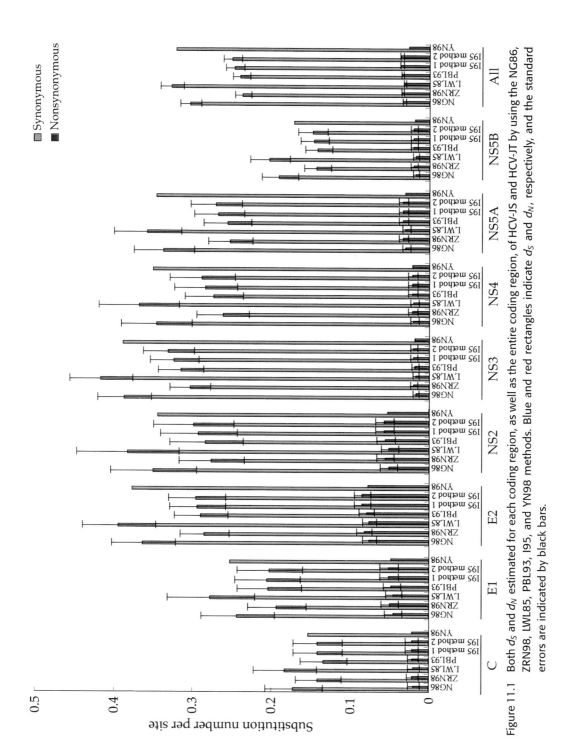

Figure 11.1 Both d_S and d_N estimated for each coding region, as well as the entire coding region, of HCV-JS and HCV-JT by using the NG86, ZRN98, LWL85, PBL93, I95, and YN98 methods. Blue and red rectangles indicate d_S and d_N, respectively, and the standard errors are indicated by black bars.

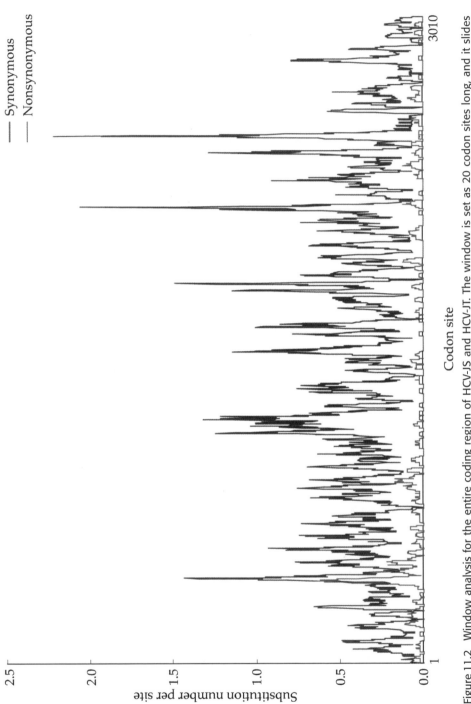

Figure 11.2 Window analysis for the entire coding region of HCV-JS and HCV-JT. The window is set as 20 codon sites long, and it slides by one codon site for each step. Within the window at each step d_S and d_N are estimated and plotted in two dimensions, in which the abscissa indicates the codon site and the ordinate indicates d_S and d_N.

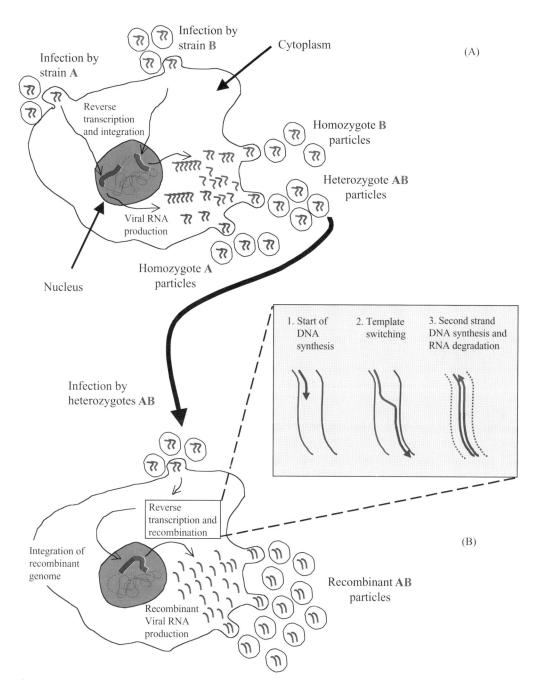

Figure 14.1 Retroviral recombination. (A) Upon super- or co-infection of a target cell by two retroviral strains (labeled A and B and colored blue and red), both strains may integrate a provirus into the host cell nuclear genome after reverse transcription in the cytoplasm. Co-expression of the different proviruses generates both homozygous and heterozygous viral particles due to the diploidy of retroviral virions. (B) When the newly generated heterozygous particles enter a second generation of target cells, recombinant proviruses may be generated through strand-transfer during the RT-step (inset). Integration and expression of recombinant provirus produce recombinant offspring virions.

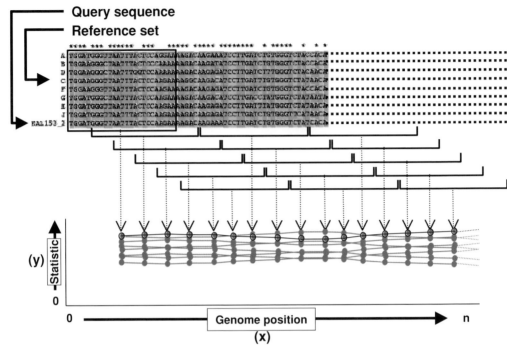

Figure 14.5 Basic principle of average-over-window scanning methods. An alignment of reference sequences (A-J) and a potential recombinant (the query sequence) are sequentially divided
into overlapping subregions (i.e., window, box, and brackets) for which a statistic/measure
is computed. This statistic/measure is then plotted (broken arrow lines) in an x/y scatter
plot using the alignment coordinates on the x-axis and the statistic/measure range on the
y-axis. Each value is recorded at the midpoint of the window and connected by a line.

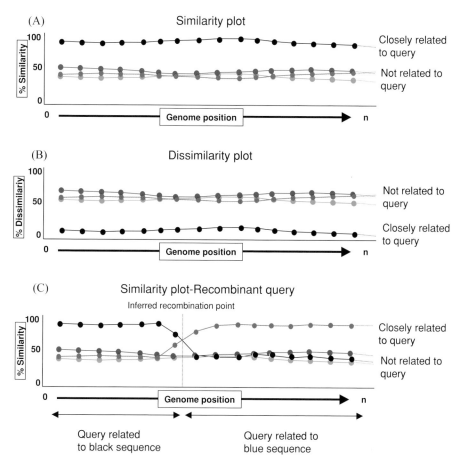

Figure 14.6 Similarity and dissimilarity methods. (A) Similarity plot. In this type of analysis, the measure recorded is the similarity value (sim) between the query sequence and each of the reference sequences. (B) The same analysis, except that the inverse value to similarity (1-sim) or the dissimilarity is plotted. In a variation of these methods, similarity/dissimilarity values corrected by an evolutionary correction model (i.e., any of the JC69, TN93, K80, F81, F84, HKY85, or GTR models (see Chapter 4 and Chapter 10)) may be used. (C) Schematic view of a plot of a recombinant sequence.

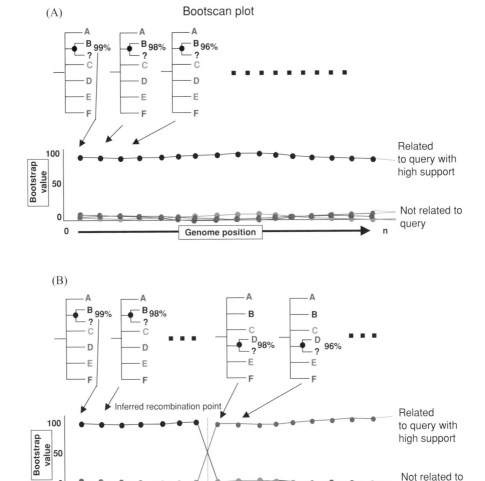

Figure 14.7 The bootscanning method. (A) In the bootscanning method, the statistic plotted is the bootstrap value of a node connecting a reference and the query (labeled "?"). (B) For a recombinant sequence, the high bootstrap support for clustering with one reference (Sequence B) suddenly switches to high bootstrap support for clustering with another reference (Sequence D).

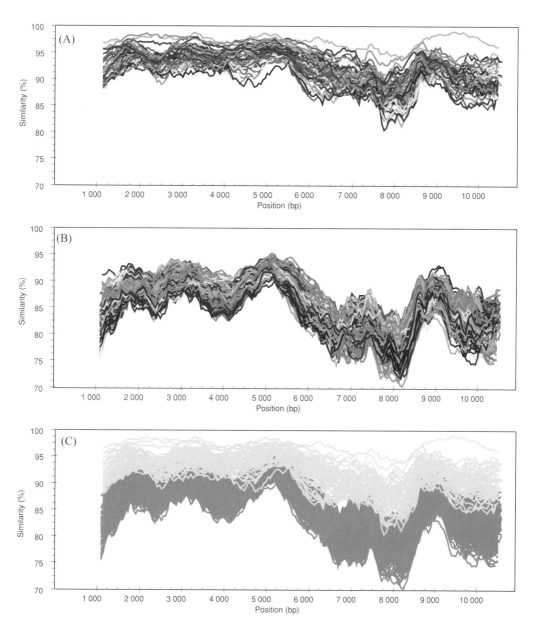

Figure 14.9 Range of HIV-1 M group intrasubtype and intersubtype variation. (A) Intrasubtype variation among the subtype reference sequences. (B) Range of intersubtype variation among the subtype reference sequences. (C) Intrasubtype and intersubtype variation superimposed (light/dark gray lines).

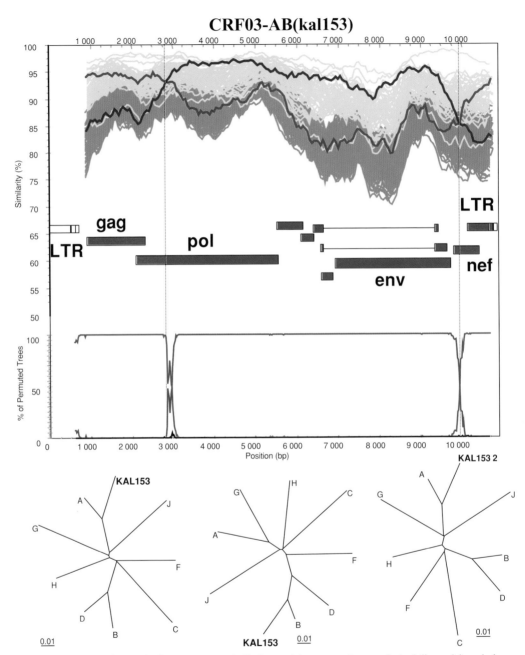

Figure 14.10 Recombinant isolate KAL153. Similarity and bootscanning analysis followed by phylogenetic analysis (K2 + NJ) of the identified segments. The similarity analysis with Subtypes
A, B (parental subtypes), and C (outgroup) is superimposed on the ranges of intra- and
intersubtype variation.

(A)

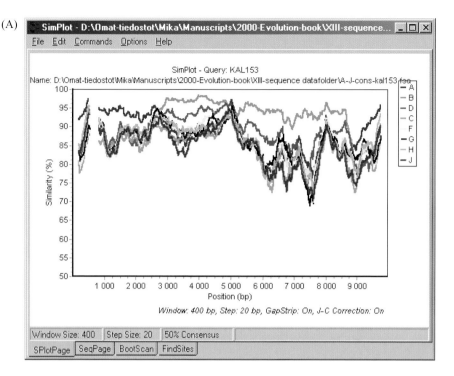

(B)

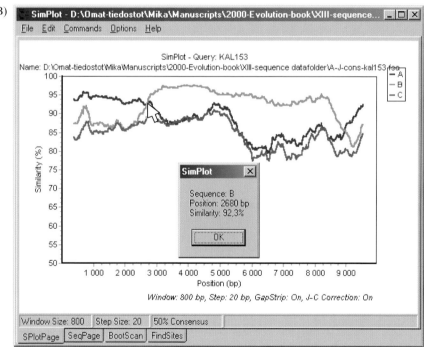

Figure 14.14 Comparing KAL153 to the reference sequences. (A) View with all reference sequences. (B) View after hiding all but the A, B, and C reference sequences. Clicking on a breakpoint brings up the location in the alignment (pop-up window). Consult the Simplot manual for details on how to zoom in/out of the plot.

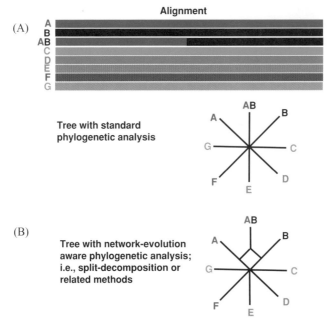

Alignment

(A)

Tree with standard
phylogenetic analysis

(B)

Tree with network-evolution
aware phylogenetic analysis;
i.e., split-decomposition or
related methods

Figure 14.8 Split-decomposition analysis. (A) When a recombinant sequence contains equal amounts of phylogenetic signal from both parental lineages (alignment schematic), standard phylogenetic methods incorrectly interpret this as a new evolutionary lineage, branching the sequence off from the center of a tree (tree). (B) Split-decomposition and other network-evolution aware related methods can indicate the dual ancestry of the sequence by showing the tree as a network rather than as a strictly binary-split tree.

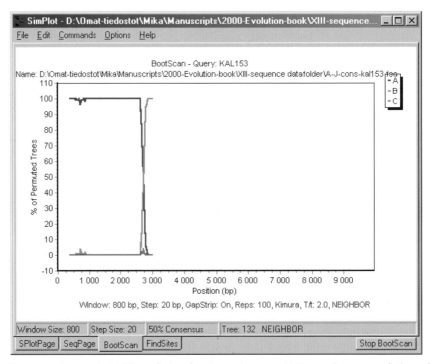

Figure 14.15 Bootscanning analysis. View during bootscanning run. Note continuous updating of the screen and run-log footer showing the status of the background run.

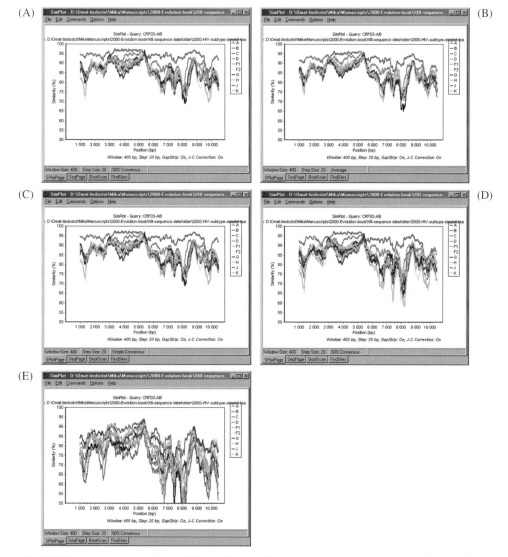

Figure 14.18 Different consensus types. Different types of consensus sequences may be calculated from the groups. The type of consensus used can be specified in the Options menu. (A) 50% consensus. (B) Average consensus. (C) Simple consensus. (D) 60% consensus. (E) 90% consensus. A detailed description on how the consensus sequences are calculated can be found in the Simplot documentation.

Tetrapod phylogeny and data exploration using DAMBE

THEORY

Xuhua Xia and Zheng Xie

13.1 The phylogenetic problem and the sequence data

One major controversy in the phylogenetic relationships among the **tetrapods** is whether birds are more closed related to crocodilians (Romer, 1966; Carroll, 1988; Gauthier et al., 1988) or to mammals (Gardiner, 1982; Løvtrup, 1985). Hedges et al. (1990) collected a set of 18S rRNA sequences to evaluate these two alternative hypotheses, and found that the bird-mammal grouping is more strongly supported – with the bootstrap value of 88% – than the bird-crocodilian grouping (in contrast with the result of the exercise in Sections 5.4 and 5.5 based on a different data set). A subset of these 18S rRNA sequences was subsequently used in a statistical test, based on the **minimum-evolution** criterion, to evaluate relative support of these alternative phylogenetic hypotheses (Rzhetsky and Nei, 1992). The nine shortest trees, including the **neighbor-joining** (**NJ**) tree, all group the bird and mammal together as a monophyletic **taxon**.

In this chapter, the subset of 18S rRNA sequences used by Rzhetzky and Nei (1992) is used to check the validity of the conclusions made by previous studies (Hedges et al., 1990; Rzhetsky and Nei, 1992) and to show the principles of data exploration in phylogenetic studies using the DAMBE program (see Section 2.8). The data set originally consisted of six 18S rRNA sequences from a human (*Homo sapiens*), a robin (*Turdus migratorius*), a snake (*Heterodon platyrhinos*), a frog (*Xenopus laevis*), a turtle (*Pseudemus scripta*), and an alligator (*Alligator mississippiensis*). However, sequences for the robin, snake, turtle, and alligator have many unresolved sites. The first three species were removed, the alligator sequence was replaced with a fully re-solved one, and sequences for a chicken (*Gallus gallus*), a tuatara (*Sphenodon punctatus*), and a rat (*Rattus norvegicus*) were added. The source of the sequences, together

Table 13.1   The six species whose 18S rRNA sequences are used in this chapter, together with GenBank LOCUS name, sequence lengths, and nucleotide frequencies

Species	LOCUS	Length[a]	A	C	G	T
(A)						
Xenopus laevis	XLRRN18S	1826	0.2371	0.2562	0.2841	0.2226
Alligator mississippiensis	AF173605	1733	0.2365	0.2516	0.2858	0.2261
Sphenodon punctatus	AF115860	1734	0.2354	0.2533	0.2864	0.2249
Gallus gallus	AF173612	1737	0.2336	0.2597	0.2875	0.2191
Homo sapiens	HSRRN18S	1869	0.2284	0.2638	0.2939	0.2139
Rattus norvegicus	RNRRNA06	1874	0.2301	0.2614	0.2928	0.2157
(B)						
Xenopus laevis			0.2033	0.3333	0.1789	0.2846
Alligator mississippiensis			0.2069	0.2586	0.1810	0.3534
Sphenodon punctatus			0.1795	0.2906	0.1966	0.3333
Gallus gallus			0.1500	0.3917	0.2083	0.2500
Homo sapiens			0.0663	0.4518	0.3313	0.1506
Rattus norvegicus			0.0877	0.4269	0.3041	0.1813

[a] The aligned sequences, after trimming the leading and trailing sequences but including gaps, are 1,790 bp. Part (A) shows the nucleotide frequencies. Part (B) shows the nucleotide frequencies of polymorphic sites only.

with descriptive information, is shown in Table 13.1. The aligned sequences can be found on the Web site **http://aix1.uottawa.ca/~xxia/research/data/18SrRNA.htm**.

13.2   Results of routine phylogenetic analyses without data exploration

The phylogenetic methods that have been employed include the ***maximum-parsimony*** method (Kluge and Farris, 1969; Fitch, 1971), ***maximum-likelihood*** method (Felsenstein, 1993; Olsen et al., 1994), and the NJ (Saitou and Nei, 1987) and ***Fitch-Margoliash*** (Fitch and Margoliash, 1967) methods, together with the following ***genetic distances***: JC69 (Jukes and Cantor, 1969), TN84 (Tajima and Nei, 1984), K80 (Kimura, 1980), F84 (Felsenstein, 1993), TN93 (Tamura and Nei, 1993), and the paralinear distance (Lake, 1994) (see Chapter 4). All of these methods generate a tree with the same topology as Tree 1 in Figure 13.1A, consistent with the findings in Hedges et al. (1990) and Rzhetsky and Nei (1992). Given the tree, it is of interest to know the relative statistical support of alternative subtrees, and resampling methods have been developed for attaching confidence limits to them. As discussed in Chapter 5, there are two kinds of commonly used resampling methods: ***bootstrapping*** and ***jackknifing***. Although they generally produce similar

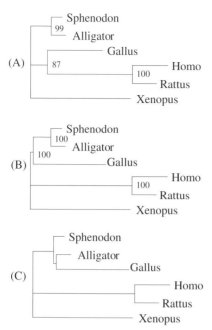

Figure 13.1   Three alternative phylogenetic trees for the tetrapod data set. The branch lengths are evaluated by the least-squares method using the TN93 distance. The numbers at the internal
nodes are percentage of bootstrapping values out of 500 resamples. (A) Best tree in Hedges
et al. (1990) and Rzhetsky and Nei (1992). Bootstrap values are based on resampling of
nucleotide sites. (B) Tetrapod phylogeny calculated according to indel-based distance. (C)
Alternative tree topology.

results, there are subtle differences. Consider a set of aligned sequences N sites long.
In the bootstrap resampling, the probability of each site being sampled is $1/N$, and
the mean number of times a site is sampled in each bootstrapping resampling is
simply one. Thus, a site that is sampled 0, 1, 2, ..., N times follows a Poisson
distribution with mean equal to 1. This implies that about 37% of the sites will
not be sampled, whereas 63% will be sampled at least once. In jackknifing, 50% of
the sites is not sampled and the other 50% is sampled just once. Thus, a jackknifed
sample is expected to be less similar to the original sample than a bootstrapped
sample. Consequently, jackknifed samples should be less similar to each other than
bootstrapped samples.

The 18S rRNA sequences were bootstrapped 500 times to generate the ***consensus
tree*** (see Figure 13.1A, Tree 1) by using each phylogenetic method discussed previously. The results from different phylogenetic methods are highly consistent, with
the cluster that groups the chicken and mammalian species having high bootstrap
values (i. e., 87% in Tree 1 in Figure 13.1). The jackknife resampling produces similar
results (not shown), which are similar to the findings in Hedges et al. (1990).

13.3   Distance-based statistical test of alternative phylogenetic trees (optional)

Statistical tests of alternative phylogenetic hypotheses are available for distance, maximum-parsimony, and maximum-likelihood methods, which are briefly reviewed in Xia (2000b). The **distance-based test** implemented in DAMBE – slightly different from the one in Xia (2000b) – is based on the fit to the original distance matrix. DAMBE reads unrooted user trees in a tree file. The user trees represent alternative phylogenetic hypotheses to be evaluated. Based on any one of many genetic distances chosen, DAMBE will evaluate tree branch lengths for each unrooted topology in the user tree file by using the least-squares method. From the evaluated branch lengths, the pairwise distances between taxa based on the evaluated branch lengths can be obtained. Thus, two matrixes are computed: the original distance matrix, designated x and used for phylogenetic reconstruction; and the reconstituted matrix, designated y and derived from the reconstructed tree. Elements in the x matrix represent "observed" values; those in the y matrix represent the estimated or expected values. If $x_{ij} = y_{ij}$ for all i and j, then the fit between the observed and the estimated values is perfect, and the tree is considered the best possible. The error variance is calculated as

$$Var E = \frac{\sum_{i=1}^{n-1} \sum_{j=i+1}^{n} (x_{ij} - y_{ij})^2}{n(n-1)/2 - m - 1} \tag{13.1}$$

where n is the number of taxa and m is the number of taxa pairs separated by a single internal node on the tree. Such taxa pairs will have $x_{ij} = y_{ij}$ when evaluated by the least-squares method and will not contribute to $VarE$. $VarE$ is equivalent to the variance of $D_{ij} (= x_{ij} - y_{ij})$ because the mean D is expected to be zero. In other words, $(D_{ij} - \bar{D}) = (D_{ij} - 0) = (x_{ij} - y_{ij})$. If $VarE = 0$, then the tree is the best possible. In the case of only two trees, an F-test can be performed to check whether one tree is significantly better than the other, with F calculated as follows:

$$F = \frac{Var E_{\text{Large}}}{Var E_{\text{Small}}} \tag{13.2}$$

For k alternative phylogenetic hypotheses, there are k $VarE$ values. Testing whether all alternative topologies are equally good can be done by the standard Bartlett's test. The test is based on $B_c = B/C$, where

$$B = \left(\ln s_p^2 \right) \left(\sum_{i-1}^{k} v_i \right) - \sum_{i=1}^{k} v_i \ln s_i^2 \tag{13.3}$$

$$C = 1 + \frac{1}{3(k-1)} \left(\sum_{i=1}^{k} \frac{1}{v_i} - \frac{1}{\sum_{i=1}^{k} v_i} \right) \tag{13.4}$$

Table 13.2 Testing the three alternative phylogenetic hypotheses, using genetic-based distances (Xia, 2000b)

Trees	Diff_Log(Var)	q	k	Prob
1 vs. 2	0.94546	2.3159	2	0.1015
1 vs. 3	2.48066	6.0764	3	<0.001
2 vs. 3	1.53520	3.7605	2	0.0078

Prob: the probability that two topologies fit data equally well.
k: the number of variances falling within the range between the two variances.

$$s_p^2 = \frac{\sum_{i=1}^{k} SS_i}{\sum_{i=1}^{k} v_i} \tag{13.5}$$

In this case, the degree of freedom v_i is the denominator in Equation 13.5, $s_i = VarE_i$, and s_p^2 is the pooled variance. The distribution of B_c is approximated by the chi-square distribution with $k - 1$ degrees of freedom, where k is the number of phylogenetic hypotheses. The Bartlett's test is an overall test of homogeneity of variance. If the null hypothesis of equal variance is not rejected, then the user should not proceed further. However, if the null hypothesis is rejected (i.e., some trees are significantly better, having smaller variances than others), then it is natural to want to know whether the best tree is better than all the others. It is not appropriate to do an F-test between $VarE$ of the best tree and each of the $(k - 1)$ $VarE_i$ values because it does not control for the experimental error rate. Various methods have been invented to control for the experimental error rate without making the test unduly conservative, and the Newman-Keuls test is perhaps one of the most widely used for significance tests involving multiple comparisons. The test statistic, q, is based on the same statistic as in Tukey's HSD test. It is calculated as follows:

$$q = \frac{\ln Var E_i - \ln Var E_j}{\sqrt{\frac{1}{v_i} + \frac{1}{v_j}}} \tag{13.6}$$

where v_i and v_j are the degrees of freedom associated with trees i and j, respectively.

The application of this method shows Tree 1 (Figure 13.1A) to be significantly better than Tree 3 (Figure 13.1C) and almost significantly better than Tree 2 (Figure 13.1B; Table 13.2). If the experimental error rate is ignored (which is not a good practice), then the first tree is significantly better than the other two alternative trees, which is consistent with the findings of Rzhetsky and Nei (1992).

13.4 Likelihood-based statistical tests of alternative phylogenetic trees

There are two slightly different tests for evaluating alternative phylogenetic hypotheses by using the maximum-likelihood method in DAMBE, both being derivatives from the ***Kishino-Hasegawa test*** (Kishino et al., 1990). The original Kishino-Hasegawa test begins by first calculating the log likelihood for each alternative topology, then the difference (D) in log likelihood between the best tree and each alternative topology, and finally the variance of the differences (V_D) estimated by resampling methods such as bootstrapping. The z-score is calculated as follows:

$$z = \frac{D}{\sqrt{V_D}} \tag{13.7}$$

The z-score is declared significant if it is larger than 1.96. Results from two different genes in a combined test may be pooled. Suppose there are two data sets of 18S and 28S rRNA genes for evaluating three alternative topologies. Let $L_{18S.i}$ and $L_{28S.i}$ be the log-likelihood values obtained for topology i ($i = 1,2,3$) from the 18S and the 28S rRNA genes, respectively. Let $V_{18S.i}$ and $V_{28S.i}$ be the corresponding variances for the difference in log likelihood between topology i and the best topology (i.e., the one with the largest log likelihood). Next, the following is computed:

$$
\begin{aligned}
L_{sum.i} &= L_{18S.i} + L_{28S.i} \\
L_{Max} &= \max(L_{sum.i}) \\
L_{diff.i} &= L_{sum.i} - L_{Max} \\
SE_i &= \sqrt{V_{18S.i} + V_{28S.i}} \\
z_i &= \left| \frac{L_{diff.i}}{SE_i} \right|
\end{aligned}
\tag{13.8}
$$

where z_i can be used to perform a standard z-test. If $z_i > 1.96$, then topology i is rejected at the 0.05 significance level. The test is also called the ***resampling estimated log-likelihood*** (**RELL**) test because V_D is estimated by resampling.

The ***likelihood-based test*** implemented in DNAML (Felsenstein, 1993) is slightly different from the RELL test; it does not estimate V_D by bootstrapping, but rather assumes that all sites are independent and performs a test similar to a paired-sample t-test. For a set of sequences N base pair long, the log-likelihood l_i for each site i is computed for each tree topology, with $\boldsymbol{L_1} = (l_{11}, l_{12}, \ldots l_{1N})$ for Topology 1 and $\boldsymbol{L_2} = (l_{21}, l_{22}, \ldots l_{2N})$ for Topology 2. If Topology 1 has a larger log-likelihood value, then $D_i = l_{1i} - l_{2i}$, where $i = 1,2,\ldots N$, is obtained. Tree 1 is considered significantly better than Tree 2 when the mean of D_i is significantly greater than zero by a standard two-tailed t-test with t calculated as follows:

$$t = \left| \frac{\bar{D}}{\sqrt{V_D/N}} \right| \tag{13.9}$$

In DNAML, N is assumed to be large, and t is treated as the z-score and declared significant if it is greater than 1.96. This test assumes that the D_i values are normally distributed. If they are not, the test becomes extremely conservative; that is, with little power to reject the null hypothesis of two topologies being equally good. The RELL test is expected to be more robust although more computationally expensive. However, the RELL test ignores the site-specific information and, therefore, may not be as sensitive to detecting the difference in likelihood between the two topologies. This can be illustrated with a simple example: suppose there is a set of sequences of five base pair long to evaluate two topologies. If the log-likelihood values for the five sites for Topology 1 is $L_1 = (1, 1, 3, 4, 5)$ and the corresponding values for Topology 2 is $L_2 = (2, 3, 4, 5, 6)$, then Topology 2 is a better tree because the log likelihood is larger by 1.2. The paired-sample t-test will reject the null hypotheses of the two topologies being equally good with $p = 0.0039$; the RELL test will not, because V_D in Equation 13.7 is quite large (5.7). In contrast, V_D in Equation 13.9 is only 0.2. The numerators in the two equations are the same.

Both tests, as presented, are valid only for comparing two topologies. The two tests can be extended to accommodate multiple comparisons involving more than two topologies by using the q statistic in the Newman-Keuls test. This extension also has been implemented in DAMBE (Xia, 2000a). For the sample data sets, the two likelihood-based tests yield similar results, with Tree 1 in Figure 13.1 being the best, but not significantly better than the others. This, of course, is not always the case.

13.5   Data exploration

The previous example is typical of many published phylogenetic studies; that is, applying a battery of tests to the data set without carefully examining whether any of them is appropriate for the data. This is not the right way of doing data analysis, but instead is an efficient way to reach a "solid" and misleading conclusion. This point is illustrated in the following sections.

13.5.1   Nucleotide frequencies

A phylogenetic study involving molecular data requires the researcher to have an understanding of the ***nucleotide substitution pattern.*** This pattern is typically summarized in a substitution model (see Chapter 4), which is characterized by two categories of parameters: frequency and rate ratio. In the tetrapod sample data set (see Table 13.1A), one pattern stands out when only polymorphic sites are considered: the three poikilothermic species (i.e., frog, alligator, and tuatara) have lower GC-content than the three homoeothermic species (i.e., rat, human, and chicken) (see Table 13.1B). The pattern applies to other sets of poikilothermic and homoeothermic vertebrates as well (Rzhetsky and Nei, 1992), which brings up an interesting question: is the nucleotide frequency an informative phylogenetic character? If it is, then the pattern would favor the grouping of the homoeotherms, consistent with the phylogenetic relationship advocated in previous studies (Hedges et al., 1990; Rzhetsky and Nei, 1992). However, nucleotide frequencies often can change rather quickly, with the consequence that distantly related species have similar nucleotide frequencies and closely related species have different nucleotide frequencies (Xia, 2000b). It might also be argued that the high GC content in the 18S rRNA sequences of the homoeotherms is due to convergent evolution. The function of rRNA molecules greatly depends on the three-dimensional (3-D) structure stabilized by the intramolecular base-pairing. Because G/C pairs have three hydrogen bonds, whereas A/T pairs have only two, an increased GC content would seem to increase the stability of the molecule in homoeothermic species. In other words, the 18S rRNA genes in the avian and mammalian species may increase their GC content as a response to increased body temperature as a consequence of convergent evolution.

The convergent increase in the GC content in the 18S rRNA sequences in the homeothermic species may bias phylogenetic reconstruction. A direct and simple way to appreciate such an effect is simply to treat the nucleotide frequencies as one locus with four alleles, calculate pairwise genetic distances, and construct a phylogenetic tree. The resulting tree has the same topology as Tree 1 in Figure 13.1A. The resulting distance matrix from the nucleotide frequency data also can be used to evaluate the relative statistical support of the three alternative phylogenetic trees in Figure 13.1. The support for Tree 1 is similar to that in Table 13.2. If the similarity in nucleotide frequencies among the four homoeothermic species is the result of convergent evolution, then the strong support for Tree 1 from nucleotide frequencies represents a bias inherent in the phylogenetic analysis using the 18S rRNA sequences. The hypothesized convergent evolution would lead to what is known as a nonstationary evolutionary process of nucleotide substitutions, which tends to bias phylogenetic estimation. Unfortunately, thus far there is no reliable way of correcting the bias. However, the result shown in Table 13.2 does suggest that previous phylogenetic analyses involving the 18S rRNA sequences (Hedges et al., 1990; Rzhetsky and Nei, 1992) may be biased.

Nucleotide frequencies, especially those at the polymorphic sites (see Table 13.1B), also can help discern evolutionary dynamics. One apparent pattern that stands out from the data in Table 13.1B is that nucleotides A and T are less frequently and G and C are more frequently involved in substitutions in mammalian species than in the poikilothermic species. The nucleotide frequencies of polymorphic sites in the chicken are similar to the mammalian species. This observation has two implications: (1) nucleotide C is a **substitution hot spot** in all species, and the homoeothermic species happen to have more C in their 18S rRNA sequences than the poikilothermic species, which results in a faster **evolutionary rate** in birds and mammals, leading to what is known as the ***long-branch attraction*** problem (see Sections 5.3 and 8.1); and (2) the difference between nucleotide frequencies of the whole sequences and those of the polymorphic sites suggests that the substitutions do not occur randomly over sites, leading to ***rate heterogeneity over sites.***

13.5.2   Substitution saturation and the rate heterogeneity over sites

The phylogenetic bias introduced by the ***similarity*** in nucleotide frequencies increases with the degree of ***substitution saturation*** (see Chapter 4). Consider, for example, four taxa with Taxon 1 and Taxon 2 being sister taxa and Taxon 3 and Taxon 4 also being sister taxa. If the sequences have experienced full substitution saturation, then it is obvious that the genetic distance calculated for any taxa pair will reflect only the similarity in nucleotide frequencies. If Taxon 1 and Taxon 3 have the most similar nucleotide frequencies, they tend to cluster together as a ***monophyletic group***, which is a wrong result.

Is there any reason to suspect that the 18S rRNA sequences of the sample data set have experienced substitution saturation? The concept is trickier than most realize; for example, it is not uncommon to come across statements such as, "Our sequences have experienced little substitution saturation because the largest percentage difference between sequences is less than 5%." The largest percentage difference in the sample data set is only 4.2%, so it would seem reasonable to ignore the problem of substitution saturation. However, the ***nucleotide substitution rate*** greatly differs over sites. It is likely that 95% of the sequences are invariant; that is, maintained by strong ***purifying selection.*** This means that a percentage difference close to 5% actually represents a high degree of substitution saturation among the noninvariant sites. An indication of substitution saturation in the sample data set is that segments in the sequences have varied so much that they cannot be aligned reliably and must be excluded from the analysis (Hedges et al., 1990).

Rate heterogeneity over sites is also a thorny problem without a good solution. All commonly used phylogenetic methods failed to retrieve the true tree when applied to sequences with high rate heterogeneity over sites (Kuhner and Felsenstein, 1994) because of two main problems. First, it implies that many ***molecular clocks*** have been ticking at different sites, consequently creating a situation more complex than modern computational tools can handle. Second, highly variable sites on which phylogenetic inferences are based often have experienced substitution saturation and contribute little else but noise to the analysis. For these reasons, it is necessary to know whether the sequences under investigation exhibit strong rate heterogeneity over sites. Heterogeneity is often measured by the shape parameter (i.e., the α parameter) of the Γ distribution, with a small α corresponding to large rate heterogeneity (see Section 4.6.1). The 18S rRNA sequences have a small α equal to 0.143, suggesting strong rate heterogeneity over sites. In conclusion, the

Table 13.3 Empirical substitution patterns in the tetrapod data set

	AG	CT	AC	AT	CG	GT
Obs	29	57	10	0	18	9
Exp	22	19	20	17	24	21

similarity in nucleotide frequencies among the homoeothermic species may play a significant role in grouping them together as a monophyletic clade, whether or not they are sister taxa in reality.

13.5.3 The pattern of nucleotide substitution

Thus far, only one aspect of the nucleotide substitution model has been explored: the frequency parameters; the other aspect is the rate ratio parameters. When obtaining the empirical substitution pattern, it is important to remember that pairwise comparisons are nonindependent (Felsenstein, 1992; Nee et al., 1995; Xia et al., 1996). For example, if there is one species that has recently experienced numerous $A \rightarrow G$ transitions and few other substitutions, then all pairwise comparisons between this and other species will each contribute one data point with a large $A \rightarrow G$ transition bias. To obtain the substitution pattern properly, it is first necessary to know the phylogenetic relationship among the taxa, construct ancestral sequences at the internal nodes, and make pairwise comparisons between neighboring nodes along the tree. This alleviates the problem of nonindependence in the resulting substitution data. It is of some comfort to know that the tree required for this purpose does not need to be the true tree, and the substitution model used for reconstructing ancestral sequences does not need to match reality exactly. As long as the tree and the model are reasonable, the resulting empirical substitution pattern will be sufficient for the purpose at hand.

In Table 13.3, pairwise comparisons are carried out between neighboring nodes along the tetrapod tree. A total of 123 nucleotide substitutions are inferred to have occurred on the tree. The value of 29 under the heading AG means that there are 29 cases in which Nucleotide A is at one site in one sequence and that G is at the *orthologous* site in another sequence. The tree and the reconstructed ancestral sequences are obtained by maximum likelihood based on the TN93 model (Tamura and Nei, 1993). One particular pattern emerging from the data is the dramatic heterogeneity among the four transversions with no $A \leftrightarrow T$ substitution (see Table 13.3). This pattern is also true for the data set used in Rzhetsky and Nei (1992). The maximum-parsimony method used in Hedges et al. (1990) cannot accommodate such a complex substitution pattern. For calculating genetic distances, Rzhetsky and Nei (1992) used the JC69 (Jukes and Cantor, 1969) model and Hedges et al. (1990) used the TN84 model (Tajima and Nei, 1984). Both models assume that the six types of nucleotide substitutions (i.e., two transitional and four transversional substitutions) all occur with equal frequencies. These models are apparently inappropriate for the substitution pattern in Table 13.3.

Although the substitution pattern in Table 13.3 could be described by the GTR model with five rate ratios, it is unsatisfying not to be able to formulate a plausible evolutionary hypothesis to explain the rarity of the A ↔ T substitutions. In addition, the unusual substitution pattern in Table 13.3 further weakens the certainty that birds and mammals are sister taxa.

13.5.4 Insertion and deletion as phylogenetic characters

Similarity in insertion and deletion patterns has been used frequently in phylogenetic reconstruction, especially in rRNA sequences. The rRNA sequences are known to have long-range correlated changes for the following reason: the function of the rRNA gene mainly depends on the 3-D structure stabilized by the intramolecular base-pairing. If a C at Site 36 pairs with a G at Site 136, then a C ↔ G change at Site 36 would favor a corresponding G ↔ C change at Site 136. Almost all substitution models used in phylogenetics assume that all sites evolve independently and cannot handle the evolutionary scenario of correlated changes.

Another problem with the rRNA sequences is that they often have many ***indels***, which may introduce systematic bias during sequence alignment. Consider the following example of three orthologous sequences with many indels occurring in the middle of Sequence 1, but not in Sequence 2 and Sequence 3. Consequently, Sequence 1 would be much shorter than the other two sequences. Given that indels are generally rare, the divergence time between Sequence 1 and Sequence 2 – or between Sequence 1 and Sequence 3 – should be longer than that between Sequence 2 and Sequence 3. However, during sequence alignment, Sequence 1 – being much shorter and necessitating many alignment gaps – will have more leeway to move the bases in order to match those in the other two sequences. In contrast, there is no flexibility for aligning Sequence 2 and Sequence 3 that have not experienced indels. After sequence alignment and the deletion of all sites with indels, the genetic distance between Sequence 1 and one of the other two sequences may well be smaller than that between Sequence 2 and Sequence 3.

The result of grouping mammals and birds as a monophyletic clade was obtained after aligning sequences of different lengths and then omitting the sites with indels. Such an approach also is expected to have systematic bias. An alternative to using nucleotide substitution data as phylogenetic characters is to use indels. A variant of maximum parsimony treats a gap as a fifth character, in which an indel of length *n* is considered equivalent to *n* indels of length 1. Because little is known about the evolutionary dynamics of indels, it is difficult to assess the validity of this approach. However, the approach does provide a simple way of using indels in phylogenetic

Table 13.4 Indel-based genetic distances between species

Species	Alligator	Sphenodon	Homo	Gallus	Rattus
Sphenodon	0.0006				
Gallus	0.0022	0.0017			
Homo	0.0302	0.0296	0.028		
Rattus	0.0318	0.0313	0.03	0.005	
Xenopus	0.0095	0.0089	0.008	0.027	0.0279

analysis, and the result obtained could at least enable a different perspective of the phylogenetic problem.

This treatment of indels also allows the estimation of a simple ***indel-based distance*** between two sequences; that is, the number of unshared gaps divided by the sequence length. If N_1 is the number of gapped sites in Sequence 1 that are not shared by Sequence 2, N_2 is the number in Sequence 2 not shared by Sequence 1, and L is the length of the aligned sequences, then the ***gap-based distance*** is simply

$$D_{12} = \frac{(N_1 + N_2)}{L} \tag{13.10}$$

A matrix of genetic distances obtained for the tetrapod data in this way is shown in Table 13.4, with the resulting tree strongly supporting the grouping of the chicken with other reptilian species rather than mammals (see Tree 2 in Figure 13.1B). Bootstrapping and jackknifing the indels produce trees that always group the chicken with the other reptilian species as a monophyletic clade.

To evaluate the relative statistical support for the three alternative topologies, with genetic distances, either the previous test or the test in Rzhetsky and Nei (1992) can be used. The result in Table 13.5 strongly supports the grouping of the chicken with other reptilian species as a monophyletic clade. In fact, even the conventional hypothesis of grouping the alligator with the bird seems better than clustering the chicken with the mammalian species (see Table 13.5).

Table 13.5 Testing the three alternative phylogenetic hypotheses using gap-based distances

Trees	Diff_Log(Var)	q	k	Prob
2 vs. 3	3.32379	8.1416	2	<0.001
2 vs. 1	4.25155	10.4141	3	<0.001
3 vs. 1	0.92777	2.2726	2	0.1080

Prob: the probability that two topologies fit data equally well.
k: the number of variances falling within the range between the two variances.
(Xia, 2000b).

In summary, this discussion began with the seemingly convincing conclusion that birds and mammals are sister groups (Hedges et al., 1990; Rzhetsky and Nei, 1992). By exploring various aspects of the data set – such as nucleotide frequencies, the pattern of nucleotide substitutions, the heterogeneity of substitution rates over sites, and the presence and absence of indels along the sequences – we gradually became critical of the original conclusion, and finally reach a more convincing one. It is important to remember that the data analysis process is not a series of blind execution of computer programs, but rather one of becoming intimate with the data.

PRACTICE

Xuhua Xia and Zheng Xie

13.6 Data exploration with DAMBE

This section discusses how to carry out data exploration with the computer program DAMBE (see Section 2.8 for a general description of the package). The aligned 18S rRNA sequences of the tetrapod data set in FASTA format are available at **http://aix1.uottawa.ca/~xxia/research/data/18SrRNA.htm**, together with the file containing the three tree topologies shown in Figure 13.1. In the following sections, it is assumed that the sequence files were downloaded and saved as `tetrapod.fas` and the tree file as `tetrapod.nhm`. For basic instructions on how to begin using DAMBE, see Sections 2.9.1 and 3.16.

13.6.1 Nucleotide frequencies

Open the file `tetrapod.fas` in DAMBE and select `Nucleotide Frequency` in the `Seq. Analysis` menu. A dialog box appears that allows the user to select which sequences to analyze. Click the `Add All` button to move all sequences in the left list to the right. Click the `Go!` button to generate results for nucleotide frequencies and dinucleotide frequencies. The data in Table 13.1A is found in the output file generated this way. To obtain nucleotide frequencies for polymorphic sites only (Table 13.1B), select `Polymorphic sites only` in the `Sequences` menu, and then repeat the steps to obtain nucleotide frequencies. To work again with the entire sequence, select `Restore Sequences` in the `Sequences` menu.

13.6.2 Basic phylogenetic reconstruction

By selecting the `Phylogenetics` menu, trees can be estimated in DAMBE according to distance, maximum-parsimony, and maximum-likelihood methods. To construct a *phylogenetic tree* with indel-based distances, select phylogenetics, then `Distance Methods`, and finally `Indel-based distance`; a dialog box appears. In the `Methods` section, click either `Neighbor-joining`, `Fitch-Margoliash`, or `UPGMA`. In the `Choose outgroup` dropdown list, click `Xenopus`, leaving all other options and checkboxes as defaults; then click Done. A phylogenetic tree is displayed in the `Tree Tool` window. Close the window with the tree to return to the `Display` window. A tree in Newick format is displayed on the screen at the end of the output (see Figure 5.5), which is in fact Tree 2 in Figure 13.1B. It is possible to display a tree in Newick format by first copying it from a text editor and then going to DAMBE and selecting `Paste tree into tree display panel` from the `Phylogenetics` menu.

To construct a tree by using the maximum-parsimony or maximum-likelihood method, select the appropriate method in the `Phylogenetics` menu. Both the bootstrap and the jackknife resampling methods are implemented in DAMBE in conjunction with phylogenetic analysis using distance, maximum-parsimony, and maximum-likelihood methods. The option for bootstrapping or jackknifing is found in the dialog box of distance-based methods or by checking the `Resam-pling` checkbox in the maximum-parsimony or maximum-likelihood dialog box. However, remember that resampling with the maximum-likelihood method can be computationally expensive (see Chapter 6). DAMBE resamples the data and produces a consensus tree that shows resampling support for the subtrees. Resampling nucleotide sites of the tetrapod sequences and using nucleotide-based genetic distances generates a topology similar to Tree 1 in Figure 13.1A. Resampling indels and using the indel-based distance leads to a topology similar to Tree 2 in Figure 13.1B.

13.6.3   Rate heterogeneity over sites estimated through reconstruction of ancestral sequences

To estimate the shape parameter α of the Γ-distribution, it is necessary to have a tree topology to reconstruct ancestral sequences. The tree can be calculated with any of the algorithms described and must be saved in Newick format. The following exercise analyzes the tetrapod sequences with the first tree saved in the file `tetrapod.nhm`.

1. Open the tetrapod data set.
2. Select `BASEML` in `Maximum Likelihood` of the `Phylogenetics` menu. In the dialog box, choose `User tree` under Run Mode. A standard `File/Open` dialog box appears to specify which file contains the user tree.
3. Select the file `tetrapod.nhm` and then click `Open`. Only the first tree in the file is used in the reconstruction. Check the `Get ancestral sequences` checkbox and click Go !
4. At the prompt, enter a file name to save the reconstructed sequences. The file will be saved in text format, and the ancestral sequences will be added to the existing sequences in order to apply further pairwise comparisons between neighboring nodes along the tree.
5. Once the ancestral sequences are constructed, select `Fit gamma distribu-tion to substitutions` from the `Seq. Analysis` menu. A standard `File/Open` dialog box appears. Enter the file name `tetrapod.rst` for saving the results.
6. The file so obtained will contain the tree topology, together with the sequences for all terminal taxa and the reconstructed sequences for all internal nodes. DAMBE also asks whether the user wants a graphic output of the empirical frequency distribution of the estimated number of substitutions per site. Click `Yes` to view the frequency distribution.

7. Quit the `Graphics` window to return to the `Display` window and check the estimated α.

DAMBE has done the following steps. First, the number of transitions and transversions are counted for each site from pairwise comparisons between neighboring nodes along the phylogenetic tree. Second, the estimated number of substitutions per site is calculated by using Kimura's (1980) two-parameter method (see Section 4.9). Finally, the shape parameter α of the Γ-distribution is estimated by maximum likelihood.

13.6.4   Empirical substitution pattern

Pairwise comparisons between neighboring nodes along an established tree also should be used to obtain the empirical substitution pattern (see Section 13.5.3). The following exercise uses the reconstructed ancestral sequences for the tetrapod data set obtained previously and saved in the file `tetrapod.rst`.

1. Select `Seq. Analysis`, then `Nucleotide Substitution Pattern` and `Detailed Output`. A dialog box appears with two lists: the one on the left shows the sequence pairwise comparisons available for selection; the one on the right lists the selected comparisons. If the input file contains reconstructed ancestral sequences and a tree topology, as in this case, the pairwise comparisons available on the left include only comparisons between neighboring nodes in the phylogenetic tree. Click `Add all` to move all sequences in the left list to the right list, and click `Done`.
2. After a few seconds, the standard `File/Open` dialog box appears. Enter the file name for saving the result or simply use the default, and click `Save`. The output file, saved in text format, will contain the results shown in Table 13.3.

13.6.5   Testing alternative phylogenetic hypotheses with the distance-based method

The following exercise shows how to test alternative phylogenetic hypotheses using the distance method with indel-based distances.

1. Open the file `tetrapod.fas`.
2. Select `Distance` method from the `Phylogenetics` menu and `Indel-based distance`.
3. Select `Fitch-Margoliash` and `User tree`, and choose `Trees from a file`. Enter the name of the text file that contains the alternative phylogenetic hypotheses (i.e., `tetrapod.nhm`). Click `Done` and the three topologies will be evaluated.

The output file has two parts. The first one is the Bartlett's test (see Section 13.3) with $B_c = 33.714$, $DF = 2$, and $p = 0.0000$. The result means that at least one topology is significantly better (or worse) than the others. The second part of the output file evaluates which topology is better than the others and shows two tests. The first test, the F-test of the difference between two variances, is valid only when two topologies are compared; it is not relevant here. The second test is the Newman-Keuls test, which is appropriate for multiple comparisons involving more than two topologies (see Section 13.4). The topology of the second tree has the smallest variance; that is, the best fit between the observed distance matrix and the reconstituted distance matrix. In fact, the topology of the second tree appears to be significantly better than the two other alternatives, which do not differ significantly.

13.6.6   Testing alternative phylogenetic hypotheses with the likelihood-based method

The RELL test with DAMBE is performed as follows:

1. Open the tetrapod data set.
2. Select `Phylogenetics`, `Maximum Likelihood`, nucleotide sequences, and BASEML.
3. In the dialog box, click the `User tree` option. A standard `File/Open` dialog box appears. After selecting the `tetrapod.nhm` file and clicking Go!, DAMBE computes the log like-lihood for each topology, the difference in log like-lihood between the best tree and all alternative trees, and the variance of the difference by bootstrapping, and finally performs the Student-Newman-Keuls test (see Section 13.4).

DAMBE also can perform the paired-sample test, which is identical to the one implemented in DNAML of the PHYLIP package.

1. Select `Phylogenetics`, `Maximum Likelihood`, nucleotide sequences, and DNAML.
2. In the dialog box, click the `User tree` option in the `Run Mode` dropdown box.
3. Select `tetrapod.nhm` to read in the tree file, and click Go! to evaluate the different topologies in the file.

REFERENCES

Carroll, R. L. (1988). *Vertebrate Paleontology and Evolution.* New York: W. H. Freeman.

Evans, M., N. Hastings, and B. Peacock (1993). *Statistical Distributions.* New York: Wiley.

Felsenstein, J. (1985). Confidence limits on phylogenies: An approach using the bootstrap. *Evolution,* 39, 783–791.

Felsenstein, J. (1992). Estimating effective population size from samples of sequences: Inefficiency of pairwise and segregating sites as compared to phylogenetic estimates. *Genetic Research,* 59, 139–147.

Felsenstein, J. (1993). *PHYLIP* 3.5 (*Phylogeny Inference Package*). Seattle, WA: University of Washington, Department of Genetics.

Fitch, W. M. (1971). Toward defining the course of evolution: Minimum change for a specific tree topology. *Systematic Zoology*, 20, 406–416.

Fitch, W. M. and E. Margoliash (1967). Construction of phylogenetic trees. *Science*, 155, 279–284.

Gardiner, B. G. (1982). Tetrapod classification. *Zoological Journal of Linnaean Society*, 74, 207–232.

Gauthier, J., A. G. Kluge, and T. Rowe (1988). Amniote phylogeny and the importance of fossils. *Cladistics*, 4, 105–209.

Gojobori, T., W. H. Li, and D. Graur (1982). Patterns of nucleotide substitution in pseudogenes and functional genes. *Journal of Molecular Evolution*, 18, 360–369.

Hedges, S. B., K. D. Moberg, and L. R. Maxson (1990). Tetrapod phylogeny inferred from 18S and 28S ribosomal RNA sequences and a review of the evidence for amniote relationships. *Molecular Biology and Evolution*, 7, 607–633.

Jukes, T. H. and C. R. Cantor (1969). Evolution of protein molecules. In: *Mammalian Protein Metabolism*, ed. H. N. Munro, pp. 21–123. Academic Press, New York.

Kimura, M. (1980). A simple method for estimating evolutionary rates of base substitutions through comparative studies of nucleotide sequences. *Journal of Molecular Evolution*, 16, 111–120.

Kishino, H., T. Miyata, and M. Hasegawa (1990). Maximum-likelihood inference of protein phylogeny and the origin of chloroplasts. *Journal of Molecular Evolution*, 31, 151–160.

Kluge, A. G. and J. S. Farris (1969). Quantitative phyletics and the evolution of anurans. *Systematic Zoology*, 18, 1–32.

Kuhner, M. K. and J. Felsenstein (1994). A simulation comparison of phylogeny algorithms under equal and unequal evolutionary rates. *Molecular Biology and Evolution*, 11, 459–468.

Lake, J. A. (1994). Reconstructing evolutionary trees from DNA and protein sequences: Paralinear distances. *Proceedings of National Academy of Sciences of the USA*, 91, 1455–1459.

Løvtrup, S. (1985). On the classification of the taxon tetrapoda. *Systematic Zoology*, 34, 463–470.

Nee, S., E. C. Holmes, A. Rambaut, and P. H. Harvey (1995). Inferring population history from molecular phylogenies. *Philosophical Transactions of Royal Society of London, Series B*, 340, 25–31.

Nee, S., E. C. Holmes, A. Rambaut, and P. H. Harvey (1996). Inferring population history from molecular phylogenies. In: *New Uses for New Phylogenies*, eds. P. H. Harvey, A. J. L. Brown, J. Maynard Smith, and S. Nee, pp. 66–80. Oxford University Press, Oxford.

Nei, M. (1972). Genetic distance between populations. *American Naturalist*, 106, 283–292.

Olsen, G. J., H. Matsuda, R. Hagstrom, and R. Overbeek (1994). fastDnamL: A tool for construction of phylogenetic trees of DNA sequences using maximum likelihood. *CABIO*, 10, 41–48.

Romer, A. S. (1966). *Vertebrate Paleontology*. Chicago: University of Chicago Press.

Rzhetsky, A. and M. Nei (1992). A simple method for estimating and testing minimum-evolution trees. *Molecular Biology and Evolution*, 9, 945–967.

Saitou, N. and M. Nei (1987). The neighbor-joining method: A new method for reconstructing phylogenetic trees. *Molecular Biology and Evolution*, 4, 406–425.

Swofford, D. L., G. J. Olsen, P. J. Waddell, and D. M. Hillis (1996). Phylogenetic inference. In: *Molecular Systematics*, eds. D. M. Hillis, C. Moritz, and B. K. Mable, pp. 407–514. Sinauer Associates, Sunderland, MA.

Tajima, E. and M. Nei (1984). Estimation of evolutionary distance between nucleotide sequences. *Molecular Biology and Evolution*, 1, 269–285.

Tamura, K. and M. Nei (1993). Estimation of the number of nucleotide substitutions in the control region of mitochondrial DNA in humans and chimpanzees. *Molecular Biology and Evolution*, 10, 512–526.

Xia, X. (1998). The rate heterogeneity of nonsynonymous substitutions in mammalian mito-chondrial genes. *Molecular Biology and Evolution*, 15, 336–344.

Xia, X. (2000a). *DAMBE 4.0 (Software Package for Data Analysis in Molecular Biology and Evolution)*. Hong Kong: University of Hong Kong, Department of Ecology and Biodiversity.

Xia, X. (2000b). *Data Analysis in Molecular Biology and Evolution*. Boston: Kluwer Academic Publishers.

Xia, X. and W.-H. Li (1998). What amino-acid properties affect protein evolution? *Journal of Molecular Evolution*, 47, 557–564.

Xia, X., M. S. Hafner, and P. D. Sudman (1996). On transition bias in mitochondrial genes of pocket gophers. *Journal of Molecular Evolution*, 43, 32–40.

Detecting recombination in viral sequences

THEORY

Mika Salminen

14.1 Introduction and theoretical background to exploring recombination in viral sequences

Recombination of viral genomes is used as the model for this chapter, primarily because viral genomes are often small and, therefore, sequence data from multiple strains are more readily available than for higher organisms (Salminen et al., 1995b; Korber et al., 1997). In addition to generating genetic variation by accumulating point mutations, many viruses have evolved other methods to increase their evolutionary potential. The most important – from a mechanistic point of view – are undoubtedly recombination and *reassortment*. Whereas reassortment, the *shuffling* of chromosomes of organisms with a segmented genome, is an important mechanism, this chapter is mostly concerned with true recombination. Nevertheless, many of the methods described here are applicable for detecting reassortment as well. For the purpose of the chapter, only *homologous recombination* is discussed. Homologous recombination is defined as the exchange of segment coding for the same genome regions in different strains of a virus. *Heterologous recombination* (i.e., recombination resulting in either the physical joining of segments in unrelated genes or gross insertion/deletion events) is not considered, although many of the methods may be applicable as well for such events.

Some viruses use DNA and some use RNA as their genetic material, and mechanisms for recombination of both types of nucleic acids have evolved. However, not much is known about the exact molecular details of all the mechanisms that lead to recombination in different viruses. It is generally assumed that viruses using a DNA genome and replicating in the nucleus use the recombination mechanisms of the host cell; that is, the same recombination mechanisms that function during normal cell division and *cross-over* between cellular chromosomes. For RNA genome viruses, other mechanisms must be used because there are no known host-provided mechanisms for **RNA recombination**. In addition, some viruses that use both an RNA and a DNA stage for their replication have evolved their own means of recombination.

For the retroviruses, which belong to the last group mentioned previously, the mechanisms of recombination have been extensively studied. Recombination is made possible by the fact that retroviruses are diploid; that is, carrying two RNA molecules within the viral particle. Retroviral recombination occurs during the reverse-transcription step, before integration, and is dependent on the copackaging of two different viral genomes. This, in turn, requires simultaneous infection of the same cell by two different strains and subsequent integration of two different parental-generation proviruses in the same nucleus (Figure 14.1A). One of the prevailing hypotheses for recombination is as follows: simultaneous expression and packaging of viral RNA in dually infected cells effectively generates a population of heterozygous first-generation virus particles (Figure 14.1A). The actual recombination event is assumed to occur upon reinfection of these first-generation heterozygote particles into new host cells, after entry into the cytoplasm, as the different RNA molecules in the particles are reverse-transcribed (Figure 14.1B). In this reaction, the growing DNA strand can jump from one RNA template to the other by a copy-choice mechanism, creating chimeric second-generation proviruses (Figure 14.1B). The viral particles produced by the second-generation proviruses will now contain recombinant RNA forms (Coffin, 1979; Goodrich and Duesberg, 1990; Vartanian et al., 1991).

14.2 Requirements for detecting recombination

To be able to detect recombination between nucleic-acid sequences, certain criteria must be met. First, there must be enough genetic variation between lineages (or alleles) to be able to distinguish exchanges between the lineages and to separate them from variation due to point-mutation accumulation. Second, in most cases, sequence data from relatively long gene segments are required. Third, a too-high accumulation of divergence after recombination will make it difficult to detect recombination because the recombination event will be masked by accumulation of point mutations. In the case of HIV-1, favorable conditions exist for the detection of recombination. The HIV-1 M-group consists of several evolutionary lineages or clades, called genetic subtypes (Figure 14.2), which show a star-like or bush-like phylogeny, and are therefore approximately equidistantly related. Furthermore, evolution of HIV is sufficiently similar in all regions of the genome, which is evidenced by a similar general topology of the whole tree and position of the individual strains in the **phylogenetic trees**, independent of the region used for the analysis (Figure 14.3).

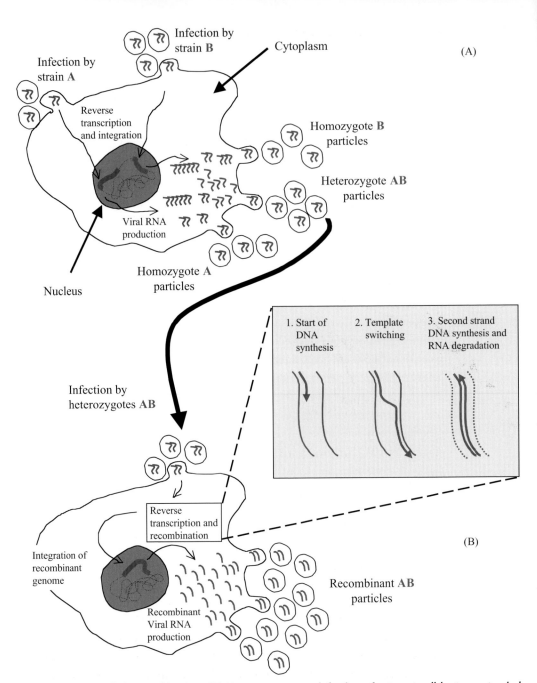

Figure 14.1 Retroviral recombination. (A) Upon super- or coinfection of a target cell by two retroviral strains [labeled A and B and colored blue and red (see color plate)], both strains may integrate a provirus into the host cell nuclear genome after reverse transcription in the cytoplasm. Coexpression of the different proviruses generates both homozygous and heterozygous viral particles due to the diploidy of retroviral virions. (B) When the newly generated heterozygous particles enter a second generation of target cells, recombinant proviruses may be generated through strand transfer during the RT-step (inset). Integration and expression of recombinant provirus produce recombinant offspring virions. (See color section.)

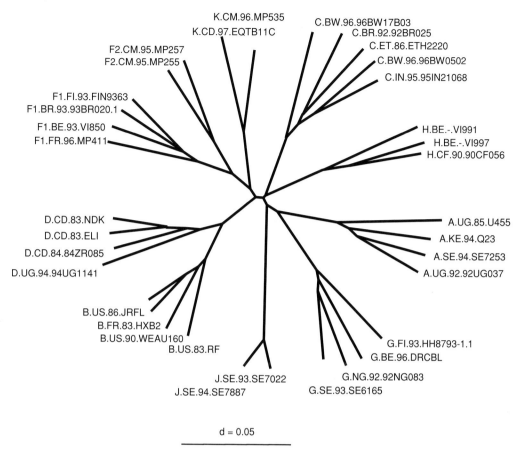

Figure 14.2 HIV subtypes. K80 model Neighbor-Joining phylogenetic tree using the 1999/2000 Los Alamos HIV database complete genome reference sequence alignment (http://hiv-web.lanl.gov). Strain-name coding: X.CC.00.YYYY -> X = Subtype, CC = two-letter country code, 00 = year of sampling, YYYY = original isolate identifier.

In such a phylogeny, a recombination event between members of the equidistantly related clades can be detected by the "jumping" of a strain between clades if different regions of the genome are used (Figure 14.4 A and B). This is a direct consequence of the horizontal transfer of genetic material from one evolutionary lineage to another, and provides a phylogenetic visualization of how separately evolved and inherited properties are transferred between lineages.

14.3   Theoretical basis for methods to detect recombination

Most methods developed to detect recombinant strains or map the positions of recombination breakpoints in gene sequences apply **distance-** or **phylogenetic topology-based** methods. In addition, many (but not all) methods apply some

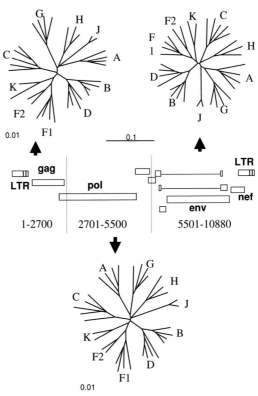

Figure 14.3 Consistent genome-wide HIV-1 M-group phylogenetic clustering. Phylogenetic analysis results in similar general clustering of HIV-1 strains irrespective of which region of the genome is used for the analysis. The trees represent clustering in the three different regions indicated by the vertical lines.

variant of an average-over-window-based scanning approach. This approach starts with a set of optimally aligned sequences of the gene region of interest, and computes a statistic or measure for a successive set of overlapping subregions (windows) of the alignment (Figure 14.5). The statistic/measure could either be for one sequence – called a *query sequence* – compared to all the others in the alignment, or a measure describing the relation of a group of sequences to the rest of the aligned sequences in that window. The statistic is then plotted on an x/y plot along the genome, so that the x-values reflect the genome position at the midpoint of the analysis window and the y-values reflect the measure/statistic calculated from the window.

   Similarity- and/or *dissimilarity-based* methods are among the fastest and theoretically least complicated of these methods. They are based on the reasonable assumption that a given query sequence is most similar to those members of a set of sequences from which it has evolved more recently by point-mutation accumulation. Therefore, if the similarity of a sequence to a given set of reference sequences is

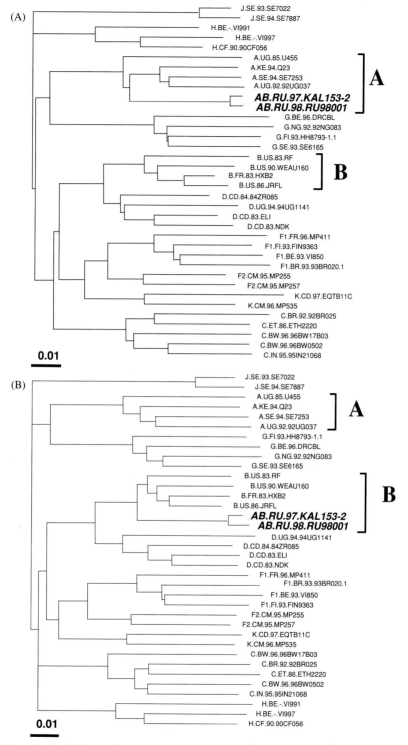

Figure 14.4 "Jumping" of recombinant isolates. Two representatives (AB.RU.97.KAL153–2 and AB.RU.98.RU98001, in boldface) of a recombinant between HIV-1 M-group Subtypes A and B were phylogenetically analyzed in two different genome regions. The trees represent a region clustering the recombinant isolates in (A) with Subtype A and in (B) with Subtype B. Reference sequences are from the 1999/2000 Los Alamos HIV-1 database.

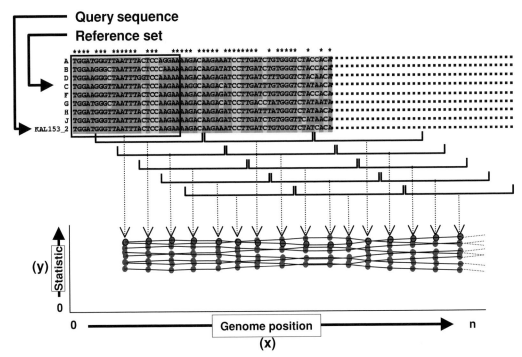

Figure 14.5   Basic principle of average-over-window scanning methods. An alignment of reference sequences (A-J) and a potential recombinant (the query sequence) are sequentially divided
into overlapping subregions (i.e., window, box, and brackets) for which a statistic/measure
is computed. This statistic/measure is then plotted (broken arrow lines) in an x/y scatter
plot using the alignment coordinates on the x-axis and the statistic/measure range on the
y-axis. Each value is recorded at the midpoint of the window and connected by a line. (See
color section.)

computed and plotted along the sequence using a sliding-window fashion (described previously), the sequence will show the highest similarity to other members
of the same evolutionary lineage and lower similarity to those of less directly related lineages (Figure 14.6A). Alternatively, computing the dissimilarity will result
in inverted plots. In this analysis, the most closely related sequences show the lowest
dissimilarity values (Figure 14.6B).

If recombination results in the crossover between two sufficiently separated evolutionary lineages, it will be reflected in a plot of the query sequence, the putative
recombinant, to a set of reference sequences by a gradual switch of the highest
similarity/lowest dissimilarity from one reference sequence to another. In a sliding-
window–based analysis, the putative recombination breakpoint lies in the position
where the two plots of similarity values cross each other (Figure 14.6C).

In addition to using similarity/dissimilarity to evaluate the genetic relatedness
between sequences, other statistics can be used. In a similar sliding-window–based
analysis method, a combination of phylogenetic analysis and **bootstrap values**

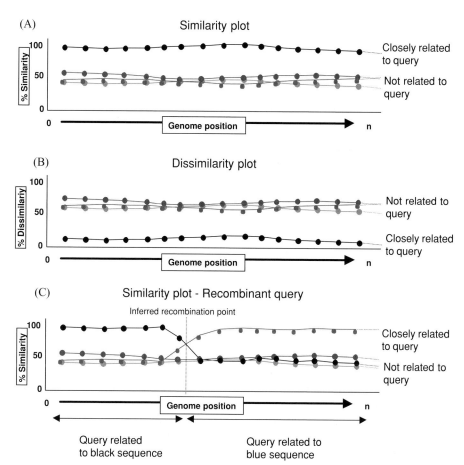

Figure 14.6   Similarity and dissimilarity methods. (A) Similarity plot. In this type of analysis, the measure recorded is the similarity value (sim) between the query sequence and each of the reference sequences. (B) The same analysis, except that the inverse value to similarity (1-sim) or the dissimilarity is plotted. In a variation of these methods, similarity/dissimilarity values corrected by an evolutionary correction model [i.e., any of the JC69, TN93, K80, F81, F84, HKY85, or GTR models (see Chapter 4 and Chapter 10)] may be used. (C) Schematic view of a plot of a recombinant sequence. (See color section.)

associated to specific clusters of sequences can be used to map recombination breakpoints (Figure 14.7A). The method has been called ***bootscanning*** (Salminen et al., 1995a). In this type of analysis, the phylogenetic relationship of the sequences in each window is calculated using ***bootstrap resampling***, and the bootstrap value of the clusters containing the query and any of the reference sequences are recorded. Bootstrap values are then plotted along the genome similar to the way the genetic similarity values are used.

The advantage of the bootscanning method is that for bootstrap values, the switch between a high and a low value is usually more pronounced than when similarity/dissimilarity values are used. Typically, in situations where recombination

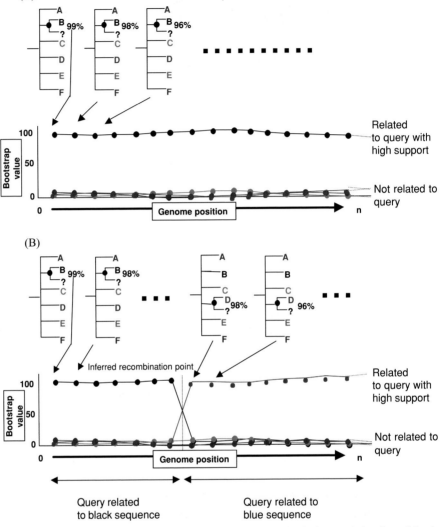

Figure 14.7 The bootscanning method. (A) In the bootscanning method, the statistic plotted is the bootstrap value of a node connecting a reference and the query (labeled "?"). (B) For a recombinant sequence, the high bootstrap support for clustering with one reference (Sequence B) suddenly switches to high bootstrap support for clustering with another reference (Sequence D). (See color section.)

has occurred relatively recently, the relative bootstrap value switches suddenly from 100 to 0% for the first parental lineage and from 0 to 100% for the other lineage (Figure 14.7B). This allows more precise mapping of the breakpoints. In addition, bootstrap support for clusters of sequences can be analyzed, allowing for systematic studies of genome-wide evolution especially of viral sequences.

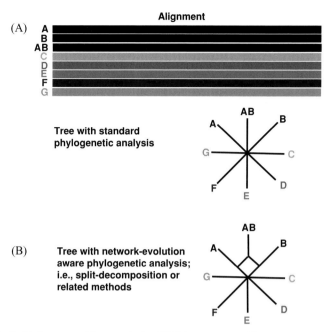

Figure 14.8   Split-decomposition analysis. (A) When a recombinant sequence contains equal amounts of phylogenetic signal from both parental lineages (alignment schematic), standard phylogenetic methods incorrectly interpret this as a new evolutionary lineage, branching the sequence off from the center of a tree (tree). (B) Split-decomposition and related network-evolution aware methods can indicate the dual ancestry of the sequence by showing the tree as a network rather than as a strictly binary-split tree. (See color section.)

Scanning analyses, like those described previously, are most useful in situations where there is preexisting suggestive evidence of recombination, such as discordant clustering in sequences from different parts of a genome/gene (see Figure 14.4). In this situation, a longer region can then be scanned and the breakpoints precisely mapped. However, it is often not clear whether a particular sequence may be recombinant or if there is a need to scan through a set of sequences for evidence of recombination. In phylogenetic trees, chimeric sequences often fall between groups because the discordant phylogenetic signals in the different parts of the sequence make it impossible for the algorithm used to place the sequence properly in the tree (Figure 14.8A). The reason for this is that ordinary phylogenetic analysis tools produce only **strictly bifurcating trees**, where there can only be one immediate ancestor to a sequence (see Chapter 12). If recombinants are included in a data set, they will have at least two immediate ancestors, which cannot be resolved using phylogenetic analysis methods that disregard recombination. Instead, methods that include the possibility of **network-like** evolution must be used in which the phylogeny or evolutionary history can be described not only as bifurcating trees,

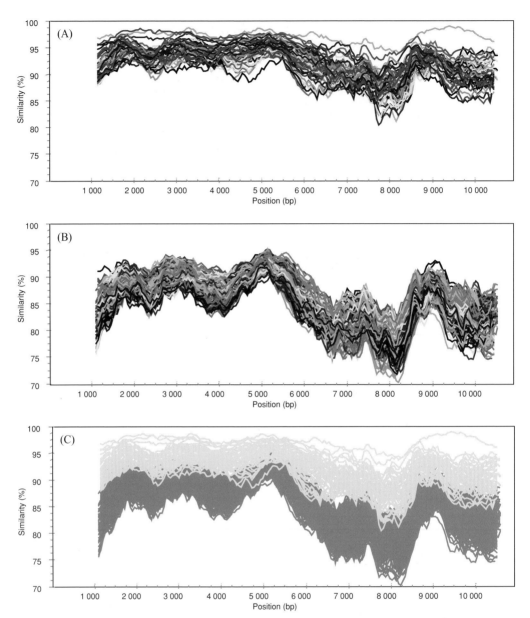

Figure 14.9   Range of HIV-1 M group intrasubtype and intersubtype variation. (A) Intrasubtype variation among the subtype reference sequences. (B) Range of intersubtype variation among the subtype reference sequences. (C) Intrasubtype and intersubtype variation superimposed (light/dark gray lines). (See color section.)

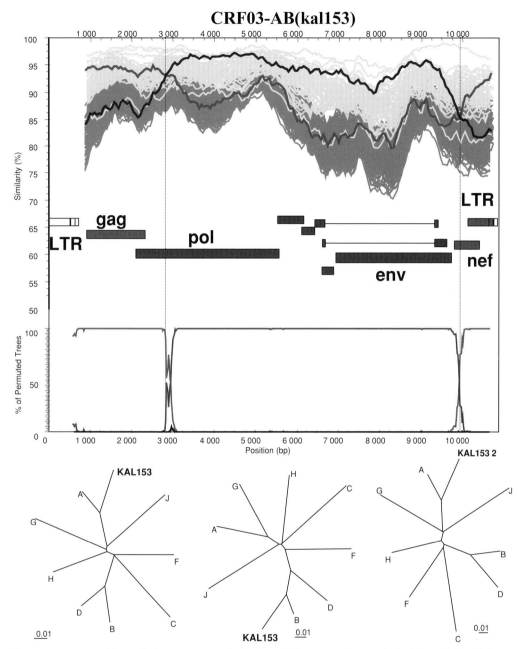

Figure 14.10 Recombinant isolate KAL153. Similarity and bootscanning analysis followed by phylogenetic analysis (K2 + NJ) of the identified segments. The similarity analysis with Subtypes
A, B (parental subtypes), and C (outgroup) is superimposed on the ranges of intra- and
intersubtype variation. (See color section.)

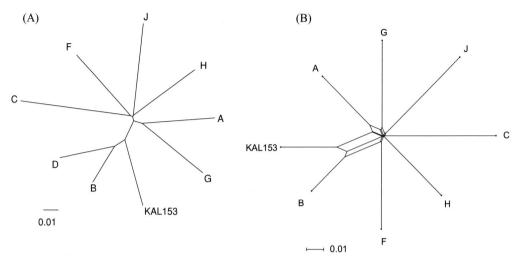

Figure 14.11 Comparison of K80-based NJ tree and split-decomposition phylogenetic analysis. Recombinant isolate KAL153 (a representative of CRF03-AB) analyzed by (A) the NJ algorithm, and (B) by the split-decomposition algorithm.

but also as **networks** of **multiple splits** (Figure 14.8B). Therefore, these methods can be used to screen an aligned set of sequences for evidence of recombination.

14.4   Examples of viral recombination

As already discussed, HIV-1 Group M has evolved multiple, approximately equidistantly related parallel lineages, which are called genetic subtypes and used for classification purposes (Robertson et al., 2000a and 2000b). In addition to the genetic lineages that evolved by accumulation of point mutations, recombination clearly influenced HIV-1 Group M evolution. When complete (or virtually complete) HIV-1 proviruses were studied using the methodology described in previous sections, many recombinant viruses were detected (Carr et al., 1996, 1998; McCutchan et al., 1996; Salminen et al., 1997; Gao et al., 1998; Liitsola et al., 1998, 2000a). The ease of detecting recombination between HIV-1 Group M subtype sequences comes from the genetic equidistance of the groups. When an intrasubtype and intersubtype similarity comparison of all the nonrecombinant strains (i.e., Subtypes A through K) (see Figure 14.2) is performed for the whole genome, it is clear that the range of intrasubtype and intersubtype distances do not significantly overlap (Figure 14.9) (Carr et al., 1996). In addition to Subtypes A through K of HIV-1 Group M, other lineages labeled CRF01 to CRF04 have been described. These are lineages that have a recombinant origin, also reflected in their names, where the letters indicate the parental lineages (Robertson et al., 2000a). Using the similarity analysis to study the structure of one strain of these circulating recombinant lineages (Kal153 of the

CRF03-AB lineage), the recombinant nature of the strain is revealed (Figure 14.10). Using `SplitsTree` (see Chapter 12) for an analysis of a recombinant genome in parallel with a standard phylogenetic analysis method reveals the potential of the **split-decomposition method** to detect recombination. In an analysis of the Kal153 AB recombinant complete genome sequence, **neighbor-joining (NJ)** only indicates that the sequence is related to Subtypes B and D, but using `SplitsTree` correctly reveals that the sequence could be a recombinant between Subtypes A and B (Figure 14.11).

Retroviruses definitely have high recombination rates due to their replication strategy and diploid virion composition, but other viruses show clear evidence of recombination as well. For example, a recently published study of complete enterovirus genome sequences shows that recombination seems to be common in this group of RNA viruses (Santti et al., 1999).

PRACTICE

Mika Salminen

14.5   Existing tools for analysis of recombination

Computer tools for analyzing recombination between gene sequences are available for multiple operating systems (e.g., several brands of UNIX/Linux, MS Windows, and MacOS systems). The most basic approach is to use any of the available multipurpose phylogenetic analysis packages (described elsewhere in this book) to explore sequence alignments for evidence of discordant branching order among a set of sequences. However, this quickly becomes impractical as longer and more sequences are analyzed; therefore, dedicated packages have been developed. Among the first attempts to build tools to scan sequence alignments for evidence of recombination was the Bootscan package, which is a set of shell scripts, GDE-menu files (a specialized script file for the Genetic Data Environment-alignment editor/sequence manipulation user interface; **http://www.ktl.fi/hiv/mirrors/pub/programs/**), and two C programs. These tools can be used from within the GDE to break an alignment into sequentially overlapping pieces and feed them to appropriate subprograms of the PHYLIP package (i.e., Seqboot and the sequence-based phylogenetic analysis programs). Finally, the results are compiled in a tab-delimited list-file, which can be imported into a plotting program (e.g., MS-Excel) to generate graphs of the bootstrap values of the nodes of interest.

The Bootscan package is available as binaries and source code for the Sun-Solaris and Linux (i.e., Red Hat 7.2 ELF binaries) operating systems. However, they should compile on any UNIX system using the ANSI C compatible GNU-C compiler gcc, but they do require use of the GDE as a front end.

Although Bootscan performs its task as intended, it has some practical drawbacks, of which probably the most significant is that it is not user friendly. Also, there are no built-in tools to graphically plot the results, so the analysis does take significant time. More recently, an application capable of performing a bootscanning analysis in the MS-Windows environment was developed: the software tool called Simplot. Written by Stuart Ray from the Johns Hopkins University in Baltimore, Maryland (USA), Simplot is one of the most versatile of the currently available recombination exploration applications (Lole et al., 1999). Simplot Version 2.5 can be downloaded from **http://sray.med.som.jhmi.edu/RaySoft/SimPlot/** and is a Windows 95/98/NT program; it can be used on the Macintosh platform using Virtual PC emulation software. Simplot is capable of reading an alignment in a wide variety of formats and performing similarity/dissimilarity plots of any of the sequences against all others in the alignment. Multiple parameters of the analysis can be varied, including window size, window overlap, use of evolutionary-distance correction, similarity/dissimilarity plot, and inclusion/disregarding of gap regions. The graphics can be exported both in bitmapped (.bmp) and Windows Metafile formats (.wmf). The program provides precise mapping of inferred re-combination breakpoints by reporting alignment positions when the user clicks anywhere on the plot. Another useful property of Simplot is the ability to export segments of an alignment into separate files, which can then be used for phyloge-netic analysis and verification of the mapped recombination breakpoints.

Simplot also can perform bootscanning by adding Joe Felsenstein's PHYLIP package (Felsenstein, 1996) (**http://evolution.genetics.washington.edu/PHYLIP/ getme.html**) to Simplot (see the Simplot Web page for installation instruc-tions). The program automatically extracts sequential alignment regions using the set window and overlap parameters, subsequently passing them over to the necessary subprograms of the PHYLIP package (i.e., Seqboot -> DNAdist -> Neighbor or Fitch -> Consense). After each window subanalysis, the program extracts the bootstrap values of the query sequence, plus each refer-ence sequence, and plots them in the Simplot bootscan window. As with the similarity/dissimilarity analysis, locations of the breakpoints can be identified by clicking on the plot and the plot can be saved as graphics files (in .bmp or .wmf).

The only known application that uses the split-decomposition method, SplitsTree developed by Huson (1998), is discussed in Chapter 12 and is used in the following discussion.

To install Simplot, go to **http://sray.med.som.jhmi.edu/RaySoft/SimPlot/** and download the zip-compressed installation file SIMPLOT25.ZIP. Place the file in a temporary file folder (e.g., the C:\temp folder found on most systems) and un-compress the file. Install Simplot using the installer file SETUP.EXE. Simplot adds itself to the Start menu as its own group. More detailed instructions can be found on the Simplot Web site.

If no Bootscanning is to be run, the program is ready to use; to run Bootscanning, the PHYLIP package must be added. If the Windows 95 ver-sion of PHYLIP is already installed on the local computer, it is not necessary to install it again. PHYLIP can be downloaded from **http://evolution.genetics. washington.edu/PHYLIP/getme.html**. The necessary files are PHYLIP95.exe and PHYLIP96.exe, which are self-expanding archive files. Create a folder named PHYLIP in a convenient location. Not all the PHYLIP programs are needed to use Bootscanning, but retaining all of the files is recommended (see Web site for de-tails). When first using Simplot for Bootscanning, the program will prompt the user to navigate to the folder into which PHYLIP is installed.

To successfully perform the exercises, two other programs are needed: `Tree-view` (see Section 5.5) and `SplitsTree` (see Chapter 12).

14.6   Analyzing example sequences to visualize recombination

Several sets of aligned sequences will be used for the exercises and are available at the Web site **http://www.kuleuven.ac.be/aidslab/phylogenyBook/datasets.htm** and described as follows:

1. File `A-J-cons-kal153.fsa`: A single recombinant HIV-1 strain (KAL153) aligned with 50% consensus sequences derived for each HIV-1 subtype (proviral LTR-genome-LTR form).
2. File `A-J-cons-recombinants.fsa`: A set of multiple recombinant HIV-1 sequences in FASTA format aligned with 50% consensus sequences derived for each HIV-1 subtype (proviral LTR-genome-LTR form).
3. File `2000-HIV-subtype.fsa`: A reference set of virtually complete HIV genome sequences aligned in FASTA format (virion RNA R-U5-genome-U3-R form). This set represents the Los Alamos International HIV database year 1999/2000 official reference set and uses the group feature of `Simplot` (see the next section).
4. Files `A-J-cons-kal153.nex` and `A-J-cons-recombinants.nex`: Two examples of `Nexus`-formatted sequence alignments to demonstrate the use of `SplitsTree`.

14.6.1   Exercise 1: Working with `Simplot`

This exercise shows the use of `Simplot` and its basic properties. To start `Simplot`, select the program from its group on the Start Menu in Windows or double-click on the `Simplot` icon in the `Simplot` folder. The program will start with a window from which it is possible to select the file menu to open an alignment file. `Simplot` can read numerous different sequence-alignment formats, including those discussed throughout this book. For this demonstration, open the file `A-J-cons-kal153.fsa` (the `.fsa` extension may or may not be visible, depending on the settings of the local computer). This file contains an alignment of the CRF03-AB strain Kal153 and 50% consensus reference sequences for Subtypes A through J. The tree-like graph shown in Figure 14.12 appears, and is similar to Windows Explorer. The sequences are the terminal branches, each contained within a group with the same name as the sequence (by default). Groups are used primarily when the user wants to quickly generate consensus sequences; they are discussed in another exercise.

At the bottom of the window is the line "`----- Hidden below this line -----`". Anything below this line will be excluded from the analyses

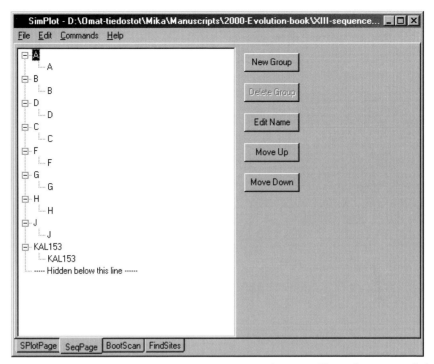

Figure 14.12 Starting `Simplot`. View after opening a sequence-alignment file. Each sequence belongs to a group, which by default is named the same as the sequence itself.

without generating any change to the original file. The line can be moved up or down and/or any of the groups/sequences can be placed above or below the line by dragging them with the mouse or using the buttons to the right. Select the KAL153 sequence and try to move it to the top of the tree by pressing the `Move up` button sequentially. Near the bottom of the window are four tabs; click on `SPlotPage` to go to the page where the *similarity plots* are performed. The other tabs, `SeqPage`, `Bootscan`, and `Findsites`, are the input page, the bootscanning-analysis page, and the phylogenetically informative site-analysis page; the bootscanning analysis is discussed later, and the `Simplot` documentation describes the Find sites.

To do the first analysis, go to the `Commands` menu and select `Query` and KAL153 on the pop-up list, which reflects the sequences in the order determined in the `SeqPage` window. This first operation determines which of the aligned sequences is compared to all the others (Figure 14.13). To perform the analysis, go to the `Commands` menu again and click on `DoSimplot`; the result should be similar to Figure 14.14A. The analysis indicates that the KAL153 strain is a recombinant of Subtypes A and B, as reflected by the legend key. The analysis was run by the `Simplot` default values; that is, a window of 400 nucleotides and a step size of 20 bases. The values can be changed by clicking on them in the lower frame of

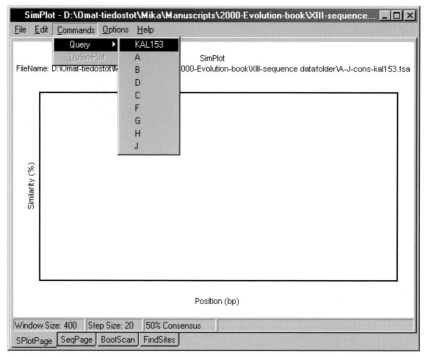

Figure 14.13 Working with `Simplot`. Selecting the query sequence in the `SPlotPage` window.

the window or from the `Options` menu. Many parameters can be changed from the `Options` menu, but one of the most useful is `AutoUpdate`, which updates the plot any time a parameter is changed. To examine some of the effects of the different options select (for example) `Jukes–Cantor correction`. Distances will increase a little, which is exactly what is expected when using the ***Jukes and Cantor correction formula*** for ***evolutionary distances*** (see Chapter 4). Change the window size to 800 bases and compare the plot to the ***composite similarity plot*** shown in Figure 14.10. Another useful option is whether to exclude gap regions from the analysis (usually recommended). Consult the `Simplot` documentation, accessible from the `Help` menu, for more information about the different options.

Return to the `SeqPage` tab and move the sequences so that only A, B, C, and KAL153 are above the "`Hidden below this line`" separator. Go back to the `SPlotPage` tab. When doing the analysis, it should be easy to pinpoint the recombination breakpoints. In fact, the user can actually click on the breakpoints (or anywhere on the curves) and a pop-up window will show at which position in the alignment the breakpoint is located (see Figure 14.14B).

The next test is `Bootscan`. Make sure at the `SeqPage` tab that only the four sequences are still included in the analysis. Click on the `Bootscan` tab; the

(A)

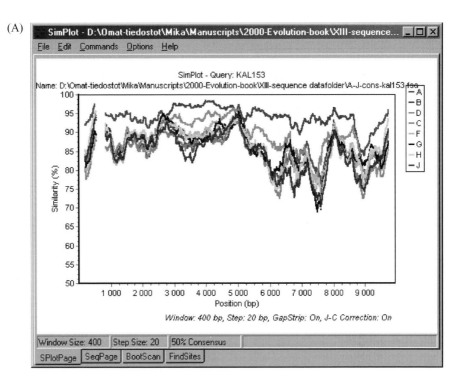

(B)

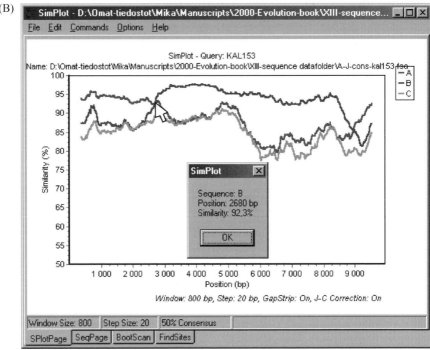

Figure 14.14 Comparing KAL153 to the reference sequences. (A) View with all reference sequences. (B) View after hiding all but the A, B, and C reference sequences. Clicking on a breakpoint brings up the location in the alignment (pop-up window). Consult the Simplot manual for details on how to zoom in/out of the plot. (See color section.)

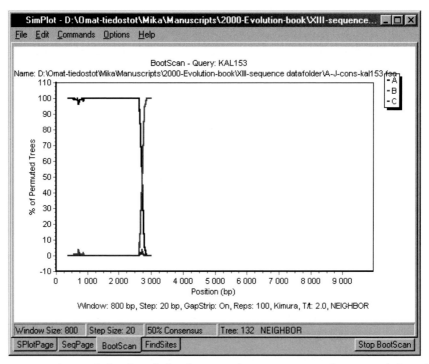

Figure 14.15 Bootscanning analysis. View during bootscanning run. Note continuous updating of the screen and run-log footer showing the status of the background run. (See color section.)

Bootscan window appears. Some of the options, all of them related to the PHYLIP package (see Chapters 4 and 5), have to be changed in the menu, which should be self-explanatory. For example, the tree algorithm *distance model*, and number of *bootstrap replicates* can be set. After setting the query sequence to KAL153, the analysis can run from the Commands menu by selecting DoBootscan. As the bootscanning begins, a running dialog appears in the bottom right corner that shows which tree is being processed; simultaneously, the plot grows on the screen (Figure 14.15). The analysis is fairly slow but can be stopped from the Stop Bootscan button in the lower right corner (it is not possible to continue the analysis after prematurely stopping it). When Bootscan is finished, the resulting plot can be compared (Figure 14.15).

14.6.2   Exercise 2: Mapping recombination with Simplot

In this exercise, the structure of four HIV-1 recombinant strains are mapped. The sequences under study are in the file A-J-cons-recombinants.fsa. Open the file A-J-cons-recombinants.fsa. Use the skills learned from Exercise 1 to map the recombination breakpoints in the sequences. First, individually analyze the recombinants using the `Hide below this line` feature. Review the subtypes of the parental sequences with all reference sequences and a relatively large window size (i.e., 600–800). Finally, map the recombination breakpoints with only the parental sequences and a smaller window size (i.e., 300–400). To verify the results, the putative recombinant regions identified by `Simplot` can be exported and analyzed running separate phylogenetic analyses with PHYLIP or PAUP*, as demonstrated previously. To export a region of an alignment from `Simplot`, first record the region of interest (i.e., map the breakpoints). Go to the `File` menu and select `Save slice of alignment as...` (see Figure 14.16; the KAL153 sequence is used as an example). In the dialogue box, indicate the region of the alignment to be saved and save it in a separate file. The format of the new sequence file can be chosen; for example, the PHYLIP interleaved format. Table 14.1 shows the correct result of the anaysis.

14.6.3   Exercise 3: Using the "groups" feature of `Simplot`

A limitation of `Simplot` is that it can analyze only 26 individual sequences at any time, partly due to memory constraint. The "group" feature is a way to specify groups of sequences within an alignment that are then used to quickly calculate consensus sequences. This is done by adding an extra string of characters to the beginning of an alignment file. The new string is put on a sequence line of its own, before all the other sequences in the alignment, and it must contain exactly as many characters as there are sequences in the aligned sequences. This group-characters sequence is simply named `Simplot`. The following is an example of a group-specifying line at the beginning of an alignment:

```
>Simplot
AAAABBBBCCCCCDDDDFFFFFFGGGGHHHJJKKEEEEXXXXYYZZZQQUU
>sequence1
TGCCTTGAGTG-CTTCAAGTAGTGTGTGCCCGTCTG-TTGTGTGACTCTG
>sequence2
TGCCTCGAGTG-CTTCAAGTAGTGAGTGCCCGTCTG-TTGTGTGACTCTG
[etc ... up to sequence 51]
```

The FASTA-formatted alignment, the beginning of which is shown in the example, contains 51 sequences because the `Simplot` groups line had exactly 51 characters. The different letters in the line reflect the groups: each letter specifies one sequence in the alignment in order of occurrence. In the example, the first four sequences of the alignment belong to Group A, the next four to Group B, and so

(A)

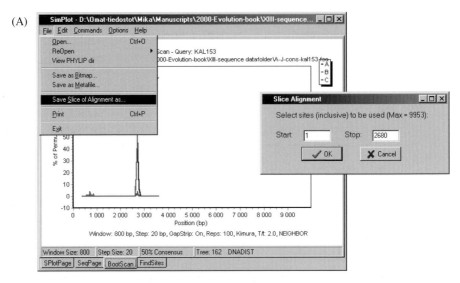

(B)

(C)

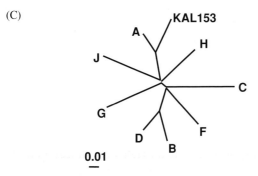

Figure 14.16 Exporting a region of the analysis for separate phylogenetic-tree analysis. (A) From the File menu, pieces (slices) of the alignment can be exported, which can be analyzed separately. The user records the coordinates of the region to be saved. (B) Exporting the slice of the alignment in the desired folder. Select the file format from the pop-up menu and name the file. (C) The result of running a PHYLIP K80-NJ analysis on the fragment used in this example (Region 1–2680 in the file A-J-cons-Kal153.fsa) is displayed by Treeview (unrooted tree).

Table 14.1   Key to recombinant-strain structures of Exercise 2

Strain	Parental subtypes	Breakpoints[a]
SE7812	CRF02-AG	2480, 3420, 4360, 5290, 5430, 5770, 6150, 6550, 8690, 9090, 9510
UG266	AD	5430, 6130, 6750, 9630
VI1035	CRF02-AG/C (AGC)	1800, 2200, 5840, 8920[a]
VI1310	CRF06-DF	2920, 3640, 4300, 5260, 6100

[a]   These are the breakpoints between CRF02-AG and the subtype C segments, depending on the window settings used there may be some variation in the exact coordinates.

on. To specify groups, all the standard letters of the ISO Latin alphabet can be used, but no other characters; hence, the 26-group/single-sequence limitation. However, there is no limitation (other than computer memory) to how many sequences could be in a group. The group sequence must be added by manual editing of an alignment text file or using the Move Up, Move Down, and New Group buttons in the SeqPage of Simplot.

To start the exercise, open the file 2000-HIV-subtype.fsa in Simplot and compare the SeqPage view to the previous example; this file contains exactly the same groups (Figure 14.17A). Scroll up and down in the window to see how the program groups all the sequences as described previously. When the user scrolls to the bottom of the window, six groups are visible, E, X, Y, Z, Q, and U. They contain reference sequences for four currently recognized circulating recombinant forms of HIV-1 (CRF01-CRF04) and two pending candidates. Because the predetermined group feature of Simplot allows only one character to designate the group, the names must be changed interactively by using the buttons to the right to reflect the real names and keep track of the sequences. Change the names of the groups as follows: E to CRF01-AE, X to CRF02-AG, Y to CRF03-AB, Z to CRF04-cpx, Q to CRF05-DF, and U to CRF06-cpx (Figure 14.17B) . The F subtype actually consists of two groups, the F1 and F2 sub-subtypes; split them into two groups. First, rename the F group F1. Next, create a new group using the buttons on the right, name it F2, and move it next to the F1 group. Select the F2 sequences one by one and move them to the new F2 group using the buttons to the right or by dragging and dropping (Figure 14.17C). Move all other CRFs except CRF03-AB below the Hide-line. Switch to the SPlotPage tab and perform a default parameter analysis with window settings of 400 nucleotides/20 steps and CRF03-AB as the query. The program calculates by default a *50% consensus* for the groups and plots the *similarity values* to that consensus. By clicking on the consensus panel in the lower part of the window, it is possible to change the type of consensus used in the analysis. Figure 14.18 shows results of analyses with different types of consensus models (see the Simplot documentation for more details).

(A)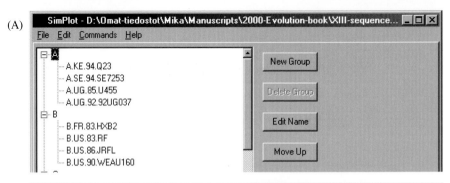

(B)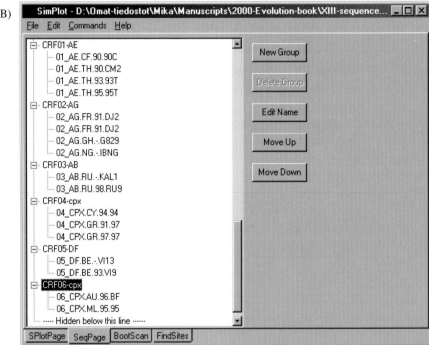

(C)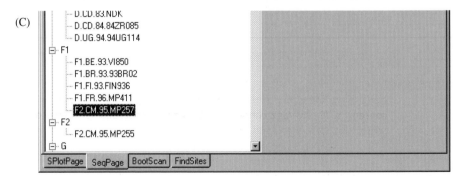

Figure 14.17 Using and rearranging groups specified in the alignment file. (A) Top view of the 2000-HIV-subtype.fsa file in which groups have been specified by the inclusion of a group line. (B) Changing the group names. (C) Creating a new group (F2), renaming group F to F1, and moving the F2 sub-subtype sequences to the F2 group.

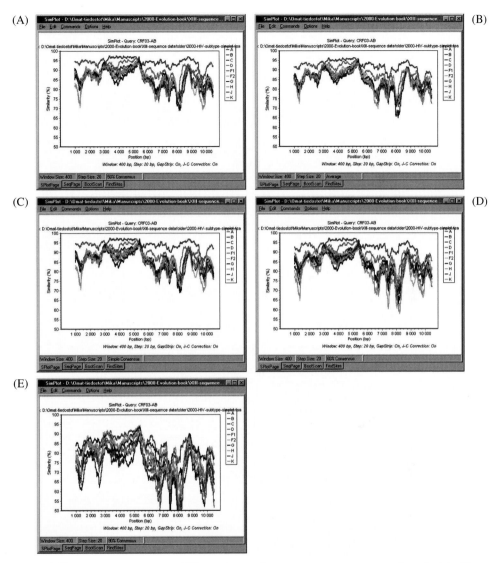

Figure 14.18 Different consensus types. Different types of consensus sequences may be calculated from the groups. The type of consensus used can be specified in the Options menu. (A) 50% consensus. (B) Average consensus. (C) Simple consensus. (D) 60% consensus. (E) 90% consensus. A detailed description on how the consensus sequences are calculated can be found in the `Simplot` documentation. (See color section.)

14.6.4 Exercise 4: Using `SplitsTree` to visualize recombination

`SplitsTree` is a program that deduces not only bifurcating evolutionary relation-ships, but also networks; that is, situations in which both evolution by accumulation of point mutations and horizontal transfer between lineages (i.e., recombination) take place (see Chapter 12). Start `SplitsTree` by clicking on the program icon and open the file `A-J-cons-Kal153.nex`.

(A)

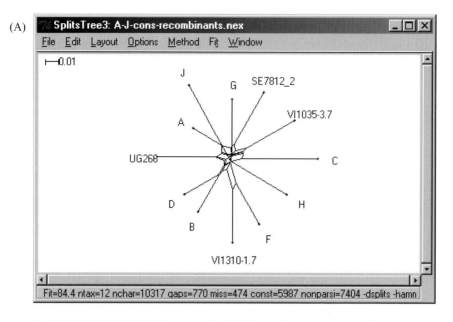

(B)

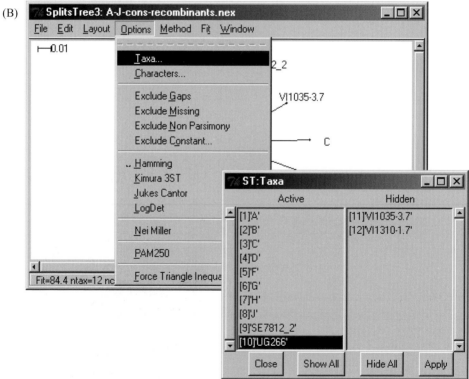

Figure 14.19 Using split-decomposition analysis. (A) The file A-J-cons-recombinants.nex opened into SplitsTree and analyzed with default parameters. The extensive network seen in the center of the tree indicates recombination between multiple sequences. (B) Selecting the taxa (individual sequences) to be included/excluded from the analysis.

(A)

(B)

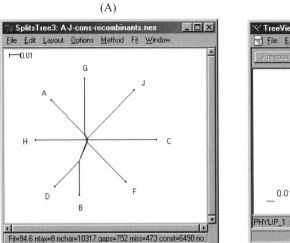
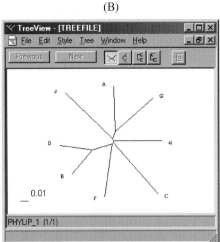

(C)

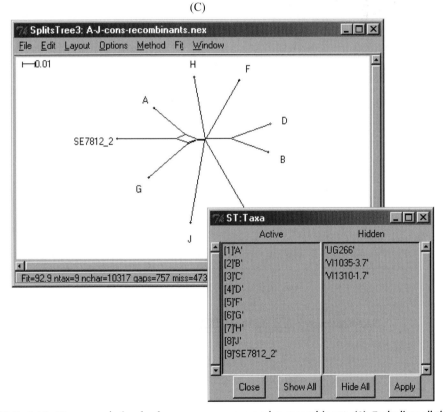

Figure 14.20 `SplitsTree` analysis of reference sequences and a recombinant. (A) Excluding all the recombinant sequences from `SplitsTree` results in a binary split tree. (B) The same data set as in (A) analyzed using the `PHYLIP` K80 + NJ algorithm and displayed with `TreeView`. (C) Including one recombinant with the reference sequences: accurate indication of the AG recombinant structure of the SE7812 sequence.

SplitsTree produces a *split-decomposition analysis* of the relationships be-tween the subtype reference sequences and the CRF03-AB recombinant strain. Compare the results with Figure 14.11. The analysis correctly indicates that the Kal153 strain is related to both Subtypes A and B of HIV-1. The relationship to Subtype D is also close, which is a reflection of the relatively recent separation of Sub-types B and D. Open the file A-J-cons-recombinants.nex, which contains many recombinant sequences with the same subtype consensus reference sequences as in the previous point. As with the Kal153 strain, the split-decomposition analy-sis indicates multiple possible branching orders, as shown by the network-like tree structure (Figure 14.19A).

With SplitsTree, it is possible to select which of the taxa to include in the analysis (see Chapter 12). Open the Options menu and select Taxa. Move all the recombinant viruses into the second panel with the heading Hidden by double-clicking on them one by one (Figure 14.19B). Then click on the Apply button to recalculate the tree using the subtype consensus reference sequences only. The tree will look quite similar to what would be generated by ordinary phylogenetic methods (Figure 14.20A and 14.20B).

Individually move each of the recombinant sequences from the Hidden panel into the Active panel and apply the analysis, which will indicate the recombinant structure of the sequences (Figure 14.20C). Compare the results with those of Exercise 2 and the key found in Table 14.1.

REFERENCES

Carr, J. K., M. O. Salminen, J. Albert, et al. (1998). Full genome sequences of human immunod-eficiency virus type 1 subtypes G and A/G intersubtype recombinants. *Virology*, 247, 22–31.

Carr, J. K., M. O. Salminen, C. Koch, et al. (1996). Full-length sequence and mosaic structure of a human immunodeficiency virus type 1 isolate from Thailand. *Journal of Virology*, 70, 5935–5943.

Coffin, J. M. (1979). Structure, replication, and recombination of retrovirus genomes: some unifying hypotheses. *Journal of General Virology*, 42, 1–26.

Felsenstein, J. (1996). *PHYLIP: Phylogeny Inference Package, Version 3.572c.* Seattle, WA: Uni-versity of Washington.

Gao, F., D. L. Robertson, C. D. Carruthers, et al. (1998). An isolate of human immunodeficiency virus type 1 originally classified as subtype I represents a complex mosaic comprising three different group M subtypes (A, G, and I). *Journal of Virology*, 72, 10234–10241.

Duesberg, P. H., D. Goodrich, and R. P. Zhou (1991). Cancer genes by non-homologous recom-bination. *Basic Life Sciences*, 57, 197–211.

Higgins, D., A. Bleasby, and R. Fuchs (1992). ClustalV: Improved software for multiple-sequence alignment. *Cabios*, 8, 189–191.

Huson, D. H. (1998). `SplitsTree`: Analyzing and visualizing evolutionary data. *Bioinformatics*, 14, 68–73.

Korber, B., B. H. Hahn, B. Foley, et al. (eds.) (1997). *Human Retroviruses and AIDS: A Compilation and Analysis of Nucleic Acid and Amino Acid Sequences*. Los Alamos, NM: Theoretical Biology and Biophysics Group, Los Alamos National Laboratory.

Laukkanen T., J. K. Carr, W. Janssens, et al. (2000). Virtually Full-Length Subtype F and F/D Recombinant HIV-1 from Africa and South America. *Virology*, 269, 95–104.

Liitsola, K., I. Tashkinova, T. Laukkanen, et al. (1998). HIV-1 genetic subtype A/B recombinant strain causing an explosive epidemic in injecting drug users in Kaliningrad. *AIDS*, 12, 1907–1919.

Liitsola, K., K. Holm, A. Bobkov, et al. (2000a). An A/B recombinant HIV-1 and its parental strains circulating in Russia: Low requirements for sequence identity in recombination. *AIDS Research and Human Retroviruses*, 16, 1047–1053.

Liitsola, K., M. Ristola, P. Holmstrom, M. Salminen, H. Brummer-Korvenkontio, S. Simola, J. Suni, and P. Leinikki (2000b). An outbreak of the circulating recombinant form AE-CM240 in the Finnish IDU population. *AIDS*, 14, 2613–2615.

Lole, K. S., R. C. Bollinger, R. S. Paranjape, et al. (1999). Full-length human immunodeficiency virus type 1 genomes from subtype C-infected seroconverters in India, with evidence of inter-subtype recombination. *Journal of Virology*, 73, 152–160.

McCutchan, F. E., M. O. Salminen, J. K. Carr, and D. S. Burke (1996). HIV-1 genetic diversity. *AIDS*, 10, S13–S20.

Robertson, D. L., J. P. Anderson, J. A. Bradac, et al. (2000a). HIV-1 nomenclature proposal [letter]. *Science*, 288, 55–6.

Robertson, D. L., J. P. Anderson, J. A. Bradac, et al (eds.) (2000b). *A Reference Guide to HIV-1 Classification*, Los Alamos, NM: Theoretical Biology and Biophysics Group, Los Alamos National Laboratory.

Salminen, M. O., J. K. Carr, D. S. Burke, and F. E. McCutchan (1995a). Identification of breakpoints in intergenotypic recombinants of HIV type 1 by bootscanning. *AIDS Research and Human Retroviruses*, 11, 1423–1425.

Salminen, M. O., J. K. Carr, D. L. Robertson, et al. (1997). Evolution and probable transmission of intersubtype recombinant human immunodeficiency virus type 1 in a Zambian couple. *Journal of Virology*, 71, 2647–2655.

Salminen, M. O., C. Koch, E. Sanders-Buell, et al. (1995b). Recovery of virtually full-length HIV-1 provirus of diverse subtypes from primary virus cultures using the polymerase chain reaction. *Virology*, 213, 80–86.

Santti, J., T. Hyypia, L. Kinnunen, and M. Salminen (1999). Evidence of recombination among enteroviruses. *Journal of Virology*, 73, 8741–8749.

Van Regenmortel, M., C. Fauquet, D. Bishop, et al. (2000). *Virus Taxonomy: Classification and Nomenclature of Viruses. Seventh Report of the International Committee on Taxonomy of Viruses*. New York, San Diego: Academic Press.

Vartanian, J. P., A. Mcyerhans, B. Asjo, and S. Wain-Hobson (1991). Selection, recombination, and G—A hypermutation of human immunodeficiency virus type 1 genomes. *Journal of Virology*, 65, 1779–88.

LAMARC: Estimating population genetic parameters from molecular data

THEORY

Mary K. Kuhner

15.1 Introduction

The **LAMARC programs** estimate parameters such as **effective population size**, **growth rate**, **population subdivision**, **migration rates**, and **recombination rate** using molecular data from a random sample of individuals from one or several populations (Felsenstein et al., 1999).

The relationship structure among sampled individuals – their **genealogy** – contains significant information about the past history of the population from which those individuals were drawn. For example, in a population that has been large for a long time, most of the individuals in the sample are distantly related; in a population that has been smaller, most of the individuals are closely related.

The mathematical theory relating a genealogy to the structure of its underlying population, called **coalescent theory**, was first developed by Kingman (1982) and expanded by Hudson and Kaplan (1988) and other researchers. However, use of the coalescent theory to estimate parameters is hampered by the fact that the sample genealogy is almost never known with certainty and is difficult to infer accurately. Population samples are less likely to yield their correct genealogy than samples from multiple species because fewer informative mutations will be available. Additionally, the possibility of **recombination** can make accurate genealogy reconstruction of a population almost impossibly daunting.

Analysis of pairs of individuals can reveal some **coalescent-based information** – the genealogy connecting two individuals is relatively easy to infer. However, considerable information is lost in reducing a genealogy to a series of pairs, and the nonindependence of the pairs also poses some problems in analysis.

The LAMARC programs work around the ignorance of the correct genealogy by considering many possible genealogies, which is accomplished using a statistical method called ***Metropolis-Hastings Markov Chain Monte Carlo sampling***. Working backward through this imposing term, Monte Carlo means that there is a random element to the sampling. Markov Chain means that the genealogies are produced using a scheme of defining transition probabilities from one state to another. Metropolis and colleagues (1953) developed (for atom-bomb research!) the specific type of sampler used today, one that starts with a plausible configuration, makes a small change, and accepts or rejects the new configuration. Hastings (1970) subsequently added statistical details to this procedure.

Markov Chain Monte Carlo (***MCMC***) ***methods*** are necessarily approximate, and they often demand considerable computer time and effort. However, methods considering the entire genealogy potentially can extract more information from the same amount of data than methods based on pairs of individuals – and they provide not only estimates of the parameters, but also approximate error bars around those estimates.

15.2   Basis of the Metropolis-Hastings MCMC sampler

Because the correct genealogy cannot be known, especially in more complex cases such as those with recombination, the estimate should be based on a good sample of possible genealogies. To make the sampling as efficient as possible, genealogies that are reasonably concordant with the data are chosen. Undirected random sampling is not efficient because the number of possible genealogies skyrockets as the number of sampled individuals increases. MCMC sampling, by imposing preferences on the random walk, allows purposeful movement through the space of possible genealogies.

There are two requirements to build up a ***Metropolis-Hastings MCMC coalescent sampler***. First, a mathematical statement of how the parameters are expected to affect the shape of the genealogy is needed. For example, in the simple case of a ***constant-size population*** without recombination or migration, Kingman's (1982) original work provides the necessary expectations, which take the general form of the probability of the genealogy G for a given value of the parameter (or parameters). In the example, the single parameter is Θ ($4N_e\mu$, where N_e is the effective population size and μ is the ***mutation rate***) and the expectation is written as $P(G|\Theta)$. Second, the relative fit of the data to the various genealogies must be assessed so that the sampler can concentrate on genealogies that explain the data well. This is the goal of phylogeny estimation as well; therefore, similar methods may be used.

Likelihood methods (see Chapter 6) (Kishino and Hasegawa, 1989) are the most appropriate in this situation because they are accurate and flexible and because they can tell not only which genealogy is better, but also by how much. The fit of data to genealogy can be expressed as the probability of the data, assuming an appropriate model of molecular evolution, with respect to any given genealogy (written $P(D|G)$). By combining the two parts, it can be shown that $L(\Theta) = \Sigma_G P(D|G)\,P(G|\Theta)$, where the likelihood of any given value of Θ is the sum – over all possible genealogies – of the probability of the data on that genealogy, multiplied by the probability of that genealogy for the given value of Θ. Unfortunately, the whole summation is not possible in any but trivial cases.

To overcome this problem, the Metropolis-Hastings sampler generates a biased set of genealogies driven by an assumed value of Θ, called Θ_0, and then it corrects for that bias in evaluating the likelihood. The result is a relative likelihood, as follows:

$$L(\Theta)/L(\Theta_0) = \Sigma_G^*(P(D|G)P(G|\Theta))/(P(D|G)P(G|\Theta_0)) \qquad (15.1)$$

Here, Σ_G^* is a sum over genealogies selected in proportion to $P(D|G)\,P(G|\Theta_0)$. If an infinitely large sample could be generated, then this approximation would provide the same results as the straightforward likelihood. In practice, a sufficiently large sample must be considered so that the region of plausible genealogies is well explored. The algorithm will efficiently explore the right region only if Θ_0, which acts as a guide, is close to the true unknown Θ. One strategy is to make short runs of the program in order to obtain a preliminary estimate of Θ, and then feed that estimate back in as Θ_0. The final run will then have Θ_0 close to Θ, and will be more efficient (and less biased) than the earlier ones.

The program generates its sample of genealogies by starting with some arbitrary or user-supplied genealogy and proposing small rearrangements to it. The choice of rearrangements is guided by $P(G|\Theta_0)$. Once a rearranged genealogy has been produced, its plausibility is assessed $(P(D|G))$ and compared to the plausibility of the previous genealogy. If the new genealogy is superior, it is accepted. If it is inferior, it still has a chance to be accepted: for example, genealogies that are ten times worse are accepted one time in ten that they occur. This behavior helps keep the sampler from walking up the nearest "hill" in the space of genealogies and sticking there, even if there are better regions elsewhere. Given sufficient time, all areas of the space will be searched, although proportionally more time will be spent in regions where $P(D|G)P(G|\Theta_0)$ is higher.

Once a large sample of genealogies has been produced, it is then used to construct a likelihood curve showing $L(\Theta)/L(\Theta_0)$ for various values of Θ, which is typically displayed as a log-likelihood curve. The maximum of this curve is the maximum-likelihood estimate (MLE) of Θ; the region within two log-likelihood units of the maximum forms an approximate 95% confidence interval. Typically, the strategy is to run 5–10 "short chains" of a few thousand genealogies each, to get a good starting value of Θ, and then one to two "long chains" to generate the final estimate.

The most difficult part of creating a Metropolis-Hastings sampler is finding a way to make rearrangements guided by $P(G|\Theta_0)$; this is particularly challenging in cases with recombination, in which the genealogy becomes a tangled graph.

15.2.1 Random sample

LAMARC assumes that the sampled individuals were chosen at random from the population(s) under study. For MIGRATE, the user may freely choose how intensively to sample each subpopulation; however, within each subpopulation, individuals must be chosen at random. Violating this assumption will lead to bias; for example, if the most diverse individuals are chosen, then Θ may be overestimated. Researchers are sometimes tempted to delete "boring" identical individuals from their data sets in favor of nonidentical ones, but doing so produces misleading results. If the data set is too large to analyze, individuals must be dropped at random.

15.2.2 Stability

LAMARC assumes that the population situation has been stable for "a long time" (i.e., longer than the history of the coalescent genealogy). COALESCE, MIGRATE, RECOMBINE, and LAMARC assume that it has been of constant size; FLUCTUATE (Kuhner et al., 1998) assumes that it has been growing or shrinking with a constant exponential rate. MIGRATE (Beerli and Felsenstein, 1999) and LAMARC assume that the **subpopulation structure** and migration rates have been stable. If these assumptions are not met, the results will be misleading. For example, two populations that currently exchange no migrants but that are recently derived from a single ancestral population will give spurious evidence for migration. These assumptions can be relaxed by constructing more complex samplers, if there is a model for how things have changed; for example, FLUCTUATE relaxes COALESCE's constant-size assumption, replacing it with a model of exponential growth.

15.2.3 No other forces

LAMARC assumes that no forces other than those modeled are acting on the **locus**. In particular, it assumes that the variation observed is **neutral** and is not being dragged by **selection** at a linked marker. Selection could bias the results in a variety of directions, depending on the type of selection. **Balancing selection** leads to overestimates of Θ; **directional selection**, in general, lead to underestimates. It is difficult to predict the effects of selection on estimates of migration or recombination rates.

15.2.4 Evolutionary model

LAMARC's results can be only as good as its model of molecular evolution. For example, analyzing human **mitochondrial DNA** with a model of DNA sequence evolution that assumes *transitions* and *transversions* are equally frequent will lead to misleading results, because mtDNA appears to greatly favor transitions. Fortunately, it is fairly easy to add more complex evolutionary models as the phylogenetic community develops them (see Chapter 10). LAMARC inherits any assumptions made by its evolutionary models; for example, a common assumption in DNA models is that all sites are independent and identically distributed. Information about the assumptions of maximum-likelihood phylogeny models is found in Chapters 6 and 7; those caveats also apply here.

15.2.5 Large population relative to sample

Coalescent theory is an approximation of the real genealogical process. To make the mathematics tractable, it is assumed that only one event happens at a time; for example, two individuals may trace back to a common ancestor at a particular point in time, but not three individuals. This assumption is safe when the population is much larger than the sample (rare violations may occur but will not affect the result). However, it can break down if the method is applied to a tiny population, most of which is being sampled. An extreme example would be a lineage of highly inbred mice, in which the size of the whole population is just a few individuals each generation. Any sample of such a population will represent a large proportion of the whole, and the coalescent approximations will not be very good. In practice, there are few stable biological populations small enough to encounter a breakdown of the coalescent approximation. Tiny populations are more likely to represent a violation of the stability assumption; they have probably not been tiny throughout their history.

15.2.6 Adequate run time

The sample of genealogies produced by a LAMARC program is meant to stand in for the entire space of possible genealogies – a tall order. Fortunately, most genealogies are such poor explanations of the data that they contribute little to the final estimate; therefore, failure to sample such genealogies is insignificant. However, the program does need to run long enough to adequately sample the high-probability regions of *genealogy space*, which is particularly difficult if it contains multiple isolated peaks rather than a single peak. For example, consider a data set that can support either high migration from A to B and low from B to A, or vice versa, but is less supportive of intermediate values. This corresponds to two peaks in the space of possible genealogies for these data. The program must run long enough to visit both peaks because an estimate based on only the first peak found may be misleading; for example, leading to a false conclusion that high migration from B to A can be rejected. Metropolis-Hastings samplers in other applications sometimes encounter cases where the "valley" between two peaks is infinitely deep and cannot be crossed, thereby stalling the sampler. It is believed that these cases do not exist for the genealogy applications, but very deep valleys are possible. Ways to detect stalled-out samplers are suggested later in this chapter, but they are not guaranteed to work. Getting good results from the LAMARC programs requires patience and a degree of skepticism.

PRACTICE

Mary K. Kuhner

15.3 The LAMARC software package

Copies of the source code and executables for various systems can be obtained from **http://evolution.genetics.washington.edu/lamarc.html** or by anonymous ftp to `evolution.genetics.washington.edu` in directory/pub/lamarc. See Table 15.1 for details of the available programs.

15.3.1 FLUCTUATE (COALESCE)

COALESCE, the first LAMARC program released, estimated Θ ($4N_e\mu$) in a single constant-size population using DNA or RNA sequence data and a Kimura two-parameter model of sequence evolution (Kuhner et al., 1995) (see Chapter 4). Its successor, FLUCTUATE, adds estimation of an **exponential growth** (or **decline**) **rate**, g. COALESCE is no longer supported because it can be easily mimicked using FLUCTUATE by simply fixing g at zero.

Estimation of Θ is quite robust and accurate even with fairly small data sets; estimation of g is more difficult. For many data sets, the **likelihood surface** resembles a long, flat ridge, so that many different values of g are almost equally likely. Each value of g has its own preferred value of Θ, but there is little to choose between various Θ/g pairs. There are two ways to overcome this problem. Using **multiple loci** can greatly improve the estimate of g. Most information about g comes from **coalescences** deep in the genealogy (toward the root), but any given genealogy has only a few such coalescences. Adding a second, unlinked **locus** nearly doubles the available information. The second option is to provide an estimate of Θ from external evidence, which allows a more stable estimate of g. If neither multiple loci nor an independent estimate of Θ are available, the estimate of g should be evaluated cautiously because it is likely to be biased upward.

15.3.2 MIGRATE

For a **subdivided population**, the MIGRATE program estimates Θ for each subpopulation and the migration rate into each subpopulation from each of the others. It also can analyze more limited **migration models**; for example, it can assume that immigration and emigration are equal or that only certain subpopulations are capable of exchanging migrants (e.g., in a **stepping-stone model**). The more parameters estimated, the less power is available for each one; thus, if a stepping-stone model is a good representation of the data, then it, rather than the most general

Table 15.1 LAMARC programs as of 2002

Program	Estimates	Data types supported
FLUCTUATE	Θ, growth	DNA, RNA
MIGRATE	Θ, migration rates	DNA, RNA, microsats, electrophoretic alleles
RECOMBINE	Θ, recombination rate	DNA, RNA, SNPs
LAMARC	Θ, migration & rec. rates	DNA, RNA, SNPs, microsats

model, should be used. The converse, of course, is that if a stepping-stone model is imposed and the biological situation is not consistent with it, estimates will be actively misleading.

MIGRATE needs multiple loci and reasonably large numbers of individuals per subpopulation – at least 20 or more – to perform well. The researcher often is aware of many subpopulations but has samples from only a few. For example, one might be studying two island populations of mink and have no samples from the adjacent mainland. In these cases, it is useful to include a third population, containing no observed individuals, to the MIGRATE analysis. Although the parameters of this unsampled population will not be estimated very accurately, its presence will allow more accurate estimation of parameters for the sampled populations.

When MIGRATE performs badly, failure of the **steady-state assumption** is often the cause. For example, if two populations recently diverged from a common ancestor but currently have little migration, MIGRATE infers high migration in an attempt to explain the shared similarities due to common ancestry. Another potential difficulty is that if two subpopulations are mixing at an extremely high rate so that they are, in effect, a single population, MIGRATE is theoretically capable of analyzing the situation and reaching correct conclusions; however, in practice, it tends to slow down.

15.3.3 RECOMBINE

The RECOMBINE program, which assumes a single population, estimates Θ and the recombination rate r, expressed as recombinations per site over mutations per site: C/μ. It also assumes that the recombination rate is constant across the region. RECOMBINE can successfully estimate recombination rate with short sequences only if Θ is high; in other words, data have to be very **polymorphic** and the recombination rate is high. If either parameter is relatively low, long sequences (thousands of base pairs) are needed to make an accurate estimate. The space of possible genealogies searched by RECOMBINE is huge, and long runs are necessary to produce an accurate estimate. One practical consideration when running RECOMBINE is whether the "short chains" used to find initial values of Θ and r are long enough. The user should examine the values of r produced from the short chains. If they fluctuate greatly from one short chain to the next, the short chains are probably too short; it may be better to run half as many short chains as before, each one twice as long.

15.3.4   LAMARC

The LAMARC program – not to be confused with the package to which it belongs – combines the basic capabilities of RECOMBINE and MIGRATE. It coestimates both migration rates and recombination rates, which is useful when both sets of parameters are of interest. It is also important when only one set is of interest – for example, a researcher trying to estimate migration rates using nuclear DNA – because acknowledging the presence of both evolutionary forces avoids bias in the results. LAMARC behaves much like a hybrid of RECOMBINE and MIGRATE, except that it does not yet have the sophisticated migration models available in the latter program and uses a narrower range of data types. Attempts to estimate both recombination and migration in complex cases demand extremely long runs; a good rule of thumb is to start by individually adding together the run length needed for each program. In the long run, the LAMARC program will be enhanced with additional evolutionary forces such as growth and selection, so that complex biological situations can be analyzed accurately.

15.4   Starting values

The closer the initial values are to the truth, the more efficiently the program will search. If a "quick and dirty" estimator is available (e.g., the **Watterson estimate**; see Section 15.8.1), it should be used to generate initial values. This estimator is provided by most of the LAMARC programs. If it is necessary to guess at an initial value, the guess should err upward rather than downward and should never be zero. If a parameter is known to be zero, it should be fixed at that value; otherwise, never use zero as an initial guess. Trying more than one set of initial values can help spot problems in the analysis. The results of multiple runs with different starting values should be quite similar. If they are not, the reason could be twofold: (1) the likelihood surface is so flat that almost any value is equally plausible; or (2) the program is not running long enough for it to reach its final solution. Examining the shape of the likelihood surface helps to distinguish between these two alternatives. If it is very flat, scenario (1) is true and the data simply do not allow a sharp estimate. If the likelihood surface is sharply peaked but varies hugely with starting values, the program is not running long enough.

15.5 Space and time

On many computers, extra memory will have to be assigned to the LAMARC programs to run them on realistic data sets. The method for accomplishing this is operating-system specific: ask the local system operator for assistance. This is worth trying even if the program does function with less memory because additional memory may allow it to run more quickly. If the program still does not run, or runs at an intolerably slow speed, consider reducing the size of the data set: randomly delete sequence regions or individuals. With the MIGRATE program, deletion of individuals can be nonrandom between subpopulations as long as it is random within subpopulations. For example, the user can randomly drop three individuals from each subpopulation. (Do not preferentially delete "uninteresting" sites or individuals because this will bias the results!)

Increasing the interval between sampled genealogies saves space and a little time. The LAMARC programs do not normally use all of the genealogies they generate in their final estimate, but rather sample every 10th or 20th genealogy. Denser sampling does not help the estimate much, because successive genealogies are highly correlated. There is significantly more information in a sample of every 100th genealogy from 100,000 than every 10th genealogy from 10,000, although the number of genealogies sampled and the amount of space used are the same. Of course, generating 100,000 genealogies is still ten times slower than generating 10,000.

15.6 Sample size considerations

If one must choose between studying more individuals, more sites, or more loci, almost always choose more loci for FLUCTUATE and MIGRATE and more sites for RECOMBINE. Users of LAMARC must decide whether they are more interested in recombination or migration. However, a minimum number of individuals is needed to get a good estimate. In general, FLUCTUATE should have at least 10–20 individuals, RECOMBINE should have at least 20, and MIGRATE and LAMARC should have at least 20 per subpopulation. Doubling or tripling these numbers is reasonable, but there are diminishing returns beyond that point. Running FLUCTUATE with 400 individuals gains almost nothing over running it with 40. DNA or RNA sequences should be at least a few hundred base pairs long for FLUCTUATE or MIGRATE, but will need to be much longer – thousands to tens of thousands – for RECOMBINE and LAMARC unless both Θ and r are very high. *Microsatellite* and *electrophoretic data* require multiple loci even more strongly than do sequences because the amount of information per locus is smaller. Adding more individuals does not necessarily help much because the first few individuals already establish the basic structure of the genealogy. After that point, additional individuals tend to join the genealogy as twigs, and each twig carries only a minimum amount of new information. Adding a second locus, in contrast, gives a new trunk and main branches. The reason that longer sequences do not much improve the estimate in FLUCTUATE and MIGRATE is that uncertainty in the *LAMARC* *estimates* comes from three sources. First, the program introduces some uncertainty by its approximate nature, which can be overcome by longer runs. Second, the data are only an approximate representation of its underlying genealogy, which can be overcome by longer sequences. Finally, the genealogy is only an approximate representation of its underlying parameters, which can be improved slightly by a larger genealogy and improved significantly by adding a second unlinked locus, thereby examining a fully independent new genealogy. After a certain point, no amount of running the program longer or adding longer sequences improves the estimate because the remaining error comes from the relationship between genealogy and parameters. This is not true in RECOMBINE because a long recombining sequence, like multiple loci, provides additional information. RECOMBINE can accommodate multiple loci, but is actually most efficient with a single long contiguous stretch of markers.

15.7   Virus-specific issues

Estimating population genetic parameters for viral data sets, especially for fast-evolving viruses like HIV or HCV, requires a few additional considerations.

15.7.1   Multiple loci

Many viruses are too small to provide multiple independent loci for analysis. RECOMBINE or LAMARC can extract additional information out of partially linked loci. Another alternative is to examine several different isolates of the virus, using the same locus from each isolate but entering them into the program as separate loci. If each isolate traces back to a common ancestor before any shared ancestry with any other isolate, then they are independent samples of the evolutionary process. However, this method assumes that the viral population size and mutation rate are the same for each isolate, which is not necessarily true.

15.7.2   Rapid growth rates

Experience from using FLUCTUATE to estimate HIV growth rates shows that no sensible estimates are produced when Θ and g are coestimated. *HIV genealogies* often resemble stars, and a star genealogy is equally consistent with any high value of Θ and corresponding high value of g. Therefore, the likelihood surface is flat along the Θ/g diagonal, and the program will wander about on this surface until the estimate becomes so huge that the program overflows. If there is a star-like genealogy, it is only possible to estimate g if an external estimate of Θ can be provided. Conversely, if g is known, then Θ can be estimated. One external source of information about these parameters is sequential sampling; that is, two or more samples taken at a reasonable time interval allow a direct estimate of the population parameters (Rodrigo and Felsenstein, 1999; Rodrigo et al., 1999).

15.7.3 Sequential Samples

It would be desirable to adapt the LAMARC programs to handle sequential samples because they should be very powerful for estimating growth in particular. Currently, however, mixing samples from different times in a LAMARC analysis results in confused estimates.

15.8 An exercise with LAMARC

The three sample data sets were examined to decide if they were suitable for analysis. The mtDNA and the bacterial (see Chapter 8) data sets could be eliminated immediately because they are clearly comparisons across species, not within a species. All of the current LAMARC programs assume a single species: the Θ parameter is not meaningful in a cross-species comparison. After removing the two simian viral sequences and the HIV-O sequence, the HIV data set seemed initially promising. However, upon examination, it has the following two troublesome aspects:

1. The sequences included are not random, but rather were selected to illustrate the various subtypes. Thus, any estimate of Θ will be grossly inflated. Because all of the other LAMARC parameters show at least some correlation with Θ, they will probably be misestimated as well.
2. It does not appear to have a ***molecular clock***. A simple ***neighbor-joining*** (*NJ*) tree (see Chapter 5) shows unequal branch lengths, suggesting that some lineages may be evolving more rapidly than others. This could be checked using the ***likelihood ratio test*** (see Section 10.11.2).

The LAMARC programs are meant to be used on random samples from a population or several subpopulations; they will not give valid results if used on a deliberately chosen sample. It would not be possible to reconstruct a random sample using published subtype frequencies because the given sequences almost surely do not contain the entire range of variability present in a real sample.

Instead, a data set from Ward et al. (1991) that contained 360 bp from the ***human mitochondrial genome*** D-loop region, sequenced in 63 members of the ***Nuu-Chah-Nulth*** people, was selected as a more appropriate example. The data originally came in the form of the 28 unique sequences only. The data were expanded into a full-sized data set with all 63 individuals, including only those individuals who

```
1
63 360
A     TTCTTT [up to 360 nt]
B     TCTTAC [up to 360 nt]
```

Figure 15.1 The first few lines of the FLUCTUATE infile.

differed from each other. Although it was not appropriate for the HIV data, this procedure can be used in this situation because it was certain that the sequences listed as identical actually were identical. A text editor was used to make an appropriate number of copies of each sequence, and each one was given an arbitrary unique name. The data set can be found at **http://www.kuleuven.ac.be/aidslab/phylogenyBook/ datasets.htm**.

Each software module of LAMARC works in the same basic way: it needs a file containing the input data – for example, aligned DNA sequences in PHYLIP format – to be placed in the same directory where the program resides. Then it produces an output file, called outfile, in text format, containing the result of some kind of analysis. By default, any application reads the data from the file named infile (no extension type!) if such a file is present in the same directory; otherwise, the user is asked to enter the name of the input file. Each program runs into a new window, from which a list of settings can be edited. The user interface looks like one of the applications of the PHYLIP package (see Chapters 4 and 5).

15.8.1 Exercise using FLUCTUATE

Copies of the LAMARC programs can be obtained from the Web site and installed on local computers. A quick run should be performed using FLUCTUATE and the infile from the package to test whether the installation was done properly (Figure 15.1). When the program starts, a window similar to the one shown in Figure 15.2 appears. Change option I if the input file is in PHYLIP-interleaved format (see Box 2.2), and type Y to start the run. Check the outfile to be sure that everything ran correctly.

The data are *mtDNA* from a single population. It should be possible to estimate Θ and growth rate, although it will be interesting to see whether the results estimated the population size of the Nuu-Chah-Nulth or some larger and perhaps older grouping. With only one population, estimating migration rates does not sound practical: it is possible to include an unknown population, but it is unlikely that there will be any power available for estimation. Finally, mtDNA of humans is not supposed to undergo recombination, but it will be interesting to see whether RECOMBINE confirms this assertion. The sequences may be too short to give a firm estimate of recombination rate, but it is expected that the confidence interval should

```
FLUCTUATE:  Hastings-Metropolis Markov Chain Monte Carlo method,
version 1.0

INPUT/OUTPUT FORMATS
   I            Input sequences interleaved?  No, sequential
   E          Echo the data at start of run?  No
   P  Print indications of progress of run?  Yes
   G                Print out genealogies?   No
   Q    Allow interactive design of output?  No
MODEL PARAMETERS
   T          Transition/transversion ratio:   2.0000
   F       Use empirical base frequencies?  Yes
   C   One category of substitution rates?  Yes
   W       Use Watterson estimate of theta?  Yes
   H       Population can change in size?  Yes
   V  Rate of change parameter for growth:  1.000000e-05
   A            Allow parameters to change?  Yes
   U      Use user tree in file "intree" ?  Yes
SEARCH STRATEGY
   S          Number of short chains to run?      10
   1              Short sampling increment?      20
   2   Number of steps along short chains?    1000
   L          Number of long chains to run?       2
   3               Long sampling increment?      20
   4   Number of steps along long chains?    15000

Are these settings correct? (type Y or the letter for one to change)
```

Figure 15.2 The FLUCTUATE menu.

include zero. The most interesting questions to explore are population size, which can be derived from Θ using an independently published estimate of the D-loop mutation rate, and growth rate. However, it is not wise to expect a good estimate of growth rate from a single locus. Generally, several unlinked loci are required, which means the analysis cannot be improved by adding additional mtDNA sequences. Consequently, this study will have only exploratory interest.

The settings for the production run are given in Table 15.2. When examining the printout of the input data from FLUCTUATE, it should be noted that all of the variable sites represent transitions. Clearly, as is common for mammalian mtDNA, transitions occur more often than transversions. There is a substantial deficit of the G nucleotide, which is also a common trait of mtDNA. It is not appropriate to use the defaults of equal nucleotide frequencies and a *transition/transversion ratio* of 2; rather, it is better to set the ratio to 10.0 and ask the program to estimate *nucleotide frequencies* from the data (options T and F, respectively). There are 360 sites in the sequences of the data set, which should allow a reasonable estimate. It is not known whether 10 is the best ratio, and PAUP* could be used to make an estimate (see Chapter 7). However, it would be time-consuming and the models in use are not highly sensitive to exact values. Is rate heterogeneity among sites expected? Quickly scanning the data shows that some of the variable sites appear

Table 15.2 Input settings for FLUCTUATE

INPUT/OUTPUT FORMATS

    I   Input sequences interleaved? No, sequential

    E   Echo the data at start of run? No

    P   Print indications of progress of run? Yes

    G   Print out genealogies? No

    Q   Allow interactive design of output? No

MODEL PARAMETERS

    T   Transition/transversion ratio: 10.0000

    F   Use empirical base frequencies? Yes

    C   One category of substitution rates? 2 categories

    R   Rates at adjacent sites correlated? No, they are independent

    W   Use Watterson estimate of theta? Yes

    H   Population can change in size? Yes

    V   Rate of change parameter for growth: $2.400000e + 02$

    A   Allow parameters to change? Yes

    U   Use user tree in file "intree" ? No, construct a random tree

SEARCH STRATEGY

    S   Number of short chains to run? 5

    1   Short sampling increment? 20

    2   Number of steps along short chains? 8000

    L   Number of long chains to run? 2

    3   Long sampling increment? 20

    4   Number of steps along long chains? 200000

---

to have mutated more than once (Figure 15.3). All four possible combinations are found at these two sites, which implies either multiple mutations or recombination; because mtDNA is thought not to recombine in mammals, this probably indicates multiple mutations. There are several such sites in the data; although this could have happened by chance, it is more likely that some sites mutate more often than others. In this case, two rate categories will be used, one ten times as high as the other. This is not accurate, but it is more accurate than assuming that all sites mutate at the same rate. It is reasonable to speculate that 5% of sites have the high rate; after all, there are only a few visible multiple-hit sites. These settings can be obtained

Name	Sequence
A	AT
C	GT
L	GC
M	AC

Figure 15.3 A probable double mutation.

as follows:

1. Choose option C and type 2 at the prompt Number of categories? in order to define two rate categories.
2. At the prompt rate for each category?, type 1 10 (remember that only relative rates matter).
3. At the prompt Probability for each category?, type 0.95 0.05.

How are starting values for Θ and g generated? For Θ, the Watterson estimate (Watterson, 1975) seems like a reasonable starting point (option W default). Because the **Watterson test** is based on an infinite-sites model, the estimate will not be perfectly accurate – the presence of multiple-hit sites in the data guarantees that – but it should be reasonably close. There is no such convenient estimate for g. It is expected to be positive because human populations tend to grow. The outcome of the default run, 240, may be used to design a search strategy. It is best to start with short chains long enough that the estimates settle down fairly quickly; if they bounce around throughout the short chains, they are probably too short. The default run can also indicate how long the run will take. On an Alpha workstation, for example, the Ward et al. (1991) data set takes about 12 minutes at defaults; however, much of that time came from parameter maximization on the first short chain. Starting with better initial values may help; longer chains also may lead to quicker parameter estimates. The production run will take at least twice as long due to the two categories of evolutionary rates. For the production run, the length of the short chains should be doubled and the length of the long chains increased tenfold, based on results of the default run (e.g., the slow maximizer suggests that the short chains may be too short and the large difference between short chains 1 and 2 suggests that the long chains could be longer). This scheme should take about 20 times as long as the default run.

Results of the production run in the outfile should be examined. There is variability from chain to chain; the first long chain had Θ = 0.038, and the second Θ = 0.049. The values of g were also quite different. The long chains are probably not running long enough. The first part of the output simply confirms the input parameters, followed by a table showing the log-likelihood surface, which will be repeated later as a graphic. The fact that the log likelihood at the maximum is much greater than 1 suggests that the estimates were still improving (i.e., the final values of Θ and g fit the trees much better than the initial values). A longer run should be tried next. The approximate standard deviations give some idea of the expected error around the estimates. The estimate of Θ is considered fairly precise; the estimate of g is much less so. The next entry is a table that shows what the growth values mean.

If there is an idea of the mutation rate of the D-loop, then there is an expectation for what kind of population growth the value of g implies. For example, if mtDNA mutates at about 10^{-6} per generation, then a modern population size of 12,282 is estimated. One thousand generations ago (about 20,000 years), the population size would have been 12,001. This is the population size of mitrochondria, not of humans. Only women count, so the mitochondrial effective population size is about one-half the human effective population size, which suggests a population size estimate of about 24,500 today. Incidentally, people often wonder how to interpret estimates of Θ for mtDNA, and ask if they should correct the estimate to reflect the fact that for mtDNA, $\Theta = 2N_f\mu$ (N_f, female population size), not $4N_e\mu$. This is not necessary. The easiest way to think of the population size factor N is as the number of reproducing gene copies in the population; it is up to the user to relate that to the number of organisms. This number is not easy to interpret, and one reason is that it is unlikely that the Nuu-Chah-Nulth behave as a single isolated population. They probably derive from a much larger Asian ancestral population; ultimately, they are part of the worldwide human population. Because population subdivisions were not allowed in this analysis, the estimate is a murky compromise between different levels of population structure. A better estimate could be obtained if information about more than one human subpopulation were available and MIGRATE or LAMARC could be used.

It is clear from inspection that if the mtDNA D-loop mutation rate is assumed to be 10^{-5} instead of 10^{-6}, strikingly different answers would be obtained: a modern size of 1,228, decreasing to 97 women just 1,000 generations ago. It is necessary to be certain of the mutation rate estimate before attaching much importance to these numbers. Conversely, two human populations could be compared for which the mutation rate is likely to be the same, considering only relative values.

It is interesting to ask whether a nongrowing population can be rejected. The log-likelihood table shows an approximate 95% confidence interval consisting of all values within 2 units of the maximum. While some extrapolation is needed, the table certainly does not suggest that any entry in the $g = 0$ column is within 2 units of the maximum value of 14.49; in fact, $g = 0$ seems to be emphatically rejected. To improve the precision of this conclusion, the program could be rerun using the Allow interactive design of output option and adjusting the table until it showed the appropriate columns.

One way to improve the estimates obtained with the exercise would be to add loci. Because mtDNA cannot offer any more unlinked loci, nuclear DNA must be used. A different value of Θ is expected for nuclear DNA because both the effective population size is greater and the mutation rate is lower. This requires choosing the Theta varies between loci? option of the menu, which is not visible on the sample menu because it is not relevant to a run with a single locus.

```
Metropolis-Hastings Markov Chain Monte Carlo method, version 1.40

STARTUP MENU
  #               Goto Data/Search Menus
  O           Save current options to file?  No
  N            Use trees from previous run?  No
  E          Echo the data at start of run?  No
  S                Save MHMC output to files?  No
  P Print indications of progress of run?  Yes
  U       Use user tree in file "intree" ?  No
  V            Number of temperatures used?  1
  H                      Infer haplotypes?  No
  L          Calculate confidence intervals?  Yes

Are these settings correct? (type Y or the letter for one to change)
```

Figure 15.4 The RECOMBINE menu.

Another improvement is to add additional populations and run the analysis using
MIGRATE or LAMARC; this is not a perfect solution because the human species
does not conform to the stability assumptions of MIGRATE, but it should be better
than the assertion that the Nuu-Chah-Nulth exist in a vacuum.

15.8.2   Exercise using RECOMBINE

It is first necessary to rename the outfile from FLUCTUATE, or transfer it to a
different directory, so that it will not be destroyed when RECOMBINE writes its own
outfile. The header of the input data file needs to be edited because RECOMBINE
uses a slightly modified PHYLIP format. The first line of the file must contain, in
the following order, the number of populations represented in the data, the number
of loci within each population (constant across all populations), and the title of the
data set as a whole. The second line contains one number for each locus, indicating
the number of bases present in each sequence of that locus. After the header, the
populations are presented one at a time, each introduced by a single line that gives
the constant number of sequences for each locus of the population, followed by
the subtitle. Finally, the sequences for that population are presented, grouped by
locus.

The RECOMBINE menu is shown in Figure 15.4. By typing a # character, a new
menu appears, similar to the one for FLUCTUATE. The same parameters used
for the FLUCTUATE run should be entered to set reasonable starting values for
the program. An initial recombination rate of 0.5 is assumed (i.e., option Z in
the SEARCH MENU), meaning that recombination is about half as common as
mutation. This is much too high for mtDNA, but it is interesting to see whether
the program will find its way to the expected value of zero. RECOMBINE runs more
slowly on the Ward et al. (1991) data than FLUCTUATE did – shown immediately
by the progress reports on-screen. If there is no access to a fast computer, the user
should try RECOMBINE with a smaller data set.

When examining the `outfile` from RECOMBINE, it is observed that the value of the recombination rate r oscillates back and forth between zero and a fairly high value – a bad sign. It suggests that some short chains produced genealogies with recombination and others did not. The program is not achieving a steady state; the short chains are almost surely too short. The other conclusion from the output file is that $r = 0$ appears to be rejected, which is surprising for mtDNA and most likely is an incorrect result driven by the too-short chains because the last long chain happened to be one of the more recombinant ones. It may be helpful to increase the length of both short and long chains fivefold and rerun the program overnight. However, the results of such an analysis still suggest that the program has not settled into a steady state. The final estimate of r is quite high and again rejects zero; the chains are not settling down on stable values (i.e., each chain has a higher value of Θ than the one before it).

Perhaps the initial strategy should be reevaluated. The data set contains only 360 base pairs, which is probably insufficient for a good estimate. It also contains 63 sequences – a large number – and the programs may be having difficulty searching its space adequately. The amount of genealogy space to search increases rapidly as the number of sequences increases. Also, RECOMBINE is rejecting most of the genealogies it is producing; therefore, it is not moving through the space very quickly. It is still possible to obtain a stable estimate from these data. One tactic is to eliminate half of the sequences at random to reduce the search space and get the program to run more quickly. However, with such short sequences, the best result is a rather flat likelihood surface, and it is probably not worth the effort. RECOMBINE needs thousands of base pairs of sequence to accurately estimate low recombination rates, and there are simply not enough data.

15.9   Conclusions

With LAMARC programs, as with phylogeny-estimation programs, it is important to know when *not* to use them. Results from inadequate or inappropriate data have no meaning and, although they look impressive, they are potentially misleading. To assess the appropriateness of LAMARC analysis, users should consider the following questions:

Is the population really likely to be in a stable state?
Are forces other than the ones analyzed likely to affect the population?
Is the data set large enough?
Were the individuals chosen at random?
Are the results stable from chain to chain? From run to run?

What is the uncertainty of auxiliary values used (e.g., as external estimates of the mutation rate)?

How wide are the confidence intervals of the estimates?

People often hesitate at the slowness of modern phylogenetic and population-genetic estimation programs; however, considering that it could take several years to collect samples, it is worthwhile to spend a few weeks making a thoughtful and careful analysis. Throwing together an analysis can greatly reduce the value of the results. There are currently no methods that can be applied mechanically that guarantee good results; therefore, a degree of good judgment is always essential.

REFERENCES

Beerli, P. and J. Felsenstein (1999). Maximum-likelihood estimation of migration rates and effective population numbers in two populations using a coalescent approach. *Genetics*, 152, 763–773.

Felsenstein, J., M. K. Kuhner, J. Yamato, and P. Beerli (1999). Likelihoods on coalescents: A Monte Carlo sampling approach to inferring parameters from population samples of molecular data. In: *Statistics in Genetics and Molecular Biology*, ed. Ed Seillier, IMS Lecture Notes-Monograph Series, pp. 163–185.

Hastings, W. K. (1970). Monte Carlo sampling methods using Markov chains and their applications. *Biometrika*, 57, 97–109.

Hudson, R. R. and N. L. Kaplan (1988). The coalescent process in models with selection and recombination. *Genetics*, 120, 831–840.

Kingman, J. F. C. (1982). The coalescent. *Stochastic Processes and Their Applications*, 13, 235–248.

Kishino, H. and M. Hasegawa (1989). Evaluation of the maximum-likelihood estimate of the evolutionary tree topologies from DNA sequence data, and the branching order in Hominoidea. *Journal of Molecular Evolution*, 31, 151–160.

Kuhner, M. K., J. Yamato, and J. Felsenstein (1995). Estimating effective population size and mutation rate from sequence data using Metropolis-Hastings sampling. *Genetics*, 140, 1421–1430.

Kuhner, M. K., J. Yamato, and J. Felsenstein (1998). Maximum-likelihood estimation of population growth rates based on the coalescent. *Genetics*, 149, 429–434.

Kuhner, M. K., J. Yamato, and J. Felsenstein (2000). Maximum-likelihood estimation of recombination rates from population data. *Genetics*, 156, 1393–1401.

Metropolis, N., A. W. Rosenbluth, M. N. Rosenbluth, A. H. Teller, and E. Teller (1953). Equation of state calculations by fast computing machines. *Journal of Chemical Physics*, 21, 1087–1092.

Rodrigo, A. G. and J. Felsenstein (1999). Coalescent approaches to HIV population genetics. In: *The Evolution of HIV*, ed. K. A. Crandall. Johns Hopkins University Press, Baltimore, MD.

Rodrigo, A. G., E. G. Shpaer, E. L. Delwart, A. K. N. Iversen, M. V. Gallo, J. Brojatsch, M. S. Hirsch, B. D. Walker, and J. I. Mullins (1999). Coalescent estimates of HIV-1 generation time in vivo. *Proceedings of the National Academy of Sciences of the USA*, 96, 2187–2191.

Ward, R. H., B. L. Frazer, K. Dew-Jager, and S. Pääbo (1991). Extensive mitochondrial diversity within a single Amerindian tribe. *Proceedings of the National Academy of Sciences of the USA*, 88, 8720–8724.

Watterson, G. A. (1975). On the number of segregating sites in genetical models without recombination. *Theoretical Population Biology*, 7, 256–276.

Index